Biological and Medical Physics, Biomedical Engineering

This series is intended to be comprehensive, covering a broad range of topics important to the study of the physical, chemical and biological sciences. Its goal is to provide scientists and engineers with monographs- both, authored and contributed volumes to address the growing need for information. The fields of biological and medical physics and biomedical engineering are broad, multidisciplinary and dynamic. They lie at the crossroads of frontier research in physics, biology, chemistry, and medicine.

Books in the series emphasize established and emergent areas of science including molecular, membrane, and mathematical biophysics; photosynthetic energy harvesting and conversion; information processing; physical principles of genetics; sensory communications; automata networks, neural networks, and cellular automata. Equally important is coverage of applied aspects of biological and medical physics and biomedical engineering such as molecular electronic components and devices, biosensors, medicine, imaging, physical principles of renewable energy production, advanced prostheses, and environmental control and engineering.

Reza Gholami · Seyed Morteza Naghib

Nanobioceramics for Bone Tissue Engineering and Regenerative Biomedicine

Tunable Biological Characteristics

Springer

Reza Gholami
Department of Nanotechnology
School of Advanced Technologies
Iran University of Science and Technology
Tehran, Iran

Biomaterials and Tissue Engineering
Research Group
Interdisciplinary Technologies Department
Breast Cancer Research Center
Motamed Cancer Institute, ACECR
Tehran, Iran

Seyed Morteza Naghib
Department of Nanotechnology
School of Advanced Technologies
Iran University of Science and Technology
Tehran, Iran

ISSN 1618-7210 ISSN 2197-5647 (electronic)
Biological and Medical Physics, Biomedical Engineering
ISBN 978-981-96-0040-3 ISBN 978-981-96-0041-0 (eBook)
https://doi.org/10.1007/978-981-96-0041-0

This Springer imprint is published by the registered company Springer Nature Singapore Pte Ltd.
The registered company address is: 152 Beach Road, #21-01/04 Gateway East, Singapore 189721, Singapore

Preface

The writing of this book began in 2019, inspired by an idea from Dr. Seyed Morteza Naghib. However, its writing faced various challenges. Due to the outbreak of COVID-19 and the resulting quarantines, access to the necessary research resources was severely limited. As a result, the pace of writing the book slowed down, but it did not stop. With the discovery of the coronavirus vaccine and the return of life to normal, the writing of this book continued and was pursued with greater seriousness. Bones are biological components in the human body that have high strength compared to other body parts. They form the body structure. In addition, bone tissues are responsible for protecting some internal organs of the body such as the heart and brain. The main constituent of bones is calcium. The use of ceramics in various ways in the treatment of bone-related diseases has a long history. Some ceramics can create similar properties to bone and be a suitable substitute for lost bones due to their similar structure to the bone and superior properties (biocompatibility, biodegradability, etc.). Also, the ease of access and engineering of manufacturing these materials has made them special materials in the treatment of these diseases. Nowadays, the progress of nanotechnology and its extraordinary properties is no secret to anyone. Nanotechnology has also found its place in the field of medical and therapeutic applications. In addition, the use of nanoscience in the design and manufacture of devices and the advancement of therapeutic methods and techniques has received great attention. The combination of nanoscience with bioceramics has led to the creation of nano-bioceramics, which impart unique properties to ceramics. These novel materials show amazing potential for use in the treatment of bone-related diseases, which are discussed in detail in this book. Although many books and articles have been published in the field of nano-bioceramics and bone tissue engineering, most of them have researched these topics only from a medical or engineering perspective. In this book, we have tried to combine these two perspectives to compile a complete and comprehensive book to be a reference for students, researchers and scientists in these fields. For this purpose, Chap. 1 of this book provides an introduction to nano-bioceramics, their application in bone tissue engineering and the challenges they face. In Chaps. 2–6, bone is thoroughly investigated from various perspectives such as types, structures, cells, extracellular matrix and basic proteins. The final three chapters (7–9) of the book also study nano-bioceramics and the properties required for their use in bone tissue engineering. It is hoped that the publication of this book will be a comprehensive and complete guide for all researchers in the field of bone tissue engineering and nano-bioceramics and can accelerate the technological progress in the field of nanomaterials. At the end, we are grateful to God the gracious the merciful for granting us the opportunity to research and write in the field of nano-bioceramics and bone tissue engineering, who despite all difficulties and challenges that arose in writing this book was always our companion so that we could finish writing the book.

Tehran, Iran

Reza Gholami
Seyed Morteza Naghib

Contents

Chapter 1
Introduction

Abstract Synthetic composite-based materials and scaffolds should be biodegradable, biocompatible, and supply sufficient structural aid for cell migration, along with oxygen, waste, and nutrient carriage to accelerate bone regeneration process and remodeling in defects. These properties may be reached by functioning tunable physical features, including absorption rate, degradation rate, modulus, porosity, and swelling by adjustments with the addition of ceramic phases and copolymers as synthetic composite scaffolds. Synthetic bioceramics, seek to imitate the natural hydroxyapatite (HA) crystal creation located in bone. These ceramics, particularly calcium phosphates, have exhibited great osteoinductivity, osteoconductivity, and biocompatibility. Lately, silicon-based glass–ceramics have been investigated as a substitution of calcium phosphates. Several members of this collection have exhibited high bioactivity, attractive mechanical strength, and have been exposed to increase cell proliferation, adhesion, and mineralization of extracellular matrix. Moreover, antibacterial properties of some nanostructured bioceramics established significant interests in avoiding implants rejection in surgery and biomedicine. To our knowledge, tunable biological characteristics of nano-bioceramics has not been considered deeply in literature yet. Therefore, we will focus on some challenges in bone tissue engineering and regenerative medicine.

List of Abbreviations

β-TCP	β-Tricalcium phosphate
ALP	Alkaline phosphatase
ECM	Extracellular matrix
HA, HAP	Hydroxyapatite
HF	Hydrofluoric acid
PLGA	Poly(lactic-co-glycolic acid)
PMMA	Poly(methyl methacrylate)
T_g	Glass transition temperature

R. Gholami and S. M. Naghib, *Nanobioceramics for Bone Tissue Engineering and Regenerative Biomedicine*, Biological and Medical Physics, Biomedical Engineering,
https://doi.org/10.1007/978-981-96-0041-0_1

1.1 Introduction

Ceramics (from the Greek *keramos*, representing burnt stuff or pottery) are combinations of non-metallic and inorganic resources and contain porcelain, clay products, porcelain enamels, abrasives, refractory materials, pottery, glasses, cements, ferroelectrics, manufactured single crystals and non-metallic magnetic materials [1]. Obviously, Old-fashioned ceramic artifacts—such as clay ware, silicates, soda-lime glasses, and Portland cement- and constructional clay artifacts (e.g., tile and brick) are not appropriate for implanting in the human body. Novel categories of ceramics, such as pure oxides, ZrO_2 and Al_2O_3, and mono-crystal Al_2O_3 are being used as an alternative. Glass–ceramics have found an application in implant productions. Ceramic nitrides (e.g., SiAlON, silicon nitride) is another contender for that purpose. Ceramic metal composites composed of typically aluminum oxide- and carbide-bonded metals (cermets) have been operated to build cutting devices, but they are not being thought of for implant applications. Plaster of Paris with chemical composition of $CaSO_42H_2O$ as a calcium sulfate is being investigated as a bone replacement [1]. In this chapter, we have presented an introduction on nano-bioceramics history and three generations of them, their developments, and challenges in bone tissue engineering.

1.2 Nano-bioceramics History

Ceramics are heat resistant polycrystalline compounds, mostly inorganic, such as carbides, metallic oxides, silicates, and different refractory hydrides, selenides and sulfides. Oxides including ZrO_2, Al_2O_3, and SiO_2 consist of metallic and non-metallic components whereas others are ionic salt like NaCl, ZnS, CsCl, etc. [1]. The first generation of materials with special applications within the human body was advanced during the 1960s and 1970s. In early 1969, the concept of using bioceramics as implants came to light by finding various bioceramics and bioactive glasses with the capability to bond with living bone [2]. These studies and developments became a foundation for the biomaterials field. The value of studying mechanical features of tissues, mostly bone, to attain good skeletal implants was recognized by some of the very first researchers like Professor Bill Bonfield. His attempts to comprehend the interaction between living tissues and biomaterials were so pioneering. The aim of almost all primary biomaterials was to accomplish an appropriate arrangement of physical characteristics with minimal toxicity in the host. Nearly all materials operated in the body as prostheses were single-phase materials. A general characteristic of the majority of those materials was biological inertness [3, 4].

The point inspiring the bulk of advancement of biomaterials between 1960 and 1980s was to decrease the biological response to the imported body. This engineering plan theory is still reliable. The interface between bioinert biomaterials and tissues is an acellular, slim fibrous capsule with smallest adhesion between the implant and its host tissue. However, Professor Bonfield noticed the necessity of finding a novel cohort of biomaterials. The inquiry team started to explore the sense of creating biocomposite materials that closely resemble the mechanical characteristics of the bones. These materials would reduce, or maybe remove, bone resorption and stress shielding at the interface between bone and prosthesis due to the difference of elastic modulus [3]. The biocomposites advanced in that time contain a new idea of bioactivity, as well as the strategic approach of combining two phases for achieving composite characteristics resembling those of living bone. Hapex, a novel biocomposite material, was the first to be successfully advanced for medical applications which made up of a spread bioactive phase of hydroxyapatite (HAP) particles and a polymeric matrix of polyethylene. Bioactive composites, like Hapex, had become essential in the replacement and repair of bones in the middle ear [3]. Researchers continued their leading assistance to the biomaterials field with the advancement of a new species of bioactive material called Si-substituted HAP, which shows great promise as a bone graft material [5].

Biomaterials achieved bioactivity in 1984 with the development of the second generation. By the mid-1980s, these materials had been used in a variety of dental and orthopedic applications. Bioactive materials evoke a regulated reaction in the physiological conditions [3]. Diverse combinations of bioactive composites, glass–ceramics, ceramics, and glasses were used in clinical experiments. To provide bioactive fixation, synthetic HAP ceramics had usually operated as coating, powders, and porous implants on a metallic prosthesis [2, 6]. The presence of HAP coatings with low solubility promotes the tissue's response known as *osteoconduction,* where bone growth occurs along the coating, resulting in the formation of a mechanically strong interface [3].

The clinical success of implants that are resorbable, bioactive and bioinert has been an essential solution to the healthcare demands of the aging population in developed countries. Although survivability assessments of most of the skeletal implants [7] display that one-third to one-half of implants fail in 10–25 years as well as man-made heart valves [8]. These failures need patients to have another surgery and it has a cost to society and the patients. Such expenses have a considerable contribution in the quickly increasing healthcare costs. It is important to know that any artificial materials cannot react to biochemical stimuli or changing physiological loads like living tissues. This understanding restricts the duration of all artificial body components [3].

Third generation of biomaterials has been considered to stimulate particular responses of cells at the level of the area of molecular biology [9]. In the first decade of the twenty-first century, the ideas of resorbable materials and bioactive materials came together and resorbable polymers have been constructed bioactive and bioactive materials have been built resorbable. Resorbable polymer systems have been engineered to undergo molecular adjustment, enabling particular interactions with cell integrins. This facilitates the differentiation and proliferation of cells as well as the organization and production of extracellular matrix (ECM). Hierarchical porous foams and third-generation bioactive glasses have been created to activate genes that stimulate living cells generation [3].

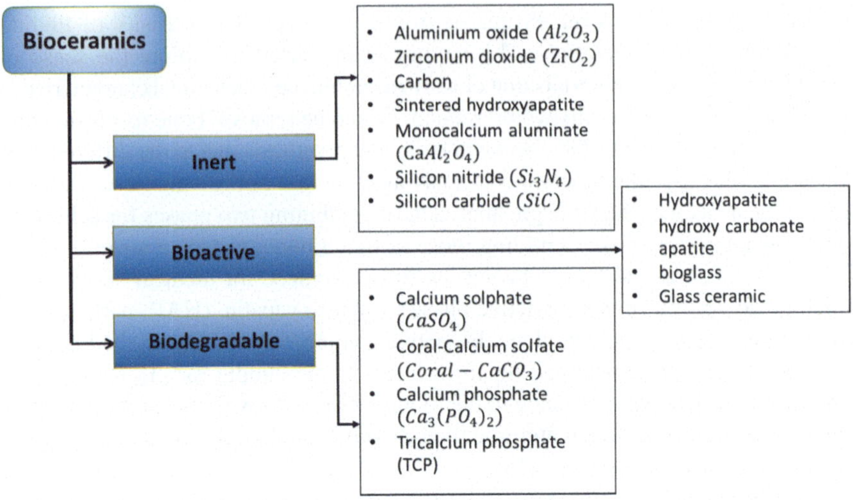

Fig. 1.1 Types of bioceramics are used for tissue engineering

Bioceramics are used to replace the bone tissues mainly divided into inert, Bioactive and biodegradable bioceramics as shown in the Fig. 1.1.

Bioinert materials (like silica ceramics, oxide ceramics and carbon fiber) should be non-toxicity and non-inflammatory [10–17]. Bioinert materials resist structural failure and are also corrosion-protective and resilient. To hinder the cracking of the material, they have a low "Young's modulus" [18]. The bioactive class of ceramics includes bioactive glass, hydroxyapatite, and β-tricalcium phosphate (β-TCP) [19]. Because of the similarity of the mineral phase of the hard tissue to hydroxyapatite ceramics, hydroxyapatite has been considered as the basic material for bone tissue engineering [1]. Bioceramics can be operated as bioactive drug delivery agents and medical implants such as drug-eluting devices and implantable pills. A medical implant can be exploited to support a biological structure that got damaged and/or to remodel a missing biological structure, to increase performance of a living biological structure. Artificial bones and teeth are comparatively regular for bioceramics applications. Surgical joint replacements and cements generally include bioceramics materials to decrease inflammatory response and wear. Bioceramics also find application in medical devices such as respirators, kidney dialysis machines, and pacemakers [18, 20].

Between nanoparticles with various material compositions, inorganic nanoparticles combined with calcium phosphate ($Ca_5(PO_4)_3OH$, hydroxyapatite) have many advantages including biocompatibility, simplicity of synthesis, potent interaction with therapeutics, and control of chemical and physical characteristics [21]. Although use of hydroxyapatite (HAP) for biomedical application has begun in a dense form, porous forms were preferred because of their better bioresorption and bioactivity as scaffolds and bone implants. Porous HAP ceramics are used as a man-made bone graft material, filler material for bone defects, and in prosthesis revision surgery within the field of bone tissue engineering. It has been used in various biomedical usages including bone tissue regeneration and cell proliferation. Owing to the accessibility of a great surface area, its high resorbability and osteoconductivity properties increase the speed of bone ingrowth. β-TCP with HAP has most widely used as block implants in orthopedics [6] and for artificial bone graft as substitution materials. The porous structure of HA shows a great affinity for bone, as the presence of pores creates a robust mechanical interaction between the material and the bone. The bone tissue exhibits robust growth within the porous HA and increases the stability of the implant. Two critical factors for excellent osteointegration are the size and morphology of pores [22–24]. Porous HA has also been utilized as scaffolds for hard tissues and loading cells [25].

Roughness of the surface for the connection of cells as well as pores interconnectivity for diffusion of osteoblast-like cells into the pores are other important demands for porous implants. Porous HA implants have limited load-bearing capacity and are mostly applied to satisfy small minor bone imperfections. Increasing the pore size of the implant significantly decreases its strength. Therefore, main viewpoints in the advancement of organ substitute materials and HA-based bone are the production of scaffolds or maintaining matrices with a proper macro- and microscopic structural morphology such as biodegradability, mechanical strength, biocompatibility, osteoconductivity, pore size and interconnectivity. Integration of fluoride ions into the composition of HA has been stated to promote the growth of bone cells and enhance the formation of new minerals in cancellous bone. Additionally, it leads to a longer resorption time and suitable integration with tissue of bone in comparison with classic calcium phosphate [25]. HA with good pore interconnectivity, controlled porosity, surface properties, and mechanical strength are required by clinical reconstructive supplies [18, 26]. Glass and ceramics have been given numerous considerations as applicants for implant materials while they have various highly attractive properties for several purposes. The key advantage is that they are inert in aqueous environments, and therefore possess high biocompatibility. The glasses, ceramics, and composites we are going to review in this book are inadequate because of the excessive cost of development and the partial information on their effects on tissue (and vice versa). One more main characteristic of glasses and ceramics is their too little strength in tension and brittleness. Hence, their application is generally restricted to compressive loading conditions, including a femoral head of a total hip joint replacement or those in an acetabular cup. A number of usages do not necessitate high loading, such as barium sulfate ($BaSO_4$) for bone cement and hydroxyapatite for artificial bone [1, 27–30].

Particular carbons have found application as implantations in such blood-interfacing purposes as heart valves. They are also being applied as a strengthening constituent for composite implant materials as a result of their biocompatibility and great specific strength as fibers. Work is also being done to discover their application in such tensile loading treatments as ligament replacements and artificial tendon. Ease of fabrication and their good biocompatibility are particularly favorable qualities for implants while their dark color could be a negative aspect in dental usages [1].

Bioceramics can be categorized on the authority of their interaction with live tissues, as bioinert or bioactive. Those ceramics (e.g., glass–ceramics and calcium phosphates) used to carry drugs or those fabricated to stimulate tissue reactions for bonding are typically bio(re)active. Most other ceramics are bioinert, which include cermets, sialon, alumina, and zirconia [1].

The most significant feature of fragile materials including glasses and ceramics is their porosity. Since it significantly intensifies and increases stress, porosity has a marvelous effect on the physical features of fragile materials. Furthermore, it is challenging to prevent creation of pores through construction of materials. The non-yielding nature of fragile materials is the major explanation for the amplified stress [1].

1.3 Nano-bioceramics Developments

The physicochemical properties or characterizations demanded of glasses and ceramics for therapeutic purposes differ widely according to the specific use. Additionally, owing to our partial comprehension of interactions between material and tissue, it is tough—if not impractical—to directly apply the physical and chemical characteristics of materials to their performance within a living organism. Nevertheless, it is essential to initiate the research of implant materials by thoroughly understanding their performance in different conditions. This must not inhibit us from thoroughly comprehending and studying the properties of materials in a controlled laboratory setting before applying them as implants [1]. There are two key features of fragile fracture: (1) determining the failure strength in ceramics is more difficult compared to polymers and metals, and (2) the fracture strength is lower than the theoretical value [1]. Nonetheless, it is possible to build these materials for "load-bearing" implants by employing a suitable approach of calculating failure including numerical methods, and creating the implant correctly preventing (tensile) stress-raisers [1].

Table 1.1 classifies the elastic modulus for different glasses and ceramics along with specific modulus, which is stated as modulus per unit density, density and Poisson's ratio. The table determines that the carbides and alumina present the uppermost specific modulus. Steels and other metallic composites are approximately 20 to 30 GPa cm^3/g which is enormously smaller than carbides or alumina [1].

Since the covalent and ionic bonds (unlike metallic bonds) are pretty localized, the explanation for their fragility is the molecular or atomic bonding characteristics. While the positioning of positive ions can occur without significant hindrance from adjacent ions, the dislocations can transfer across the metallic bonds with minimal resistance. This is in contrast to ionic bonds, where the movement of positive ions is restricted by negative ions, controlling the number of potential locations. To put it another way, metallic bonds possess more slip systems compared to ionic bonds. Covalent bonds also exhibit high directionality, and need fragmentation and restructuring, hence restricting the mobility of displacements. Consequently, making materials such as carbides and diamond fragile [1].

Table 1.1 Moduli, specific modulus, density and Poisson's ratio for various glasses and ceramics [1]

Material	Shear modulus, G (GPa)	Elastic modulus, E (GPa)	Specific modulus, (GPa cm^3/g)	Density, (g/m^3)	Poisson's ratio, ν
Zirconia (ZrO_2).cubic	58	152	27	5.56	0.32
Magnesia (MgO)	76	207	58	3.58	0.36
Alumina (Al_2O_3)	154	390	100	3.9	0.27
Spinel ($MgAl_2O_3$)	–	284	80	3.55	–
Borosilicate glass	26	62	28	2.23	0.20
Soda-lime glass	22	69	28	2.48	0.23
Titanium carbide (TiC)	–	462	94	4.92	–
Silicon carbide (SiC)	174	414	120	3.22	0.19
Silicon nitride (Si_3N_4)	123	304	88	3.44	0.24
Fused silica (SiO_2)	32	75	34	2.2	0.16
Cement, Portland	–	45	19	2.4	–

Several materials can be constructed to proceed toward their theoretical strength. Metal whiskers exhibit a rise in strength as their diameter decreases, suggesting that a smaller diameter results in fewer displacements and therefore superior strength. Tremendously fine glass fibers newly pulled from the melt. After all, the fibers' strength reduces significantly when they come into contact with the air, even momentarily. This reveals that the strength of the fiber is exceedingly reliant on the surface circumstance. These investigational outcomes suggest that the dissimilarity between observed and theoretical strength is because of surface conditions and structural irregularities [1].

Fragility of glasses is triggered by an absence of plastic deformation, which is seen in the 3D network arrangement. The absence of slip systems in glasses and ceramics avoids displacement generation and motion, causing a material that is brittle and hard, as pointed out previously. This is fairly owing to covalent bonding of C and Si and the ionic nature of the oxides bonding. Slip can happen only in diagonal directions (not horizontally) owing to repulsion of like-charged ions [1]. The strength of fragile materials is higher in compression than in tension, as declared formerly. These trial remarks directed Griffith and others to explore the connection between microflaws and strength in fragile materials [1]. One approach operated to amplify the strength of fragile materials is the omission of Griffith flaws. Fire polishing is a process that eliminates surface imperfections by warming the material slightly over its glass transition temperature (T_g), allowing it to flow and smooth out crack tips or close them. Chemical etching has the ability to refine the fracture points, therefore decreasing the stress intensity. Hydrofluoric (HF) acid is commonly utilized due to its rapid dissolution of silicates [31].

While applied weight must dominate, compressive force ahead of tensile force can take charge, in order for a ceramic material to better resist tensile stress, it can be enhanced by applying compressive force to the surface layers relative to the inside. Surface compression can be initiated through surface crystallization, quenching, and ion exchange. Ion exchange can be achieved by electrical-migration or diffusion methods. Essentially, bigger ions are replaced with littler ions (e.g., K^+ for Na^+), making the surface compressive owing to lattice straining. Since electric fields allow for the presence of external ions on the surface at smaller temperatures than what can be achieved by diffusion process alone, utilization of them can be beneficial [1]. Surface crystallization of glass–ceramics can be achieved by modifying the specific or molar volume of layers of the surface. Much the same outcomes can be attained by fast refrigeration (quenching), as the internal part of the ceramic or glass has a tendency to get smaller but is controlled by the firm surface layer chilled by quenching. This causes the interior to be under tension and the surface to be under compression [1]. Addition of an additional crystalline phase might enhance the strength by generating particles of a second phase, which can effectively impede the propagation of cracks. Matrix materials are generally more flexible materials, including metals and polymers. Glasses and ceramics are exploited efficiently for constructing composite materials owing to their high strength and modulus [1].

As pointed out before, the strength of fragile materials rests on the dimension of defects dispensed all over the material. Griffith's theory of fracture in tension posits that the biggest crack or flaw has the most influence on the crash of a material. Since flaw dimension is restricted to the extent of cross-section of specimen, strength moreover be influenced by the volume of a specimen. Hence, the minor the specimen (such as fibers), the superior the fracture strength [1]. Fatigue can also arise from dynamic (cycle) loads. This is typically shown by plotting stress against logarithmic cycles (N) or logarithmic time. The number of cycles or time of failure rests on the temperature, type of load and the magnitude, and the test environment [1]. Since materials located in the body experience loading and unloading cycles, conducting fatigue tests on materials in a simulated body atmosphere allows for a more accurate evaluation of the material's performance. Thus, a fatigue test is valuable for evaluating the operation of different implants under specific testing circumstances. Nevertheless, it is not possible to mimic the complex conditions of loading and unloading that an implant experiences inside a living organism [1]. Similar to toughness, impact strength refers to the total energy that a material can absorb when subjected to an impact force. Impact strength of a sample can be quantified by subjecting it to the force of swinging pendulum. The alteration of amplitude of the pendulum swing is an amount of the energy absorbed by the sample. Consequently, it is possible to calculate the impact energy or strength. An impact analysis typically needs a huge quantity of specimen for the reason that there is a great volume of difference in the results. Glasses and ceramics have comparatively little impact strength, which change extensively [1].

Hardness is a quantity of plastic deformation and is described as the force per unit area of penetration or indentation, and is therefore expressed in the dimensions of stress. The stress amounts achieved during the hardness test are 2.5 to 3 times greater than the tensile stress experienced by a malleable material. This is owing to impingement of the elastic material around the plastic indentation, which limits free expansion of the indented area. Rockwell test, Knoop test, and the Vickers regularly used in hardness tests. The main differences between them lie in the applied weight, their inner material, and its arrangement. It is fairly challenging to utilize an old-fashioned hardness test for glasses and ceramics because of their non-malleable nature (no plastic deformation) [1]. Wear plays a crucial role in selecting implant materials, particularly for joint prosthesis. There are various kinds of wear. Wear cannot be considered without some comprehension of the abrasion among two materials. The wear characteristics of an implant material are essential, particularly for different joint substitutes. When two solid materials get in touch, they contact just at the tips of their uppermost roughness. Recent findings indicate that the effective region of interaction rises with applied weight (P) for malleable materials but not for elastic materials similar to diamond and rubber. Hence, the visible surface area is much larger than the actual contact zone. Pressure welding is possible with ductile materials owing to creation of plastic junctions. The plastic junctions are the major supply of adhesive abrasion when two materials are moving smoothly along each other with or without a lubricating layer [1]. Adhesive wear is a procedure in which particles adhere to a surface after being removed from another through sliding. Afterward, the particles may be lost or loosened. If tough particles are avoided from the sliding surfaces (some hard particles also come off from the firm surfaces) and the surfaces are smooth, this type of wear can be reduced. This kind of wear is the most critical procedure in implant materials and compared to all other types of wear, this can be investigated more simply [1]. Chemical activity on both or only one of the sliding materials can lead to corrosive wear. The creation of subsurface or surface cracks tracked by big chunks breaking apart under frequent sliding and loading cycles is the source of surface fatigue wear. The sliding action eliminates the creation of corrosion, which would defend the surface from extra strike, leading to quicker corrosion [1].

The abrasion and wear properties alter considerably when lubrication is applied among two in touch surfaces. It is essential to recognize the lubrication procedure because various kinds of lubricants are present in the majority of implant appliances. There are typically two kinds of lubrication. Boundary lubrication occurs when an extremely thin layer of lubricant or a smoother substance is presented between two firmer materials. This kind of lubrication is further essential for man-made joints. The application of lubricant significantly reduces the direct connection between two moving surfaces, hence decreasing wear and friction. Fluid lubrication can be attained by a layer of some gas or liquid thick adequate to entirely distinct surfaces of two solids. This kind of lubrication is attained more simply at elevated sliding surface speed and is chiefly reliant on the characteristics of the lubricating fluids [1]. Implants are usually used nowadays in dental and orthopedic surgical treatment, for instance restoration of lost teeth, bone repair, and knee and hip joint substitution. The implants are usually manufactured from long-lasting bioinert metals, for instance tantalum (Ta), stainless steel, titanium (Ti) and its alloys (e.g., Ti–6Al–4V), and cobalt-chromium (CoCr) alloy. They are described by outstanding mechanical features such as fatigue resistance and elevated yield strength, as well as excellent resistance to corrosion. These qualities are crucial for their application in load-bearing anatomical areas. The implants' excellent corrosion resistance indicates their outstanding biocompatibility, which elicits a partial foreign body response [32]. The exceptional corrosion resistance arises from the formation of a protective nanoscale oxide layer, which is generated spontaneously, through a chemical reaction or by a thermal treatment [33].

The abovementioned metallic implants possess a successful and confirmed history in dental and orthopedic surgical treatment. A review of five common hip replacement brands (built of titanium alloy, CoCr or steel, implanted between 1987 and 2007) specified that the rate of revision due to aseptic loosening after 18 years was as low as 8.4% (95% confidence intervals: 7.7–9.1, $n = 24{,}728$) [34]. Nevertheless, there is a continuous ambition to attain more dependable and stronger incorporation of the implant with host bone (osseointegration). Osseointegration has been stated from different perspectives, consisting of numerical calculation of the interfacial strength, ultra-structures of the bone-implant edge, and the explanation of clinical outcomes. Virtually, it refers to direct functional and structural association between the implant and living bone [35]. The primary characteristics of effective osseous implantation are the creation and preservation of suitable tissue integration with the implants. Bioinert metals usually do not create a straight connection with living bone. Alternatively, the interface between bone and Ti implants is generally packed with a fine acellular collagenous capsule [32, 36], which has been discovered to have a thickness ranging from 500 to 1000 nm for implants of Ti–6Al–4V [36]. This non-osseous interfacial section compromises the time-consuming stability, directing to the implants loosening. Bioactive materials including various calcium phosphates and bioactive glasses usually bring on straight chemical bonding between the surface of the implant and bone deprived of an interfering fibrous connective tissue layer. However, this occurrence is rare when it comes to metallic implants [33].

Although poly(methyl methacrylate) (PMMA) cement does not directly attach to bone, it has been introduced to stabilize bioinert implants [37, 38]. Instant stabilization can be attained quickly after the completion of surgical treatment. There are growing worries about the time-consuming stability of implants attached to PMMA cement based on the stated high rates of revision while PMMA is successful in various outlooks [39–41]. The PMMA cement's inferior mechanical characteristics have been demonstrated to cause problems, as well as the PMMA wear debris leading to unfavorable biological reactions over time [42–44]. The enlargement of little wear particles of PMMA neighboring the implant often triggers an unpleasant immunological reaction, leading to osteolysis [45, 46].

In order to tackle the issue of osteolysis caused by PMMA wear debris, techniques for implant cementless fixation have been expanded [47, 48]. The operational approach is to adjust the surface of metallic implant, supplying it with novel features which can promote osteogenesis proximal to the surface of implant and attain straight bone bonding. Several surface adjustment methods have been explored, usually covering the implant with a bioactive material or aiming the manufacture of a specific surface topography. Nanotechnology has a significant role in both methods [33].

Metallic orthopedic implants are regularly fixed into the skeletal system of the human body as components of reconstructive tools (such as knee or hip joint substitute) or fracture fixation supplies (such as nails, screws, and plates). The biocompatibility and chemical and mechanical stabilities of the implant materials in the background of body fluids and tissues are of essential significance for the effective healing of bone substitutes and bone fracture [49]. The plan of these implants is limited by the physiology and stated by the anatomy of the skeletal structure of the human body [50]. If an orthopedic implant has to be deleted from the body ahead of time, it is deliberated to have failed [51].

Regrettably, several lack of successes of stainless steel implants have been stated and associated with corrosion. Clinical experiment has exposed that such implants are receptive to pitting and crevice corrosion in the human body, triggering discharge of metallic ions into the adjacent tissue. This might cause systemic effects or local inflammations and, in some situations, elimination of implants is required. Furthermore, allergic reactions occasionally happen as well. Co–Cr alloys do infrequently release ions and produce particles to their environments by means of wear and corrosion while they are usually acknowledged to have outstanding corrosion and wear resistance in vivo. Cobalt can be cancer-causing and might initiate inheritable harms in humans. Thus, it is hard to express an advantage of Co–Cr or titanium alloys over stainless steel for in vivo usages. Pure Ti and Ti–6Al–4V alloy are identified to be more sensitive to wear than stainless steels, hence producing larger quantities of metallic particles for a loose functional implant. These deficiencies restrict the application of Co–Cr and titanium alloys for implant products [49].

Corrosion is the main trouble influencing the lifetime of orthopedic implants [52, 53]. In manufacturing, usual performances to manage this phenomenon consist of using electrical current, using protective layers, decreasing temperature, altering the composition of the electrolyte and its pH, the supplement of inhibitors, etc. Unfortunately, since the atmosphere of the body is stable and cannot be changed without negative biological consequences, most of these methods cannot be operated for orthopedic implants. The use of cathodic or anodic protections which include applied currents is unrealistic for medical implants. The substitute ways to decrease corrosion of orthopedic implants are adjustment of surface by protective layers and selection of materials [54, 55]. With regard to materials choice, the number of probable materials is somewhat restricted due to biocompatibility necessities and mechanical strength. One of the biggest challenges facing materials technology and science nowadays is the advance of a new generation of biomaterials to operate successfully in the body [56].

Bioactive materials which are resorbable propose an alternate to the foundation of a stable tissue/implant edge. This idea is based on the management of the material surface chemistry. A bioactive material is one that produces a certain biological response at the interface, which gives rise to the establishment of a bond between the material and tissues. A bioactive implant chemically reacts with the body fluids in a way that is harmonious with the healing process of the tissues. The material does not reduce in strength since the chemical reactions are limited to the surface. Creation of the fibrous capsule is inhibited by adhesion of repairing tissues. The key point is that failure does not happen at the interface [51].

The attention in calcium phosphates originates from the truth that teeth and bones have a high percentage of mineralized calcium phosphate. The resemblance between the powder X-ray diffraction patterns of the basic and the in vivo mineral calcium phosphate, HAP $[Ca_{10}(PO_4)_6(OH)_2]$, was detected first in 1926 by De Jong [57]. Basic calcium phosphates deposited from aqueous mixtures from a non-stoichiometric series, fairly due to surface effects and replacements by other ions, but also because of tremendously small crystallite sizes grown by this low solubility mineral. The crystallographic and chemical resemblance of HAP to bone mineral permits it to establish extremely strong biological bonds [58]. Hydroxyapatite is a bioactive material and has been displayed to have a mimicking influence on construction of bone, also acknowledged as osteoinduction. The biological efficiency of HAP layers is based on a quick osteointegration with a pronounced growth in the load carrying capability of the implant. Hydroxyapatite ceramics noticeably exhibit desirable characteristics including lack of intervening fibrous tissue, lack of toxicity, the opportunity to accelerate bone growth, and the probability of establishing a straight contact with bone. Their weak mechanical properties and fragile nature, however, restrict their application directly as implantation materials. This has made it easier for the advancement of HAP coatings on regularly used Ti and its alloys, Co–Cr alloys, and type 316L stainless steel, mostly for bone substitution [59–62].

Currently, for all those experimental purposes where load-bearing characteristics are essential, the majority of the implants operated are metallic, with critical and subsequent issues. Another possibility is to cover the metallic implants with bioceramics. This method is being operated at the present time, both for hip joint prosthesis and dental implants. There is still a long way to go, but numerous metallic implants with bioceramics coatings are commercially existing by this time, and the study in problem solving is ongoing [63]. The bioceramics coating procedure on a metallic substrate is pretty complicated, and quite a lot of approaches are accessible in this sense. Another benefit of bioceramics coating is its ability to decrease the release of ions from the metal alloy. The clinical accomplishment heavily relies on this coating, since the durability and quality of the connection between the surfaces significantly rest on surface morphology of the substrate, layer thickness, chemical composition of the coating, particle size, and the purity. The bioceramics coating denotes a really operative obstacle that hinders the metallic ion kinetics of discharge toward the living body [51].

Bone itself is a composite containing HAP nanorods inserted in the collagen matrix. Human teeth are composed of HAP nanorods (usually less than 100 nm) placed in lamellae and connected to collagen [64, 65]. Nanosized HAP particles are attractive when biocompatibility is considered [66]. Because of their nanosize, HAP crystals have an extremely expansive surface area, facilitating uniform resorption by osteoclasts [67]. In the last few years, an understanding of HAP property has directed to various novel approaches of benefiting the nanometer HAP with the quick advancement of nanometric HAP materials [68]. The form of HAP crystal can influence several properties of HAP, for instance bioactivity, surface properties, and so on [69]. Moreover, nanometric biomaterials are in favor of cellular proliferation and adhesion, the deposition of calcium comprising minerals on their surface and synthesis of alkaline phosphatase [70]. Hence, nano-HAP demonstrates advantage in the subject of orthopedic implants for its enhanced biomechanical and biological characteristics [71, 72]. Furthermore, recent advancements in the manufacture of nanoscale hydroxyapatite particles have directed to development of various novel usages. For instance, nanoparticles of hydroxyapatite can hinder proliferation of cancer cells and be worked as an effective agent for drug delivery. Nanoparticles of hydroxyapatite are shown to have a suppressive influence on the growth of tumor cells [73]. Seemingly various properties of HAP particles, including crystallinity, size and morphology, have exhibited various biological results [74].

The outstanding characteristics and disadvantages of metallic implants and bioceramics have directed to the advancement of an arrangement of both, along with nanotechnological adjustment of metallic surfaces, HAP or both. This has directed to the development of a new generation of materials for instance biomimetics and graded structures, tissue engineering, scaffolds and surface coatings [51].

Nanotechnology has opened up a new age of technology, chiefly the advancement of nanoscale engineering and nanostructuring of the biomaterials. The fascinating features of nanobiomaterials are investigated in quite a lot of ways. Metallic nanoparticles have obtained considerable uses in an extensive field of biomedical utilities including gene targeting, drug delivery, sensing and imaging [75–78]. Carbon-based nanomaterials such as graphene, nanodiamonds, nanotubes and fullerenes exhibit excellent potential as a carrier for photothermal therapy, bacterial inhibition, drug delivery, imaging, pharmaceutical agents, biosensors, and components for nanoformulations used in the delivery of therapeutic molecules [79–81]. Nano-bioceramics have been employed widely in cranio-maxillofacial and orthopedic purposes, and polymeric nanoparticles have been operated as drug carriers [82, 83]. In fact, nanotechnology has grown quickly from the innovation phase to the function phase; its capacity effect on healthcare is massive and has spread through several aspects of the medical area [84, 85].

Nano-bioceramics such as bioglasses, calcium phosphates, zirconia, and α-alumina involve several various shapes, which vary from nanorods and nanofibers (rods and fibers with diameters in the range of 1–100 nm), to nanopowders (bulk materials containing enormously fine powders with mean particles dimensions in the range of 1 and 100 nm), nanocoatings (single or multilayer surface coatings with widths between 1 and 100 nm), nanocrystals (crystalline solid with grain sizes between 1 and 100 nm), and nanostructured (basic structural units between 1 and 100 nm) [84].

Close to bioceramics, nano-bioceramics have low electrical and thermal conductivities and are favorably corrosion-resistant, stable in the physiological situation, and biocompatible. Owing to the distinctive chemical structure, nano-bioceramics are additionally susceptible to chemical and surface adjustments. Therefore, nano-bioceramics maintain most of the intrinsic characteristics of bioceramics. At the nanoscale, however, a number of the features of nano-bioceramics become extremely changed from their normal equivalents, and this directs nano-bioceramics to expose several unique performances relative to bulk materials, which are mostly exposed in the mechanical, surface, and physical characteristics [84]. The dimension of nano-bioceramics is less than one-fourth of the light wavelength; thus they do not disperse light, which originates transparency. The transparent feature rising from degraded microstructure of nano-bioceramics can be utilized for aesthetic advantages in dental purposes. Nano-bioceramics include larger extent of grain boundaries than regular types, supplying distinctive surface characteristics including wettability, roughness, energy, and topography that potentially apply a considerable effect on behavior of the cell [69, 86–92].

One of the most noticeable benefits of a nanostructured surface is its extraordinarily superior specific surface area when compared to a surface with microscale characteristics. Nanostructured surface has an amplified number of crystal grains and atoms [92], which would become more operating in terms of recrystallization and dissolution processes. This permits nano-bioceramics to operate as a conceivable drug delivery system for many treatments including genes, growth factors, proteins, anti-inflammatory and anti-tumor drugs, and antibiotics [93, 94]. Moreover, nano-bioceramics have a superior surface area-to-volume ratio than regular microscale bioceramics, which would present more accessible sites for surface adjustment, drug loading, and protein adsorption. Furthermore, the elevated surface area in nano-bioceramics modifies the related surface energy for protein adsorption, which would in turn smoothen the quantity of cellular interactions [84]. Fragility is a most important disadvantage in bioceramics for fixing large bone flaws. Though, based on Hall–Petch relation, when grains are perfected to the nanoscale system, it gives rise to outstanding increase in the yield strength and hardness of nano-bioceramics. Nevertheless, when grain dimension is under 20 nm, nanocrystalline grains comprise huge quantities of triple grain boundary junctions, which decreases the strength. The decrease in strength can be clarified by numerous mechanisms which inspire deformation in nanoceramics by the improved diffusion alongside microcracking, grain boundary, and grain boundary sliding [95, 96]. Usually, the distribution and size of the serious defects in ceramics influence the fracture strength. By keeping a nanocrystalline microstructure for bioceramics, flaw sizes are decreased, the homogeneity of the microstructure is enhanced, the coefficient of thermal expansion divergence among microstructural phase accumulation, and thermal residual stress arise out of anisotropic crystal structure are similarly dropped. Mutually, all these effects cause a considerable superior stress for crack propagation, which result in an upper fracture toughness and fracture strength at the best dimensions of nano-bioceramics [84].

Generally, nanotechnology has much more to deal with than constructing the molecular assemblies from the atomic level and just simple miniaturization. In the last 20 years, several functions and innovations of nanotechnology have dramatically altered various disciplines of technology, science and medicine. Specifically, the use of nanotechnology in biomaterial science has turned into a new edge for bone repair, suggesting a hopeful scope to advance orthopedic care, treatment, and inhibition of bone diseases. Presently, enormous exertions are ongoing to improve characteristics and the structure of nano-bioceramics, to imitate composition and the nano- and microstructural elements of bone to achieve advantageous properties. All these efforts have directed investigation to attain the stage with primary clinical trials. In the foreseeable future, with better identification of cell connections, extensive effort will be concentrated on the distribution of the agglomerated nano-bioceramics with the aim of reveal its valuable and unique characteristics for required biological or mechanical influence, and enhancing angiogenesis within the nano-biomaterial procedure. Thus, it is predicted that nano-bioceramics will allow the advance of innovative systems that imitate the complicated hierarchical structure of the natural tissue to a superior level [84].

1.4 Nano-bioceramics' Challenges in Bone Tissue Engineering

While recent studies seem to maintain the effect of nano-bioceramics in the field of bone science, major progress is required to understand their extensive potential in medical application. The following critical matters and challenges are needed to be focused in order for nano-bioceramics to shift on the later developmental phase [84].

One of the main challenges is the discrepancy of the processing technologies of nanostructured materials, which hamper the scale-up of nanofabrication procedures. Nanofabrication methods combine a variety of advanced materials, tools and processes, which depend on bottom-up and top-down nanoscale methods. Industrial system issues in the design of production, control and design of process, quality and reliability, stream chain administration and shop floor strategies have yet to be improved, which are the crucial obstacles in developing nano-bioceramics from the research laboratory into high volume manufacture [84].

An extra challenge is nanotoxicology, which is presently a quite new area. In spite of consideration of the molecular and cellular basis that rules the contact between cells and nanostructured materials, there are still vast gaps in our comprehension about the nature of the communications of the human system with nanoparticles, particularly, the safety sides of nanoparticles stay unaddressed. Lately, an article has specified possible harmful consequences of carbon nanotubes on cells and their capability to produce genetic damage, irritation, oxidative stress, and enduring pathological influences [97]. This may pose a risk to human health as there may be conditions when the communication of nano-bioceramics with biomolecules in vivo or nano-bioceramics in their accumulation state may interact in an undesirable way, generating inflammatory responses. Moreover, the idea of dosimetry, dose metric exposure evaluation, risk description and hazard identification of nano-bioceramics are still far from convincing state. Thus, the effects of nano-bioceramics on human health are not well comprehended yet. Therefore, considerable research is certainly required to estimate the stability of nano-bioceramics in a range of test systems so as to completely clarify its application [84].

The primary obstacle in the field of biomaterials lies in the fact that all implant devices substitute alive tissues that have evolved over thousands of years to possess exceptional physical characteristics, as well as the ability to repair, regenerate, and grow. Therefore, all artificial biomaterials that are modified for the purpose of restoring or repairing the body signify a compromise. The relative failure or success of a biomaterial exposes the engineering and scientific wisdom operated in attaining this compromise. The interaction of lots of complicated technological, clinical, biological, and physical aspects should be deliberated [4].

For instance, deliberate the following properties of a native tooth that should be met to attain a effective tooth implant: an interface that biologically bonded to bone, gingival tissues, and epithelial skin cells, which leads to a dissimilarity in elastic moduli of more than 10^3 through the different interfaces connected to a tooth; a tensile strength of 15,000 to 20,000 psi when subjected to flexure; and an addition structure (the periodontal ligament) that alters compressive stresses put on the tooth into tensile stress within the jawbone. No material identified currently can consistently reach the stable interfacial attachments needed to imitate a natural tooth while several materials have the prerequisite flexural strength [4].

The most important obstacle in the field of materials science is achieving precise control over the interface of tissue biomaterials. By sensitive choice of polymers, ceramics or metallic materials, either individually or in specifically engineered arrangements, it is possible to closely replicate the physical characteristics of most tissues, within the constraints of engineering. The criterion for the biomaterial to be non-toxic to its host tissues can be accomplished quite simply by screening the materials using temporary implantation or tissue culture experiments. However, attaining the required gradient or match in physical characteristics through the interface between living and nonliving substances poses a remarkable scientific obstacle. The challenge lies in the ongoing development of the scientific understanding of the adhesion of biological interfaces. The poor understanding of tissue adherence to implant devices will persist until biochemists and cell biologists identify the specific molecular species that regulate cell bonding [4].

Chiefly, mechanical features including fracture toughness and yield strength can be improved by adjusting characteristics at the nanoscale level. Osteointegration is essential for enduring success of implant in bone tissue particular purposes. Nanoscale engineering accomplishes this through inhibiting fibrous capsule formation and elevating cell adhesion. A promising response can be achieved by designing nanostructures at the cellular level. This would be valuable for regulating and maintaining the osteogenic expression of cells that are proliferating on the surface of implants [98].

Typically, metal-on-polymer is the preferred choice for articulating implant surfaces. These surfaces, however, have substantial rates of wear that can cause foreign body response. It is essential to enhance the hardness of the moving surfaces in order to decrease the rate of wear. This can be accomplished by utilizing ceramics. Alumina is commonly chosen for this purpose because of its relatively high strength, chemical inertness, and exceptional hardness. Actually, alumina-alumina surfaces can attain friction constants as low as 0.05 in comparison to cobalt chromium-ultra high molecular weight polyethylene (CoCr - UHMWPE), which calculated 0.094 [99]. Nevertheless, alumina exhibits brittleness and is susceptible to the development of cracks and fractures because of its low ductility. To address the issue, it is suggested that reducing the grain sizes to less than 20 nm can enhance the plasticity, hardness and strength of the material, as predicted by the Hall–Petch equation (Eq. 1.1).

$$\sigma_y = \sigma_0 + \frac{k}{\sqrt{a^3}} \tag{1.1}$$

Theoretically, the Hall–Petch equation proposes that grain size (d) reduces as yield strength (σ_y) rises. Nevertheless, it is important to acknowledge that when the size is less than 10 nm, the Hall–Petch effect becomes irrelevant as defects and flaws would compromise the nanomaterial's characteristics [100]. Considering this restriction, it is feasible to manufacture nano-bioceramics that possess both high fracture toughness and high hardness, hence ensuring a minimal rate of wear [98].

Regularly, the inadequate bonding of the implantation to the juxtaposed bone is the reason for orthopedic implant failure. Osteointegration (also known as osseointegration) is a critical factor in fixing the insert in place, avoiding micromotion from hurting the nearby tissues, establishing a straight functional and structural bond between the implant and bone that operates to dispense stress and load equally all through the tissue. The inadequate osteointegration can be ascribed to the biological reaction elicited by the biomaterial. Inadequate bonding can lead to several problems, including aseptic loosening caused by extreme micromotion, and total migration of the whole implant. A bioinert material would cause no integration to the host body and cells connecting to the surfaces therefore cannot be originated. Moreover, an excessive secretion by inflammatory cells will lead to the creation of a fibrous capsule, since they recognize the material as an unknown entity. On the other hand, a substance that undergoes excessive degradation in vivo may not pose a threat to the structural integrity; however, it may liberate microparticles that stimulate macrophages to secrete cytokines, thereby speeding up the process of bone resorption. This occurrence, comprehended as osteolysis, is the primary reason for post-surgical bone loss. To avoid these disadvantageous reactions, one can promote osteointegration by utilizing nanostructured materials with certain topological characteristics that promote proliferation and cellular adhesion, while preventing the creation of fibrous tissue. An approach involves surface adjustments like coatings where fine films of nano-bioceramics can be placed on the surface of implants to enhance cytocompatibility, wettability and roughness. The enhanced surface area of bioceramics placed at the nanolevel allows for greater surface energy, which is essential for the surface adsorption of proteins and the initiation of cellular attachment [98].

Functional alterations in the near nano/micro environment encompassing the cells contribute to numerous biochemical processes related to cellular regulation, differentiation, and proliferation. It is widely recognized that cellular biomechanics operate at the nanoscale level. Several cellular signaling pathways can be prevented or promoted when a set of integrins are involved in a specific arrangement or sequence. These integrins operate at the nanoscale level, therefore, in order to induce a particular response, biomaterials should also be engineered at the nanoscale [98].

The influence of nano-topological characteristics was exhibited in a report including a range of nanoscale geometries varying from nanopitarrays, nanopost, and nanogroove to investigate the influences of cellular contact regulation. The report revealed that while cells exhibited a predominantly spherical shape on smooth surfaces, epithelial cells underwent elongation and alignment along the ridges and grooves patterns, which had feature diameters as tiny as 70 nm [101]. Indeed, these topological cues offer a chance for precise manipulation of a bone scaffold at the nanoscale. By aligning nano-bioceramic crystals along the c-axis of the bone, it becomes possible to facilitate the attachment of osteoblast cells and promote the production of extracellular matrices. This, in turn, would increase the integrity and strength of the implanted scaffold. By this time Kim and coworkers have presented a concept in which cellular proliferation and adhesion was enhanced by using nanocomposite of hydroxyapatite (HA) nanocrystals embedded in a nanofiber composed of gelatin [102]. The adherence of osteoblasts is contingent upon surface topology, and research has demonstrated that nanophase surfaces facilitate superior adhesion compared to materials on a larger scale [86]. The investigations demonstrated a 46% and 30% augmentation in osteoblast adhesion on alumina (with a grain size of 23 nm) and nanophase titania (with a grain size of 32 nm), respectively, in comparison to typical bioceramics [69]. Actually, nanotopology on its own may explain 48–51% of the overall adhesion of osteoblasts, as presented in an investigation by Palin and colleagues. They indicated that poly(lactic-co-glycolic acid) (PLGA) designed after nanophase titania exhibited a comparable percentage rise in osteoblast adhesion in comparison to nanophase titania [103]. Furthermore, when osteoblasts were cultured on nanophase bioceramics, ECM calcium content and alkaline phosphatase (ALP) were notably elevated, in comparison with common microscale ceramics [87, 98].

1.5 Conclusion

Currently, implants are commonly utilized in dental and orthopedic surgery, such as for the replacement of missing teeth, repair of bones, and substitution of knee and hip joints. The initial iteration of materials with specialized applications within the human body was developed in the 1960s and 1970s. Bioactivity in biomaterials was first achieved in 1984 with the introduction of the second generation. By the mid-1980s, these materials had been utilized in many dental and orthopedic applications. Bioactive compounds elicit a controlled response under physiological conditions. The third generation of biomaterials has been designed to specifically induce cellular responses at the molecular biology level. Bioceramics can be classified based on their interaction with living tissues, either as bioinert or bioactive. Ceramics like glass–ceramics and calcium phosphates, which are utilized to transport medications or to promote tissue reactions for bonding purposes, are commonly considered bioactive. The primary characteristic of delicate materials, such as glassware and ceramics, is their porosity. Porosity has a remarkable impact on the physical characteristics of delicate materials as it greatly enhances and amplifies stress. The surface roughness and interconnectivity of pores are crucial requirements for porous implants, since they facilitate the connection of cells and allow for the diffusion of osteoblast-like cells into the pores. Nanotechnology has ushered in a new era of technology, primarily focused on the progress of engineering at the nanoscale and the structuring of biomaterials at the nanolevel. A nanostructured surface has a significantly higher specific surface area compared to a surface with microscale features, which is one of its most prominent advantages. Among nanoparticles of different material compositions, the combination of inorganic nanoparticles with calcium phosphate (hydroxyapatite) offers several advantages, such as biocompatibility, ease of manufacturing, strong interaction with medicines, and precise control over chemical and physical properties. Nano-bioceramics, including bioglasses, calcium phosphates, zirconia, and α-alumina, come in many forms such as nanorods, nanofibers, nanopowders, nanocoatings, nanocrystals, and nanostructured materials. The required physicochemical qualities or characteristics of glasses and ceramics for therapeutic applications vary significantly depending on their unique application. Fragile fracture has two primary characteristics: firstly, the determination of failure strength in ceramics is more challenging in comparison to polymers and metals, and secondly, the fracture strength is lower than the theoretical value. Conducting fatigue testing on materials in a simulated body atmosphere provides a more precise assessment of the material's performance, as the materials in the body undergo loading and unloading cycles. Impact strength, like toughness, is the measure of the maximum amount of energy a material can absorb when it experiences an impact force. Glasses and ceramics exhibit relatively low impact strength, which varies significantly. Hardness is a measure of how much a material can be deformed without breaking. It is defined as the force applied per unit area during penetration or indentation, and is therefore measured in terms of stress. Utilizing an old-fashioned hardness test for glasses and ceramics is difficult due to their non-malleable nature, which means they cannot undergo plastic deformation. The choice of implant materials, especially for joint prostheses, is heavily influenced by wear considerations. Frequently, the insufficient adherence of the implant to the adjacent bone is the cause of orthopedic implant failure. The primary challenge in the field of materials science lies in attaining meticulous regulation of the interface between tissue and biomaterials. Nanotoxicology presents an additional difficulty. Despite studying the molecular and cellular mechanisms that govern the interaction between cells and nanostructured materials, there are still significant gaps in our understanding of how the human body communicates with nanoparticles. Specifically, the safety aspects of nanoparticles remain unexplored.

References

1. Park, J. (2009). Bioceramics: properties, characterizations, and applications (Vol. 741). Springer Science & Business Media.
2. Hench, L. L. (1991). Bioceramics: From concept to clinic. *Journal of the American Ceramic Society, 74*(7), 1487–1510.
3. Hench, L. L., & Thompson, I. (2010). Twenty-first century challenges for biomaterials. *Journal of the Royal Society Interface, 7*(suppl_4), S379–S391.
4. Hench, L. L. (1980). Biomaterials. *Science, 208*(4446), 826S31.
5. San Thian, E., et al. (2006). The response of osteoblasts to nanocrystalline silicon-substituted hydroxyapatite thin films. *Biomaterials, 27*(13), 2692–2698.
6. Hench, L. L. (1993). *An introduction to bioceramics* (Vol. 1). World Scientific.
7. Hench, L. L., & Wilson, J. (1996). *Clinical performance of skeletal prostheses*. Springer.
8. Schoen, F. J., Levy, R. J., & Piehler, H. R. (1992). Pathological considerations in replacement cardiac valves. *Cardiovascular Pathology, 1*(1), 29–52.
9. Hench, L. L., & Polak, J. M. (2002). Third-generation biomedical materials. *Science, 295*(5557), 1014–1017.
10. Kalantari, E., et al. (2019). Biocomposites based on hydroxyapatite matrix reinforced with nanostructured monticellite ($CaMgSiO_4$) for biomedical application: Synthesis, characterization, and biological studies. *Materials Science and Engineering: C, 105*, 109912.
11. Kalantari, E., et al. (2018). Nanostructured monticellite for tissue engineering applications—Part I: Microstructural and physicochemical characteristics. *Ceramics International, 44*(11), 12731–12738.
12. Kalantari, E., et al. (2018). Nanostructured monticellite for tissue engineering applications—Part II: Molecular and biological characteristics. *Ceramics International, 44*(12), 14704–14711.
13. Rahmanian, M., et al. (2019). Multifunctional gelatin–tricalcium phosphate porous nanocomposite scaffolds for tissue engineering and local drug delivery: In vitro and in vivo studies. *Journal of the Taiwan Institute of Chemical Engineers, 101*, 214–220.
14. Askari, E., et al. (2021). Local delivery of chemotherapeutic agent in tissue engineering based on gelatin/graphene hydrogel. *Journal of Materials Research and Technology, 12*, 412–422.
15. Dehghani, N., et al. (2023). Chitosan/silk fibroin/nitrogen-doped carbon quantum dot/α-tricalcium phosphate nanocomposite electrospinned as a scaffold for wound healing application: In vitro and in vivo studies. *International Journal of Biological Macromolecules, 238*, 124078.
16. Salehi, S., et al. (2023). Smart stimuli-responsive injectable gels and hydrogels for drug delivery and tissue engineering applications: A review. *Frontiers in Bioengineering and Biotechnology, 11*.
17. Khakpour, E., et al. (2023). Graphene-based nanomaterials for stimuli-sensitive controlled delivery of therapeutic molecules. *Frontiers in Bioengineering and Biotechnology, 11*, 1129768.
18. Das, R. (2016). Application of nanobioceramics in bone tissue engineering. *Nanobiomaterials in hard tissue engineering* (pp. 353–379). Elsevier.
19. O'Donnell, M., et al. (2009). The effect of phosphate content on the bioactivity of soda-lime-phosphosilicate glasses. *Journal of Materials Science: Materials in Medicine, 20*(8), 1611–1618.
20. Kassinger, R. (2003). *Ceramics: From magic pots to man-made bones*. Twenty-First Century Books.

21. Loo, S., et al. (2010). Biomedical applications of hydroxyapatite nanoparticles. *Current Pharmaceutical Biotechnology, 11*(4), 333–342.
22. Yoshikawa, H., & Myoui, A. (2005). Bone tissue engineering with porous hydroxyapatite ceramics. *Journal of Artificial Organs, 8*(3), 131–136.
23. Gauthier, O., et al. (1998). Macroporous biphasic calcium phosphate ceramics: Influence of macropore diameter and macroporosity percentage on bone ingrowth. *Biomaterials, 19*(1–3), 133–139.
24. Le Huec, J., et al. (1995). Influence of porosity on the mechanical resistance of hydroxyapatite ceramics under compressive stress. *Biomaterials, 16*(2), 113–118.
25. Sopyan, I., et al. (2007). Porous hydroxyapatite for artificial bone applications. *Science and Technology of Advanced Materials, 8*(1–2), 116.
26. Nishikawa, M., et al. (2004). Bone tissue engineering using novel interconnected porous hydroxyapatite ceramics combined with marrow mesenchymal cells: Quantitative and three-dimensional image analysis. *Cell Transplantation, 13*(4), 367–376.
27. Ansari, M., et al. (2011). Synthesis and characterisation of hydroxyapatite-calcium hydroxide for dental composites. *Ceramics-Silikáty, 55*(2), 123–126.
28. Naghib, S. M., et al. (2012). Biofunctionalization of dextran-based polymeric film surface through enzyme immobilization for phenylalanine determination. *International Journal of Electrochemical Science, 7*(1), 120–135.
29. Morteza, N. S., et al. (2012). Development and characterization of 316 L stainless steel coated by melt-derived and sol-gel derived 45S5 bioglass for orthopedic applications. *Ceramics-Silikáty, 56*(1), 89–93.
30. Naghib, S. M., et al. (2012). Bioactivation of 304 stainless steel surface through 45S5 bioglass coating for biomedical applications. *International Journal of Electrochemical Science, 7*(4), 2890–2903.
31. Griffith, A. A. (1921). VI. The phenomena of rupture and flow in solids. *Philosophical Transactions of the Royal Society of London. Series A, Containing Papers of a Mathematical or Physical Character, 221*(582–593), 163–198.
32. Brunette, D. M., et al. (2012). *Titanium in medicine: Material science, surface science, engineering, biological responses and medical applications*. Springer Science & Business Media.
33. Liu, Y., Rahaman, M. N., & Sonny Bal, B. (2014). Chapter 8: Nano-bioceramics as coatings for orthopedic implants and scaffolds for bone regeneration. In *Tissue regeneration: Where nano-structure meets biology* (pp. 343–391). World Scientific.
34. Espehaug, B., et al. (2009). 18 years of results with cemented primary hip prostheses in the Norwegian Arthroplasty register: Concerns about some newer implants. *Acta Orthopaedica, 80*(4), 402–412.
35. Albrektsson, T., & Johansson, C. (2001). Osteoinduction, osteoconduction and osseointegration. *European Spine Journal, 10*(2), S96–S101.
36. Johansson, C., et al. (1989). Ultrastructural differences of the interface zone between bone and Ti 6Al 4V or commercially pure titanium. *Journal of Biomedical Engineering, 11*(1), 3–8.
37. Jaeblon, T. (2010). Polymethylmethacrylate: Properties and contemporary uses in orthopaedics. *JAAOS-Journal of the American Academy of Orthopaedic Surgeons, 18*(5), 297–305.
38. DiMaio, F. R. (2002). The science of bone cement: A historical review. *Orthopedics, 25*(12), 1399–1407.
39. Mulroy, R., Jr., & Harris, W. (1990). The effect of improved cementing techniques on component loosening in total hip replacement. An 11-year radiographic review. *The Journal of Bone and Joint Surgery. British Volume, 72*(5), 757–760.
40. Mulroy, R., Jr., & Harris, W. (1990). Failure of acetabular autogenous grafts in total hip arthroplasty. Increasing incidence: a follow-up note. *The Journal of Bone and Joint Surgery. American Volume, 72*(10), 1536–1540.

41. Katz, R. P., et al. (1995). Results of cemented femoral revision total hip arthroplasty using improved cementing techniques. *Clinical Orthopaedics and Related Research, 319*, 178–183.
42. Ioannidis, T. T., et al. (1997). Detrimental effect of aging on the endurance of bone cement: An in vivo and in vitro study in rabbits. *Acta Orthopaedica Scandinavica, 68*(sup275), 115–118.
43. Fernandez-Fairen, M., & Vazquez, J. (1983). The aging of polymethyl methacrylate bone cement. *Acta Orthopaedica Belgica, 49*(4), 512–520.
44. Looney, M., & Park, J. (1986). Molecular and mechanical property changes during aging of bone cement in vitro and in vivo. *Journal of Biomedical Materials Research, 20*(5), 555–563.
45. Willert, H.-G., Bertram, H., & Buchhorn, G. H. (1990). Osteolysis in alloarthroplasty of the hip. The role of bone cement fragmentation. *Clinical Orthopaedics and Related Research, 1990*(258), 108–121.
46. Jasty, M. J., et al. (1986). Localized osteolysis in stable, non-septic total hip replacement. *The Journal of Bone and Joint Surgery. American Volume, 68*(6), 912–919.
47. Ries, M. D. (2008). Review of the evolution of the cementless acetabular cup. *Orthopedics, 31*(12 Suppl 2), 285–290.
48. Rothman, R. H., & Cohn, J. C. (1990). Cemented versus cementless total hip arthroplasty. A critical review. *Clinical Orthopaedics and Related Research, 1990*(254), 153–169.
49. Kamachi, M., & Baldev, R. (2008). Corrosion science and technology: Mechanism, mitigation and monitoring.
50. Williams, D. (1987). Advanced applications for materials implanted within the human body. *Materials Science and Technology, 3*(10), 797–806.
51. Sridhar, T. (2010). Nanobioceramic coatings for biomedical applications. *Materials Technology, 25*(3–4), 184–195.
52. Kumar, S., et al. (2010). Fretting corrosion behaviour of thermally oxidized CP-Ti in Ringer's solution. *Corrosion Science, 52*(3), 711–721.
53. Eliaz, N., & Nissan, O. (2007). Innovative processes for electropolishing of medical devices made of stainless steels. *Journal of Biomedical Materials Research Part A: An Official Journal of The Society for Biomaterials, The Japanese Society for Biomaterials, and The Australian Society for Biomaterials and the Korean Society for Biomaterials, 83*(2), 546–557.
54. Kumar, S., et al. (2009). Thermal oxidation of CP-Ti: Evaluation of characteristics and corrosion resistance as a function of treatment time. *Materials Science and Engineering: C, 29*(6), 1942–1949.
55. Kumar, S., & Narayanan, T. S. (2008). Corrosion behaviour of Ti–15Mo alloy for dental implant applications. *Journal of Dentistry, 36*(7), 500–507.
56. Kamachimudali, U., Sridhar, T., & Raj, B. (2003). Corrosion of bio implants. *Sadhana, 28*(3), 601–637.
57. De Jong, W. (1926). The mineral substance in bone (translated from French). *Rec Trav Chim Pays-Bas, 45*, 445–448.
58. Gilbert, J. L., & Jacobs, J. J. (1997). The mechanical and electrochemical processes associated with taper fretting crevice corrosion: a review. In *Modularity of orthopedic implants*.
59. Karthega, M., Tamilselvi, S., & Rajendran, N. (2006). Effect of pH on the corrosion behaviour of Ti–6Al–4V alloy for dental implant application in fluoride media. *Trend Biomater Artif Org, 20*, 31–34.
60. Raman, V., Tamilselvi, S., & Rajendran, N. (2007). Electrochemical impedance spectroscopic characterization of titanium during alkali treatment and apatite growth in simulated body fluid. *Electrochimica Acta, 52*(26), 7418–7424.
61. Eliaz, N., et al. (2009). Electrochemical processes of nucleation and growth of calcium phosphate on titanium supported by real-time quartz crystal microbalance measurements and X-ray photoelectron spectroscopy analysis. *Journal of Biomedical Materials Research Part A: An Official Journal of The Society for Biomaterials, The Japanese Society for Biomaterials, and The Australian Society for Biomaterials and the Korean Society for Biomaterials, 89*(1), 270–280.
62. Sridhar, T., et al. (2002). Electrophoretic deposition of hydroxyapatite coatings and corrosion aspects of metallic implants. *Corrosion Reviews, 20*(4–5), 255–294.

63. Sridhar, T., & Rajeswari, S. (2009). Biomaterials corrosion. *Corrosion Reviews, 27*(Supplement), 287–332.
64. Fratzl, P., et al. (2004). Structure and mechanical quality of the collagen–mineral nano-composite in bone. *Journal of Materials Chemistry, 14*(14), 2115–2123.
65. Sanosh, K., et al. (2009). Synthesis of nano hydroxyapatite powder that simulate teeth particle morphology and composition. *Current Applied Physics, 9*(6), 1459–1462.
66. Chen, Y., et al. (2008). In vitro biological performance of nano-particles on the surface of hydroxyapatite coatings. *Applied Surface Science, 255*(2), 375–378.
67. Kim, H.-M. (2003). Ceramic bioactivity and related biomimetic strategy. *Current Opinion in Solid State and Materials Science, 7*(4–5), 289–299.
68. Qiu, C., Xiao, X., & Liu, R. (2008). Biomimetic synthesis of spherical nano-hydroxyapatite in the presence of polyethylene glycol. *Ceramics International, 34*(7), 1747–1751.
69. Webster, T. J., Siegel, R. W., & Bizios, R. (1999). Osteoblast adhesion on nanophase ceramics. *Biomaterials, 20*(13), 1221–1227.
70. Bow, J.-S., Liou, S.-C., & Chen, S.-Y. (2004). Structural characterization of room-temperature synthesized nano-sized β-tricalcium phosphate. *Biomaterials, 25*(16), 3155–3161.
71. Prabakaran, K., & Rajeswari, S. (2009). Spectroscopic investigations on the synthesis of nano-hydroxyapatite from calcined eggshell by hydrothermal method using cationic surfactant as template. *Spectrochimica Acta Part A: Molecular and Biomolecular Spectroscopy, 74*(5), 1127–1134.
72. Xu, J., et al. (2004). Preparation and characterization of nano-sized hydroxyapatite powders produced in a radio frequency (rf) thermal plasma. *Materials Science and Engineering: A, 374*(1–2), 101–108.
73. Fu, Q., et al. (2005). Effects of nano HAP on biological and structural properties of glass bone cement. *Journal of Biomedical Materials Research Part A: An Official Journal of The Society for Biomaterials, The Japanese Society for Biomaterials, and The Australian Society for Biomaterials and the Korean Society for Biomaterials, 74*(2), 156–163.
74. Laquerriere, P., et al. (2003). Importance of hydroxyapatite particles characteristics on cytokines production by human monocytes in vitro. *Biomaterials, 24*(16), 2739–2747.
75. Minard-Basquin, C., et al. (2005). Gold-nanoparticle-assisted oligonucleotide immobilisation for improved DNA detection. In *IEE Proceedings-Nanobiotechnology*. IET.
76. Rosi, N. L., et al. (2006). Oligonucleotide-modified gold nanoparticles for intracellular gene regulation. *Science, 312*(5776), 1027–1030.
77. Ghosh, P., et al. (2008). Gold nanoparticles in delivery applications. *Advanced Drug Delivery Reviews, 60*(11), 1307–1315.
78. Murphy, C. J., et al. (2008). Gold nanoparticles in biology: Beyond toxicity to cellular imaging. *Accounts of Chemical Research, 41*(12), 1721–1730.
79. Mohanty, N., & Berry, V. (2008). Graphene-based single-bacterium resolution biodevice and DNA transistor: Interfacing graphene derivatives with nanoscale and microscale biocomponents. *Nano Letters, 8*(12), 4469–4476.
80. Sun, X., et al. (2008). Nano-graphene oxide for cellular imaging and drug delivery. *Nano Research, 1*(3), 203–212.
81. Yang, K., et al. (2010). Graphene in mice: Ultrahigh in vivo tumor uptake and efficient photothermal therapy. *Nano Letters, 10*(9), 3318–3323.
82. Uhrich, K. E., et al. (1999). Polymeric systems for controlled drug release. *Chemical Reviews-Columbus, 99*(11), 3181–3198.
83. Makadia, H. K., & Siegel, S. J. (2011). Poly lactic-co-glycolic acid (PLGA) as biodegradable controlled drug delivery carrier. *Polymers, 3*(3), 1377–1397.
84. Lim, P. N., et al. (2017). Overview of nanobioceramics. *Nanobioceramics for healthcare applications* (pp. 1–22). World Scientific.
85. Raffa, V., et al. (2010). Progress in nanotechnology for healthcare. *Minimally Invasive Therapy & Allied Technologies, 19*(3), 127–135.
86. Webster, T. J., et al. (2000). Specific proteins mediate enhanced osteoblast adhesion on nanophase ceramics. *Journal of Biomedical Materials Research: An Official Journal of The*

Society for Biomaterials, The Japanese Society for Biomaterials, and The Australian Society for Biomaterials and the Korean Society for Biomaterials, 51(3), 475–483.
87. Webster, T. J., et al. (2000). Enhanced functions of osteoblasts on nanophase ceramics. *Biomaterials, 21*(17), 1803–1810.
88. Webster, T. J., Siegel, R. W., & Bizios, R. (1999). Design and evaluation of nanophase alumina for orthopaedic/dental applications. *Nanostructured Materials, 12*(5–8), 983–986.
89. Webster, T. J., et al. (2001). Mechanisms of enhanced osteoblast adhesion on nanophase alumina involve vitronectin. *Tissue Engineering, 7*(3), 291–301.
90. Kay, S., et al. (2002). Nanostructured polymer/nanophase ceramic composites enhance osteoblast and chondrocyte adhesion. *Tissue Engineering, 8*(5), 753–761.
91. Gutwein, L. G., & Webster, T. J. (2004). Increased viable osteoblast density in the presence of nanophase compared to conventional alumina and titania particles. *Biomaterials, 25*(18), 4175–4183.
92. Webster, T. J., & Ahn, E. S. (2006). Nanostructured biomaterials for tissue engineering bone. *Tissue Engineering, II*, 275–308.
93. Ginebra, M.-P., Traykova, T., & Planell, J. A. (2006). Calcium phosphate cements as bone drug delivery systems: A review. *Journal of Controlled Release, 113*(2), 102–110.
94. Kim, K., & Fisher, J. P. (2007). Nanoparticle technology in bone tissue engineering. *Journal of Drug Targeting, 15*(4), 241–252.
95. Chokshi, A., et al. (1989). On the validity of the Hall-Petch relationship in nanocrystalline materials. *Scripta metallurgica, 23*(10), 1679–1683.
96. Richter, V., & Ruthendorf, M. (1999). On hardness and toughness of ultrafine and nanocrystalline hard materials. *International Journal of Refractory Metals and Hard Materials, 17*(1–3), 141–152.
97. Kunzmann, A., et al. (2011). Toxicology of engineered nanomaterials: Focus on biocompatibility, biodistribution and biodegradation. *Biochimica et Biophysica Acta—General Subjects, 1810*(3), 361–373.
98. Feng, J., & San Thian, E. (2013). Applications of nanobioceramics to healthcare technology. *Nanotechnology Reviews, 2*(6), 679–697.
99. Grimm, M. J. (2002). Selection of materials for biomedical applications. In *Handbook of materials selection* (pp. 1165–1194). John Wiley & Sons.
100. Catledge, S. A., et al. (2002). Nanostructured ceramics for biomedical implants. *Journal of Nanoscience and Nanotechnology, 2*(3–4), 293–312.
101. Bettinger, C. J., Langer, R., & Borenstein, J. T. (2009). Engineering substrate topography at the micro-and nanoscale to control cell function. *Angewandte Chemie International Edition, 48*(30), 5406–5415.
102. Kim, H. W., Song, J. H., & Kim, H. E. (2005). Nanofiber generation of gelatin–hydroxyapatite biomimetics for guided tissue regeneration. *Advanced Functional Materials, 15*(12), 1988–1994.
103. Palin, E., Liu, H., & Webster, T. J. (2005). Mimicking the nanofeatures of bone increases bone-forming cell adhesion and proliferation. *Nanotechnology, 16*(9), 1828.

Chapter 2
Bone

Abstract Bone is a remarkable organ of the human body that has the ability to undergo self-renewal. Also, this renewal may require a long time depending on the type of fracture. Bone tissue engineering products have the potential to accelerate the healing process of different bones. However, in order to design a bone tissue engineering device that can effectively function within the body, it is essential to possess comprehensive knowledge regarding the specific characteristics of bone tissue. Bone can be considered at the tissue and organ levels. Furthermore, it can be classified into other categories. In nanoscale, they are composed of structural proteins and polysaccharides. Osteogenesis, the process of bone formation, and the elongation of long bones begins during endochondral ossification, the second phase of long bone growth. The process continues from fetal stage to postnatal stages until skeletal maturity is reached. It begins when chondrocytes, which are experiencing proliferation within the template, change into a non-proliferative, hypertrophic state. Intramembranous bone regeneration occurs in many cases of skeletal healing. It is essential in several clinical scenarios. In bone tissue engineering, traditional methods aim to replicate the process of intramembranous ossification in order to create a matrix that closely resembles bone. This chapter presents a comprehensive overview of bone physiology, bone biology, osteogenesis, and bone development mechanisms.

Abbreviations

3d	Three-dimensional
ALP	Alkaline phosphatase
BMPs	Bone morphogenetic proteins
BMU	Basic multicellular unit
BTE	Bone tissue engineering
ECM	Extracellular matrix
HSCs	Hematopoietic stem cells
MSCs	Mesenchymal stem cells
PG	Polyglycolide

R. Gholami and S. M. Naghib, *Nanobioceramics for Bone Tissue Engineering and Regenerative Biomedicine*, Biological and Medical Physics, Biomedical Engineering,
https://doi.org/10.1007/978-981-96-0041-0_2

POC	Primary ossification center
SOCs	Secondary ossification centers
TCP	Tricalcium phosphate
VEGF	Vascular endothelial growth factor

2.1 Introduction

To design a proper implant for bone tissue engineering we need to know some basic information about its host tissue. In this chapter and the next two, we presented a full review of features and characteristics of bone and related tissues, physiology and biology, proteins, procedures to use materials as an implant and advantages and disadvantages of them. Bone is a complicated alive connective tissue that supplies flexibility, mechanical maintenance, and structural framework to the body as well as blood pH adjustment, homeostasis, and mineral storage. The skeleton is composed of over 206 distinct bones, which range from the flat bones in the skull and sternum, short bones in the wrist and ankle, and long bones found in our limbs, to irregular bones including the vertebrae and pelvis [1]. Structure of bone usually contains cancellous and cortical bone. The exclusive composition of organic and inorganic materials reveals its mechanical characteristics to the bone.

Bone flaws and their healing is the most general trouble in the world acquiring bone in the role of a second most transferred tissue type followed by blood. Surgery, fractures, trauma, congenital malformation, tumor resection, or diseases such as arthritis and osteoporosis are the foremost origin of bone imperfections. A number of clinical situations such as large bone defects, skeletal reconstruction or compromised regenerative treatments including osteoporosis, atrophic non-unions and avascular necrosis also need bone related transfers. The substitutions or healing of such broken or injured bone tissue is accomplished by accepted methods such as bone transport, distraction osteogenesis or various bone grafting approaches including bone graft replacements, allografts, autografts or by utilizing growth factors. The first marketable material as bone graft was announced in 1993 as Interpore's coral obtained Pro-Osteon® [2]. Autografts have accomplished several steps of success in curing bone defects. Prolonged rehabilitation and the donor site morbidity, however, raise danger of serious infection and limited accessibility restrains its potential usages. Bone allografts have relatively solved restrictions of transplantable bone samples, however, with possible dangers of immunological rejections, viral infection, transmissible diseases, efficacy and cost efficiency. Because of bone porous and avascular nature, osteocytes survive by diffusion of nutrients that restricts their use in case of host viability and size of bone defect. In addition, there are no synthetic or heterologous bone replacements existing currently which are better or with comparable mechanical or biological characteristics like natural bone. Therefore, an effective and alternative healing approach for regeneration of bone is essential.

There are several experimental motivations to progress bone tissue engineering substitutes, such as the demand for orthopedic implants that are mechanically more appropriate to their biological conditions and the demand for superior filling materials that can be exploited in the reform of large orthopedic defects [3]. Latest replacements of bone which can substitute typical bone grafts have exhibited potential opportunity. Usage of stem cells, gene therapy, osteoinductive matrix, and usage of osteogenic growth factors including bone morphogenetic proteins (BMPs) have exhibited their capability in bone tissue engineering (BTE) [4]. BTE may possibly present another solution that has superior mechanical characteristics than those operated nowadays. This can reduce vascular damage from the implant to the bone and trigger less-stress shielding, possibly reducing the incidence of implant-related osteopenia and following fractures. A related usage is the utilization of the surface of a tissue-engineered implant by coating the prosthesis with tissue or cells to enduringly stabilize implants earlier than implantation. This could be very helpful in reconstructive orthopedic surgical treatment that possibly have high failure rates secondary to defects of large bones [3].

Bone defects are a problematic area of orthopedic and considerable investigations have been undertaken to find a cure. An approach to decrease the quantity of autologous bone and enhance the osteogenesis of bone grafts is instantly needed. Additive manufacturing technology for BTE has significantly supported the development and research of bone regeneration. The most notable additive manufacturing methods comprise selective laser sintering, fused deposition modeling, stereolithography, and three-dimensional (3D) printing. Biocompatibility, biodegradability and bioactivity are the foremost characteristics required in an ultimate tissue healing material. Bioceramics have social value and potential market in the manufacture of a reliable and effective material to mend bone abnormalities. Hence, utilizing tissue engineering equipment for bone regeneration is quite advantageous. Bioceramics can be exploited directly in the human body or in treatments associated with the human body, and are operated in biochemical, medical, biological and other fields of investigation. Bioceramics present several promising characteristics, including good chemical and physical stability, bactericidal effects, anti-thrombus effects, excellent surface compatibility, mechanical compatibility and biocompatibility. All bioceramics have various pros and cons; so, perfect bone implants can be constructed by adjusting the scaffold microstructures or by exploiting two or more raw materials. CaP bioceramics, which present brilliant osteoconductive characteristics because of their chemical resemblance to natural bone, are extensively used to manufacture porous scaffolds for BTE. The characteristics of CaP scaffolds can be adjusted by exploiting surface coating procedures or additives to improve osteogenesis and angiogenesis. Lately, biodegradable ceramics have been tailored operating different other materials, and a new scaffold selection has been exploited for regeneration of bone. Ma et al. made a bioceramics scaffold with a uniformly self-assembled CaP/Polydopamine nanolayer surface for bone defects healing after bone tumors resection [5]. In recent years, Yang et al. [6] built a scaffold composed of tricalcium silicate with a nanosurface structure. Liu et al. [7] constructed a scaffold from akermanite that had a better rate of new bone growth than the pure tricalcium phosphate (TCP) scaffold and superior mechanical characteristics. Between the scaffolds aforementioned, it was detected that osteoblasts or mesenchymal stem cells (MSCs) of bone marrow could proliferate, migrate and adhere in or on scaffolds that have suitable biocompatibility.

Bone deformities initiated by disease, tumors or trauma typically exhibit a range of variations. There are only inadequate supplies of traditional autogenous bone. The shape and size of autogenous bone cannot always correspond to those of bone defects and the cycle of production is time-consuming, leading to unacceptable clinical results. Precision and individualization are the current guidelines of medical advancement. The tailoring of synthetic and natural materials, innovative arrangements of distinct ceramic materials, and advancement of new bioceramics materials are foremost study directions for regeneration of bone [8]. The capability of the host bed and the extent of the defect can restrict usage of bone grafts since they are reliant on diffusion and are avascular. Moreover, maintenance of the new bone volume can be challenging because of erratic resorption of bone. The grafts can be absorbed by the body before bone formation is accomplished in large defects. Distraction osteogenesis procedures are often long-lasting and arduous treats that are preserved for the most inspired patients. Grafts which are vascularized need a major microsurgical operational technique involving an advanced infrastructure. Allografting presents the danger of infection and/or disease; it can trigger an absolute loss or diminishing of the inductive factors of bone. Not only is the performing time needed for harvesting autografts costly, but the donor tissue is often rare, and there can be considerable morbidity at the donor site combined with hematoma, pain and infection. Bone cement fillers made in the operating room are another approach of repair of bone defects but they may be vulnerable to infection [3]. Bone marrow substitution is a different promising tissue engineering purpose for the healing of patients after treatment with radiation and/or high-dose chemotherapy. Marrow can be exploited in tissue-engineered cultures and particularly as a source for bone marrow growth. The bone marrow achievement needs the sterilized aspiration of the marrow from the posterior iliac crest. The progenitor cells can be chosen and cultured as required [3]. Regeneration of bone needs four elements: a morphogenetic signal, sensitive host cells that will react to the signal, an effective transporter for this signal that can transport it to particular locations and then operate as a scaffold for the expansion of the sensitive host cells, and a feasible, well-vascularized host bed [3]. It is helpful to comprehend two ideas of bone regeneration for tissue engineering fabricates, particularly osteoinduction and osteoconduction, before deliberating the required characteristics of possible tissue engineering materials. Osteoconduction maintenance is growing of cells and capillaries from the host into a 3D construction to develop bone. Osteoinduction is stated as the capability to affect pluripotent cells, from non-osseous environs to distinguish into osteoblasts and chondrocytes, climaxing in bone development. Osteoconductive materials direct healing where usual curing will happen if left unprocessed. Osteoinductive materials allow healing in areas that would typically not repair if left unprocessed [3]. Materials exploited as BTE scaffolds can be rigid or injectable, the first needing an effective implantation technique. Accordingly, the fields of materials exploration can be mostly split into cellular and acellular, with overlapping drug delivery in both subjects [3].

The materials usually exploited in all methods are composites, polymers and ceramics. The polymers and ceramics are either absorbable or non-absorbable, and polymers can be synthesized materials or naturally derived materials. BTE systems have comprised titanium, dental plaster, hydroxyapatite, polylactide-polyethylene glycol, poly(lactide-co-glycolide), polylactide, calcium phosphates, fibrin, collagen composites and demineralized bone matrix [3]. Proliferation of osteoblasts is sensitive to mechanical stimuli like strain and surface topography. Energy of the surface could have a role in absorbing specific proteins to the material surface and, sequentially, this will influence the attraction of the cells to the material. Furthermore, the aspect ratio of a material affects the fibroblastic or bony differentiation of a tissue. Cells are most clearly irritable to surface energy, topography, and chemistry, and can distinguish even the slightest variations in topography. Thus, surface roughness, shape, and particle size influence phenotype, proliferation, and adhesion of cells. Such surface characteristics are specifically attractive when taking an absorbable material into account, because it is a dynamic material and always provides a new surface [3]. Obviously, the similar shape of pore may not be perfect for all probable applications since bones have a very special structure depending on their location and function. Size of the pore is also extremely cell-type individual. The most favorable ceramic pore size has been suggested 300–400 μm by Tsurga et al. [9] that maintains ectopic bone formation. Holmes also submitted that the best range of pore size is 200–400 μm and the mean human size of about 223 μm [10]. Table 2.1 presents a digest of some serious issues in the fabrication of a BTE material.

Synthetic materials which are injectable are finding attraction as BTE scaffolds. This is partly due to their negligibly aggressive implantation. Matrices which are synthetic have also been considered as acellular BTE materials [3].

Photocrosslinkable polyanhydrides are new materials that offer specific benefits in orthopedic uses. Primary mechanical investigations reveal that these polymers exhibit improved mechanical integrity. They absorb via surface destruction

Table 2.1 Serious issues in design of a BTE scaffold [3]

Required merits of a BTE scaffold	
Absorbable with biocompatible constituents	Suitable bony apposition
Capability to sterilize without loss of characteristics	Accurate physical and mechanical characteristics for usage
No harmful results to nearby tissues because of processing	Most bone growth owing to osteoconduction and/or osteoinduction
Average size of pores in the region of 200–400 μm	Adjustable to erratic wound site, flexible
Does not stimulate growth of soft tissues at interface of implant/bone	Absorbs in expectable way harmoniously with growth of bone
Stimulates bone ingrowth	Accessible to surgeon on short notice

and consequently are not sensitive to load discarding in delivery purposes or unexpected losses in mass. The photopolymerizable component not only increases the microfabrication capability of porous scaffolds, but may also permit an injectable material that can be consequently crosslinked [3]. Ceramic composites with polymeric reinforcements have also been explored. Composite systems can be designed using TCP to enhance mechanical integrity. The injectable nature of the materials permits them to fill abnormal bone defects, and the leachable components provide space for internal growth of bone. The polymerization of materials itself can be altered to trigger foaming during the emission of CO_2, thus constructing a porous scaffold. These specific materials also have capability as a drug delivery system.

The attraction of the calcium phosphates depends mostly on their biocompatibility. The inorganic, absorbable materials that have been studied consist of $Ca_3(PO_4)_2$(beta - whitlockite. a form of TCP), $CaSO_4.2H_2O$(plaster of Paris), and $CaCO_3$(argonite). The most extensively investigated calcium phosphate ceramics are HA with chemical formula of ($Ca_{10}(PO_4)_6(OH)_2$), TCP and the latest tetracalcium phosphate. Negligible systemic toxicity, foreign body responses or immunologic reactions have been stated with their application since they are free of protein. The inorganic ceramics naturally have a significant capacity to bind directly to bone as well as osteoconductive capability while they have not exposed osteoinductive capability. Hydroxyapatite is an extremely gently degradable material and hence can be formed into a suitable structure and manufactured as a bone flap which is vascularized. Almost at the time when Urist et al. [11] first produced the decalcified bone matrix, they suggested that its osteoinductive characteristics were initiated by a unique feature. Many years later, Strates and Urist nominated this feature bone morphogenetic protein (BMP) and, together with other colleagues, purified BMP of rabbits and rats [12–14]. From that time, numerous researchers have presented that BMPs are in fact a group of proteins involved in various actions in postnatal skeleton and embryogenesis.

BMPs are of specific attention to the subject of orthopedics as they may support the important functions presented in bone repair, osteoinduction, and embryological bone formation as well as the probability that they may support the substitution of the regular autogenous bone graft [3]. It is clear that BMPs could be precious in orthopedic reconstructive surgery while the application of BMPs may demonstrate to be essential to orthopedic operation as a whole. BMP has been transferred to located sites with the aim of bone defects healing and non-unions in various investigational models. Delivery systems have comprised titanium, dental plaster, hydroxyapatite, polylactide-polyethylene glycol, poly(lactide-co-glycolide), polylactide, calcium phosphate, fibrin, collagen composites, and demineralized bone matrix. A perfect delivery system should permit a prolonged release of the BMPs, be supportive of angiogenesis and cell proliferation, rapidly absorbed, and immunologically and biologically inert. It would also have adequate firmness to endure deforming energies until absorbed. In conclusion, it would be sterilized, handled, and simply stored [3].

For long bones like the tibia, the current treatment possibilities for large segmental defects are typically restricted to multi-staged amputation and/or reconstruction.

Some authors have shown that BMP 7 and BMP 2, delivered to experimental and clinical bone defects up to 17 cm in length, had a favorable and significant effect on their capability to repair these defects [3]. Researchers have investigated the probability of exploiting synthetic polymers as agents to permit the targeted delivery of BMPs, while selected synthetic polymers like PLLA can also cause design concerns because of extended absorption times or inflammatory responses [3]. The collagen materials also function as cellular scaffolding systems. Engineering adjustments could be effective to supply a firmer polymer to aid in the transmission of force from bone through the regenerative phase of curing as collagen has no intrinsic structural mechanical characteristics. Polyglycolide (PG) fibrous, non-woven mesh is a different tissue engineering contender. Polyglycolide mesh exhibits quite low mechanical integrity in vitro and, consequently, is unsuitable as a BTE construct. This material, which is clinically famous and has the benefit of rapid absorption, is operated in almost all subjects of tissue engineering study, containing bone. A stable construct can be designed by mixing these materials with reinforcing materials. This was formerly achieved by exploiting a PL solution to bond the meshes.

Investigators have lately assumed the complex assignment of restructuring an entire joint since joint is the convergence of various kinds of tissue, like bone. This demands co-culture with a custom polymer section for particular types of cells. Primary investigations exploited PL-stabilized PG mesh co-cultured with periosteal osteoblasts, tendon cells, and chondrocytes in a try to create a finger joint [15]. The distinct parts were stitched together with absorbable sew, and inserted subcutaneously into nude mice. The basic ideas can be directly employed to osteochondral defect healing while the ultimate purpose is complicated. New synthetic polymers are also extremely interesting as possible cellular scaffolds. These materials possess rigidity and high mechanical strength, and their compressive modulus varies between 10 and 60 MPa. Additionally, they benefit from more promising mass loss as they are absorbed by surface erosion. Clinically, it is always very interesting to progress techniques of bone renewal with the purpose of decreasing the surgical trauma and costs to the patient. The capability to exploit autogenous bone-forming cells connected to mechanically sound, substitutable scaffolds impregnated with a biologically active BMP would be perfect. This would tolerate implantation of the cellular construct with no supplementary surgical treatments to implant a structural device or harvest a graft. As an agreement, a comparable cellular construct could be exploited alongside typical fixation procedures. Both methods would be dependent on a practical bed of host tissue, so there is not any dense avascular scar tissue or local tissue necrosis [3].

Sol–gel treating is another attractive method that may merge organic/inorganic constituents at the scale of nano (for instance constructing a lattice from inorganic silica chains and biological or synthetic polymers) [16]. However, it is difficult to reconstruct the same degree of nanoscale order obtained in vivo with the composition of organic and mineral components [1]. The synergistic arrangement of cell therapy and biomaterials is very interesting. In fact, since the 1980s, the capability of MSCs in bone regeneration has been underlined [17]. BTE operating cells varying from bone marrow MSCs to primary mature osteoblasts (bone cells) and biomaterials have obtained several accomplishments in animal models. Most of the research, however, is done in rodents and very few describe orthopedic uses (for instance in a bone defect) in bigger animals. In spite of some medical achievements, translation into human applications suffer from the inadequate prognostic capability for experimental results of the ectopic model in rodents [1]. Vascularization of clinically sized tissue-engineered fabricates stays a limitation in both animal to human system transmission and in vitro to in vivo tissue engineering transmission. By advancing well-controlled and favorably interconnected porosity into the scaffolding of the material, it is possible to facilitate the diffusion of nutrients and oxygen and the following permeability, along with the formation of a three-dimensional (3D) vascular network. Further approaches for "prevascularization" in vitro are developing. For instance, a 3D multicultural system including endothelial cells (cells that cover blood vessels), differentiated mature cells, and progenitor cells can produce systemic endothelial blood vessels through engineered tissue fabricates, as lately established for artificial muscle implants [18]. Exogenous administration of effective angiogenic agents like vascular endothelial growth factor (VEGF) may also promote the formation of blood vessels. Nevertheless, use of such effective biologics and actually other growth factors involved in bone formation, including BMPs (for example BMP 2 and BMP 7, which are now in medical usage), is not without disadvantages. Exogenous administration of BMPs proceeds the possibility of triggering heterotopic ossification (for example establishment of bone outside the skeleton) and is expensive. Gene therapy methods propose the probability of localized and sustained gene expression from transgenic cells, but the delivery carriers require further development. Different growth factor delivery approaches might aid in getting control of currently suboptimal release kinetics, as well as the necessity for non-physiologically high growth factor concentrations and their challenging small half-life. It should be noted that these methods do not accurately replicate the complicated temporal sequence and mixture of growth factors required for stable and safe formation of bone in vivo [1].

2.2 Bone Biology

At the tissue level, bone has the capacity to withstand deformation from impact loading, while concurrently demonstrating the ability to absorb or disperse energy through shape alteration without experiencing fracture. Bone possesses elastic qualities that enable it to absorb energy through reversible deformation under load (elastic deformation). Fatigue damage can occur as a result of repetitive loading at physiological levels. To mitigate this damage, it is necessary to have a damage detection mechanism, such as osteocytes, and a healing function, such as remodeling, that can limit its accumulation and prevent fatigue-related failure [19]. On the organ level, bones can be classified into four separate envelopes based on their anatomical location: subperiosteal, endocortical, cancellous, and intracortical. The aforementioned division indicates that these envelopes possess distinct morphology. They have distinct functions in bone biology throughout both healthy and diseased states, and vary in their reaction to alterations in the mechanical surroundings [19]. The process of modeling involves the shaping and adaption of bones to various loading or external forces. Modeling can lead to alterations in morphology, size and mass of bones. The influence of cortical modeling on cortical thickness and bone diameter is observed at the endosteal or periosteal surfaces. Remodeling refers to the ongoing and perpetual process of bone repair and renewal that persists over the entirety of an individual's adult lifespan. The rate of bone turnover is elevated throughout the process of bone modeling and acquisition during puberty and growth, but thereafter declines to its lowest point around the age 40. Menopausal women see a significant surge in turnover, while men undergo a more gradual increase in turnover that is associated with aging [20].

Remodeling of bone is enhanced during middle age of adults and occurs in four steps: (1) activation and employment of osteoprogenitor cells (activation), (2) osteoprogenitor cell resorption by osteoclasts (resorption), (3) temporary phase from resorption of bone to formation of bone (reversal), and (4) synthesis of matrix by osteoblasts (formation) [21, 22]. In summary, through resorption of bone, osteoclasts are performed by eliminating packets of "old bone"; then, a new synthetic matrix is generated, in company with the mineralized tissue. Osteoblasts (engaged in formation of bone) and osteoclasts (the promoters of resorption of bone) are the major intermediary of this procedure. Degradation and formation of bone are closely kept in equilibrium during life of humans by remodeling and homeostasis of the bone. The basic multicellular unit (BMU) is responsible for maintaining this balance, which consists of blood vessels, nerves, connective tissues, osteoblasts and osteoclasts [21, 23].

2.3 Bone Physiology

Bone physiology comprises the harmonized adaptation of a multitude of biological procedures that direct repair, homeostasis, and tissue development upon trauma. Bone re-forming progressions can consequently be very complex to imitate, partly because of the multifactorial and various signs influencing the adjustment of its ideal position. Those contain the complicated constitution of tissue and complex related paths including viewpoints such as the function of microRNAs, mechanical stimulation of cell-ECM and cell–cell interactions, configuration and regeneration of insoluble proteins in the extracellular matrix (ECM) and glycoprotein, or soluble microenvironment of bone. The comprehension of the coordinated and particular function of all interventions on physiology of bone tissue, along with their interconnected activities, has motivated the creation of an excess of bioinspired and/or biomimetic bone regeneration methods [23]. However, utilizing principles derived from nature to replicate the physiology and structure of natural bone needs precise and insightful methods. The implementation of efficient reproducible treatments is most often reliant on attaining an equilibrium among appropriate complication to guarantee perform, in the company of regulatory compliance and speed/comfort of fabrication while imitating function and structure of bone based on its physiology is an appealing conception [23].

Remarkably, though, an increasing body of knowledge including both applied and fundamental investigations concentrating on responses of bone to engineered signs have definitely evidenced that bone is a dynamic and complex structure, in which various structural properties and biological processes play supplementary roles to the effective maintenance and regeneration of healthy performance of bone [23]. Some investigations have investigated various factors related to bone renewal, such as the influence of physical features like viscoelasticity, stiffness and bioactivity of biomaterials, the combinatorial function of ECM proteins, interaction between various types of cells (comprising vascular cells, immune cells, reabsorption cells, and mineral-forming cells), and the influence of external mechanical forces acting on the natural bone (such as shear stress and compression). While several fundamental features of bone physiology have been extensively researched and have directed the progress of treatments, there are still numerous sides that are not well known and not been thoroughly investigated as potential added-value assets to improve renewal. Various instances comprise the impact of hematopoietic stem cells (HSCs) on reformation processes, the interaction between immune cells and bone resident cells, and the precise regulation of coordinated biophysical and biochemical signaling in terms of location and time [24].

Human bones can be categorized in five major classes, which contain flat (aimed at protection of inner organs; for instance, scapulae, ribs, mandible, sternum and skull), short (presenting stability and movement; for instance, carpal and tarsal), long (targeted at maintaining body weight; for instance, ulnae, fibulae, metatarsals, humeri, femurs, phalanges, tibiae, metacarpals, raddi, and clavicles), sesamoid bones (embedded at tendons; for instance, patella) and irregular (for instance hyoid, sacrum, coccyx and vertebrae) [25]. Through formation of tissue, two forms of bone can be distinguished: (1) the primary/woven bone, which present through the fracture repair and embryonic development. It is made up of collagen fibers with osteoid (non-mineralized extracellular matrix) that demonstrate minor or non-determined orientation in 3D space, accompanied by an accidental distribution of cells. The structure is temporary and is eventually substituted by fully developed lamellar bone; (2) the secondary/lamellar bone, which is composed of well-structured layers of mineralized osteon and comprises the adult skeleton. This construction exhibits reduced elasticity, greater strength and rigidity compared to the woven bone [26–28].

Bone tissues are organized either in a trabecular configuration (cancellous bone) or a compact configuration (cortical bone) [1]. Lamellar bones are found by both cancellous/trabecular and compact/cortical bone. The first one is found in the inner part of bone and possesses a porous construction with interconnected spaces. In contrast, cortical bone is positioned in the outermost layer of the tissue and is characterized by its solid and dense structure. Both cancellous and cortical bones consist of osteons. The proportion of both kinds of bone differs in accordance with the anatomical site (for instance radial diaphysis with 95:5; vertebra with 25:75; femoral head with 50:50; and generally, the human skeleton with an 80:20 cortical to trabecular ratio) [21]. From a structural outlook, the trabecular bones consist of spacious areas, with a honeycomb like arrangement of trabecular sheets. The matrix is composed of a three-dimensional network of thin columns that interconnect to create the trabeculae. This leads to the formation of a bone that is porous and light, which makes it resistant to multi-directional forces and is important for allowing the body to move. Bone marrow has filled the spaces between trabeculae. The outer surface of cortical bone is enveloped by a layer of fibrous connective tissue known as periosteum. Meanwhile the interior section is covered by a membrane structure known as endosteum, which extends over both trabecular and cortical bone surfaces. Both endosteum and periosteum have stimulated the progress of two-dimensional membranes to imitate these anatomical structures and support processes of tissue renewal. The latter structure is likewise in direct communication with the bone marrow cavities, vascular channels and is composed of osteoclasts, osteoblasts and blood vessels.

In cortical bones, lamellae with width of approximately 3 μm form concentrically and surround a vertical Haversian canal, which comprises nerves and blood vessels. This complete structure is the performing unit of bone and is defined as the Haversian system or the osteon. The system is created through the process of osteoclasts reconstructing preexisting bone, resulting in tubular cavities. These cavities are then filled by osteoblasts, which generate lamellae around the Haversian canal [23]. From a nanoscale viewpoint, bone is composed of polysaccharides and structural proteins, which collagen fibrils with up to 1 μm length, and diameters in range of 35 to 60 nm are key components, and prearranged with 40 nm gaps and a 67 nm periodicity. These fibrils are mineralized by HAP crystals, which are anisotropic and very firm inorganic constituents placed in the gaps of collagen. The simulation of the osteon configuration is reliant on a nanometer limitation of material distribution while the micro- and macrostructure of bone carefully repeated operating porous scaffolds. Actually, the advancement of biomaterials that accurately mimic the structure of bones relies on an exact arrangement of nano- and micrometric features of the bone structure. The task of nanostructured biomaterials as sign suppliers directing bone renewal should not be neglected. Platforms displaying, for instance, pro-mineralization, pre-differentiation and topographic signs through methods as advance of nanocomposites, electrospinning and nanopatterning had given rise to some of the most attractive yields in the subject. Through natural bone healing upon injury and fetal progress of mammals, bone development is attained via two procedures: endochondral and intramembranous ossification. The initial structure of these two processes is the immature hollow or woven bone, which is promptly substituted by secondary/lamellar bone (parallel fibrils lay down in reverse directions). This structure is not limited to embryonic life, but occurs whenever a bone experiences an insignificant damage. The arrangement of woven bone happens at a much faster rate than that of lamellar bone [23].

2.4 Osteogenesis

The process of osteogenesis initiates when chondrocytes that are undergoing proliferation within the template transition into a non-proliferative, hypertrophic condition. These hypertrophic chondrocytes release substances that promote bone formation and blood vessel development, such as alkaline phosphatase (ALP) and vascular endothelial growth factor (VEGF). Simultaneously, the tissue undergoes invasion by vasculature, which facilitates the transportation of hematopoietic cells, endothelial cells, osteoclasts, and MSCs to the diaphysis, resulting in the formation of a primary ossification center (POC). Osteoclasts are thought to eliminate hypertrophic chondrocytes that are experiencing programmed cell death from the template, according to traditional beliefs. MSCs subsequently undergo differentiation into osteoblasts in order to generate bone matrices. The presence of supplementary blood vessels in close proximity to the ends of the bone will give rise to one or more secondary ossification centers (SOCs), which are involved in the development of the articular cartilage and bony epiphyses [29].

Osteoblasts deposit bone in close proximity to infiltrating blood vessels, resulting in the development of a bone collar. The transportation of additional osteoblasts into the central region of the condensation is facilitated by the blood vessels. These osteoblasts then deposit bone onto the mineralized ECM "scaffold" that is generated by hypertrophying chondrocytes. The deposition of bone occurs in a wave-like manner, away from the central region of the diaphysis toward the periphery (cortex) and ends (epiphyses) of the long bone. The process of bone ossification involves the convergence of primary and secondary ossification centers, resulting in the formation of growth plates (physes). These growth plates play a crucial role in driving differential limb bone growth and limb bone elongation [30].

2.5 Bone Development Mechanisms

At the organ level, bone is composed of the mineralized cortical and trabecular bone structures, the marrow space, calcified cartilage of the growth plate (during skeletal growth only), and cartilaginous joints (Fig. 2.1) [19]. Differentiation of osteoblasts can take place through two separate mechanisms during embryonic advancement. The mesenchymal progenitor cells undergo direct differentiation into osteoblasts in the condensations anticipating the mandibles, the clavicles, and particular skull bones. The procedure referred to as intramembranous ossification does not include cartilaginous patterns. Conversely, each future skeleton, with the exception of the mandibles, the clavicles and particular skull bones, is prefigured using a cartilage pattern encased by a bone collar for the complete future bones. When the cartilage template is invaded by vascularization, the chondrocytes, which are the cells composing the cartilage, undergo apoptosis and are subsequently substituted by

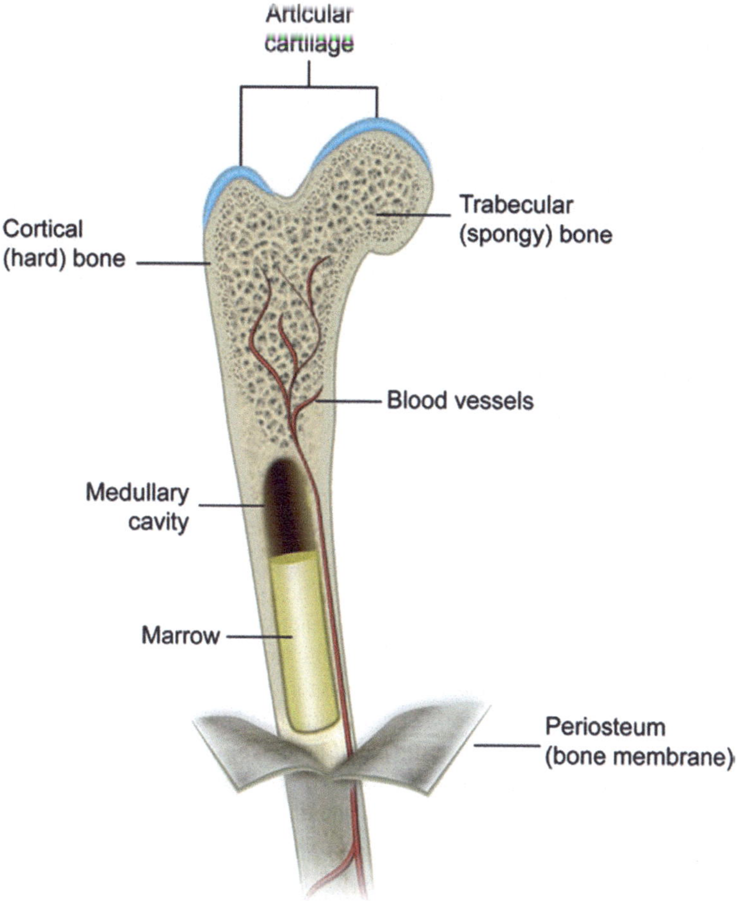

Fig. 2.1 Bone envelopes: On the macroscopic level, long bones consist of the marrow cavity, porous cancellous bone, the cartilaginous joints, and the more compact, dense cortical bone. The subperiosteal surface of long bones is enveloped by a thin fibrocellular membrane known as the periosteum, which is highly innervated and vascularized. Reprinted with permission from Gasser and Kneissel [19]

osteoblasts that migrate from the bone collar. The term used to describe this procedure is endochondral ossification [31].

2.5.1 Intramembranous Bone Formation

Intramembranous bone regeneration happens in several instances of skeletal repair. It is crucial in various clinical situations, including stress fracture healing, allograft implantation, distraction osteogenesis, dental implant surgery, joint replacement, fracture healing, and healing of skeletal abnormalities resulting from trauma or tumor removal. After undergoing dental implant surgery and joint replacement surgery, the implant's correct attachment to the surrounding tissue is ensured through osseointegration, which is facilitated by the intramembranous ossification process. During the healing of long bone fractures, a coordinated series of events occur where endochondral ossification takes place at the central callus and intramembranous ossification by the periosteum. These processes work together to ensure correct union at the location of the fracture [32].

The regeneration of intramembranous bones has been thoroughly investigated through the use of animal models. These studies indicate that many biological processes mimic the events that take place during the embryonic development of membranous bones. Furthermore, multiple potential skeletal stem cell populations have been identified, which contribute to the process of intramembranous bone regeneration. Researchers have investigated the molecular mechanisms underlying the process of intramembranous bone regeneration by studying rodent models that mimic most of these circumstances. These findings demonstrate that signaling pathways including Notch, VEGF, FGF, TGF-β/BMP, and Wnt are activated, reminiscent of embryonic formation of membranous bone. The use of mouse genetic models has allowed for the discovery of different types of skeletal stem cells/progenitors. This has also given us valuable information on the possible sources of postnatal intramembranous bone repair [32]. Recent research suggests that the strong distinction previously made between intramembranous bone regeneration and endochondral may not be as definitive as previously believed. In the calvarial defect model, the use of scaffolds or cells to promote bone regeneration can lead to the repair of the defect location through the mechanism of endochondral ossification. In the models of stress fracture healing and marrow ablation, there is a temporary upsurge in the expression of certain "cartilage" genes, such as aggrecan, preceding the up-regulation of genes related to bone formation. However, no cartilaginous template is observed histologically [32].

In terms of evolutionary classification, intramembranous bones can be classified into two categories: dermal bones and membrane bones. These two types are considered components of the exoskeleton and endoskeleton, respectively. In some cases, dermal bones can arise through metaplasia in fibrous connective tissue that is devoid of osteoblasts. However, intramembranous bones are usually generated through the differentiation of osteoblasts. The presence of RUNX2 in the osteoblast progenitor is essential for the onset of osteoblast differentiation. Subsequently, the fundamental group of factors involved in the differentiation of osteoblasts, namely Sp7, Col I, Bglap (formerly referred to as osteocalcin) and Spp1 (formerly referred to as osteopontin) are expressed in a sequential manner. The expression of late markers, such as dentin matrix protein (Dmp)1, is initiated by osteoblasts when they become enclosed by matrix. Eventually, they exhibit the osteocyte marker sclerostin, indicating their maturity as osteocytes. The promotion of osteoblast differentiation during the early stages is facilitated by RUNX2, and its role is of utmost importance in the first progenitors and Sp7 expressing cells [33]. Intramembranous bones do not utilize chondrocyte precursors, unlike endochondral bones where osteoblasts develop within and around a cartilaginous template. During the initiation phase in chicks and mice, the osteoblasts express several cartilage markers, including Col II and Sox9. As the differentiation process takes place, they also produce Col II and IX mRNAs. The simultaneous expression of cartilage markers in the intramembranous bones is observed in less advanced vertebrates such as zebrafish and frogs. In these species, the expression also includes the marker for hypertrophic cartilage, type X collagen (Col10a1) [33].

2.5.2 Endochondral Ossification

Conventional approaches in bone tissue engineering (BTE) involve the emulation of the embryological mechanism of intramembranous ossification to facilitate direct development of a matrix resembling bone. Nevertheless, the implementation of these clinical techniques for bone healing is hindered by inadequate bone regeneration and restricted vascularization following in vivo implantation. One potential approach to address these limitations involves the development of cartilaginous constructs through the replication of embryonic processes involved in endochondral ossification. These constructs demonstrated a distinctive capacity to endure hypoxic environments while also promoting ossification and neovascularization [29]. The process of bone formation, referred to as osteogenesis, and the elongation of long bones commence during the endochondral ossification (the second phase of long bone growth), which persists from the fetal to postnatal stages until skeletal maturity [30]. The endochondral ossification process involves various cellular mechanisms that have the ability to regulate the proportions and lengths of long bones throughout development, within a species (such as phalanx vs. tibia), among individuals (with variation at the population level), and eventually between species (evolutionary allometry). Both temporal and spatial variation in the dynamics and structure of the condensations, as well as the growth plates and their resident chondrocytes, are involved in these mechanisms [30]. The dimensions of the Anlage before the initiation of endochondral ossification differ depending on the bone, resulting in much smaller condensations in a phalanx or metacarpal compared to the more proximal limb bones. During endochondral ossification, which is sometimes referred to as allometric growth, the development of bones may differ in terms of both extent and rapidity. However, even before ossification occurs, the fetal and embryonic limbs already exhibit proportions that are similar to those seen in mature and postnatal stages [30]. The developmental processes that contribute to variations in proportions may undergo changes over long periods of evolution as a function of the strength of selection for differences in shape and size of limb bones. More pronounced differences in limb proportions involve the accumulation of evolutionary changes in multiple cellular factors of endochondral ossification [30]. The longitudinal growth driven by chondrocyte hypertrophy process initiates with endochondral ossification itself, building upon the events that occur during condensation and development of hypertrophic and proliferative chondrocyte populations [30].

2.6 Conclusion

Bone tissue engineering may potentially offer an alternative option with mechanical qualities that surpass those now in use. This might mitigate vascular damage caused by the implant to the bone and minimize the occurrence of stress shielding, potentially lowering the risk of implant-related osteopenia and subsequent fractures. Biocompatibility, biodegradability, and bioactivity are the primary attributes necessary for an optimal tissue healing material. Bioceramics possess societal significance and a promising market potential in the production of a dependable and efficient material for repairing bone defects. However, in order to create a suitable implant for bone tissue engineering, it is essential to possess fundamental knowledge regarding the host tissue. Bone is a complex living connective tissue that provides the body with flexibility, mechanical support, and structural framework. It also plays a role in regulating blood pH, maintaining homeostasis, and storing minerals. The primary causes of bone flaws include surgical procedures, fractures, trauma, congenital malformations, tumor removal, as well as diseases like arthritis and osteoporosis. There are multiple reasons for conducting experiments to advance bone tissue engineering substitutes. These include the need for orthopedic implants that are better suited to the mechanical requirements of the body, as well as the need for improved materials to fill large orthopedic defects. Bone physiology encompasses the coordinated adaptation of numerous biological processes that govern the repair, maintenance of balance, and growth of tissue following injury. Bone homeostasis is maintained by three types of cells: osteoblasts, osteoclasts, and osteocytes. At the tissue level, bone is capable of withstanding deformation caused by impact loads, while also being able to absorb or distribute energy through shape alteration without fracturing. Bones can be categorized into four distinct layers based on their anatomical position at the organ level: subperiosteal, endocortical, cancellous, and intracortical. The process of bone remodeling is intensified during the middle age of adults and consists of four distinct stages: (1) activation, (2) resorption, (3) reversal, and (4) creation.

References

1. Stevens, M. M. (2008). Biomaterials for bone tissue engineering. *Materials Today, 11*(5), 18–25.
2. Kenley, R. A., et al. (1993). Biotechnology and bone graft substitutes. *Pharmaceutical Research, 10*(10), 1393–1401.
3. Burg, K. J., Porter, S., & Kellam, J. F. (2000). Biomaterial developments for bone tissue engineering. *Biomaterials, 21*(23), 2347–2359.
4. Kashte, S., Jaiswal, A. K., & Kadam, S. (2017). Artificial bone via bone tissue engineering: Current scenario and challenges. *Tissue Engineering Regenerative Medicine, 14*(1), 1–14.
5. Ma, H., et al. (2016). 3D printing of biomaterials with mussel-inspired nanostructures for tumor therapy and tissue regeneration. *Biomaterials, 111*, 138–148.
6. Yang, C., et al. (2017). 3D-printed bioactive Ca_3SiO_5 bone cement scaffolds with nano surface structure for bone regeneration. *ACS Applied Materials Interfaces, 9*(7), 5757–5767.

7. Liu, A., et al. (2016). Three-dimensional printing akermanite porous scaffolds for load-bearing bone defect repair: An investigation of osteogenic capability and mechanical evolution. *Journal of Biomaterials Applications, 31*(5), 650–660.
8. Wen, Y., et al. (2017). 3D printed porous ceramic scaffolds for bone tissue engineering: A review. *Biomaterials Science, 5*(9), 1690–1698.
9. Tsuruga, E., et al. (1997). Pore size of porous hydroxyapatite as the cell-substratum controls BMP-induced osteogenesis. *The Journal of Biochemistry, 121*(2), 317–324.
10. Holmes, R. E. (1979). Bone regeneration within a coralline hydroxyapatite implant. *Plastic Reconstructive Surgery, 63*(5), 626–633.
11. Urist, M. R., et al. (1967). 24 the bone induction principle. *Clinical Orthopaedics Related Research, 53*, 243–284.
12. Urist, M. R., Mikulski, A., & Lietze, A. (1979). Solubilized and insolubilized bone morphogenetic protein. *Proceedings of the National Academy of Sciences, 76*(4), 1828–1832.
13. Urist, M. R., & Mikulski, A. J. (1979). A soluble bone morphogenetic protein extracted from bone matrix with a mixed aqueous and nonaqueous solvent. *Proceedings of the Society for Experimental Biology Medicine, 162*(1), 48–53.
14. Urist, M. R., & Strates, B. S. (1971). Bone morphogenetic protein. *Journal of Dental Research, 50*(6), 1392–1406.
15. Isogai, N., et al. (1999). Formation of phalanges and small joints by tissue-engineering. *JBJS, 81*(3), 306–316.
16. Pereira, M. M., et al. (2005). Preparation of bioactive glass-polyvinyl alcohol hybrid foams by the sol-gel method. *Journal of Materials Science: Materials in Medicine, 16*(11), 1045–1050.
17. Friedenstein, A., Chailakhyan, R., & Gerasimov, U. (1987). Bone marrow osteogenic stem cells: In vitro cultivation and transplantation in diffusion chambers. *Cell Proliferation, 20*(3), 263–272.
18. Levenberg, S., et al. (2005). Engineering vascularized skeletal muscle tissue. *Nature Biotechnology, 23*(7), 879–884.
19. Gasser, J. A., & Kneissel, M. (2017). Bone physiology and biology. In *Bone toxicology* (pp. 27–94).
20. Walsh, J. S. (2015). Normal bone physiology, remodelling and its hormonal regulation. *Surgery, 33*(1), 1–6.
21. Clarke, B. (2008). Normal bone anatomy and physiology. *Clinical Journal of the American Society of Nephrology, 3*(Supplement 3), S131–S139.
22. Raggatt, L. J., & Partridge, N. C. (2010). Cellular and molecular mechanisms of bone remodeling. *Journal of Biological Chemistry, 285*(33), 25103–25108.
23. Lopes, D., et al. (2018). Bone physiology as inspiration for tissue regenerative therapies. *Biomaterials, 185*, 240–275.
24. Yin, T., & Li, L. (2006). The stem cell niches in bone. *The Journal of Clinical Investigation, 116*(5), 1195–1201.
25. Standring, S. (2015). *Gray's anatomy e-book: The anatomical basis of clinical practice*. Elsevier Health Sciences.
26. Tzelepi, V., et al. (2009). Bone anatomy, physiology and function. *Bone metastases* (pp. 3–30). Springer.
27. Shapiro, F. (2008). Bone development and its relation to fracture repair. The role of mesenchymal osteoblasts and surface osteoblasts. *European Cells and Materials, 15*(53), e76.
28. Reznikov, N., Shahar, R., & Weiner, S. (2014). Bone hierarchical structure in three dimensions. *Acta Biomaterialia, 10*(9), 3815–3826.
29. Fu, R., et al. (2021). Bone defect reconstruction via endochondral ossification: A developmental engineering strategy. *Journal of Tissue Engineering, 12*, 20417314211004212.
30. Rolian, C. (2020). Endochondral ossification and the evolution of limb proportions. *Wiley Interdisciplinary Reviews: Developmental Biology, 9*(4), e373.
31. Ducy, P., Schinke, T., & Karsenty, G. (2000). The osteoblast: A sophisticated fibroblast under central surveillance. *Science, 289*(5484), 1501–1504.

32. Ko, F. C., & Sumner, D. R. (2021). How faithfully does intramembranous bone regeneration recapitulate embryonic skeletal development? *Developmental Dynamics, 250*(3), 377–392.
33. Galea, G. L., et al. (2021). Making and shaping endochondral and intramembranous bones. *Developmental Dynamics, 250*(3), 414–449.

Chapter 3
Bone Cells

Abstract The process of bone formation is crucial for different processes including bone remodeling, bone mineralization, and the longitudinal growth of bones. The composition of bone includes progenitor cells, osteoblasts, osteoclasts, and osteocytes. The presence of contaminant blood-derived cells, endothelial cells, and pericytes should not be disregarded as potential sources of osteoprogenitor cells. Osteoblasts have a vital function in the synthesis of extracellular matrix components. Osteoclasts are derived from the macrophage/monocyte lineage. Osteocytes are unique cells that express genes from both the hematopoietic and mesenchymal lineages. To effectively employ osteoprogenitor cells for novel treatments, it is imperative to actively extract them from tissue samples, specifically from skeletal muscles. Furthermore, several transcription factors are involved in the processes of differentiation and inhibition of bone cells. Programmed cell death also has a crucial function in cellular processes. In order to design appropriate implants, it is crucial to know the reasons and mechanisms of cell death. The interactions between bone tissue engineering materials and these cells should be considered. A comprehensive understanding of bone components is invaluable when designing and producing bone tissue engineering devices. This chapter will comprehensively consider these cells and the mechanisms that regulate them.

Abbreviations

AR	Androgen receptor
As	Androgens
BMPs	Bone morphogenetic proteins
BRUs	Bone remodeling units
Cbfa1	Core binding factor α1
CLPs	Common lymphoid progenitors
CMPs	Common myeloid progenitors
CSF1R, also identified as c-FMS	Colony-stimulating factor 1 receptor
DAP12	Dnax activation protein 12

R. Gholami and S. M. Naghib, *Nanobioceramics for Bone Tissue Engineering and Regenerative Biomedicine*, Biological and Medical Physics, Biomedical Engineering,
https://doi.org/10.1007/978-981-96-0041-0_3

DPSCs	Dental pulp stem cells
EGF	Epidermal growth factor
ER	Estrogen receptor
ERK	Extracellular signal-regulated kinase
Es	Estrogens
ESCs	Embryonic stem cells
FcRγ	Fc receptor common gamma subunit
GCs	Glucocorticoids
HSCs	Hematopoietic stem cells
IFNs	Interferons
IL	Interleukin
JAK	Janus kinases
L-arg	L-arginine
Macs	Macrophages
MCP	Monocyte chemotactic proteins
M-CSF	Macrophage colony-stimulating factor
MEPs	Megakaryocyte–erythrocyte progenitors
MPPs	Multipotent progenitors
M@@SCs	Mesenchymal stem cells
MSTN	Myostatin
NF-κB	Nuclear factor Kappa-B
NO	Nitric oxide
NOS	Nitric oxide synthase
OCPs	Osteoclast precursor cells
OCs	Osteoclasts
OCy	Osteocytes
OP/OA	Osteoporotic/Osteoarthritic
OPG	Osteoprotegerin
Osx	Osterix
PDB	Paget's disease of bone
PFF	Pulsatile fluid flow
PI3K	Phosphatidylinositol 3-kinase
PTHrP	Parathyroid hormone-related protein
RA	Rheumatoid arthritis
RANKL	Receptor activator of NF-κB ligand
S1P	Sphingosine-1-phosphate
SDF-1	Stroma-derived factor 1
sema3A	Semaphorin-3A
SSCs	Skeletal stem cells
TGF-α	Transforming growth factor alpha
TGF-β	Transforming growth factor beta
TNFs	Tumor necrosis factors
TRAIL	TNF-related apoptosis-inducing ligand

3.1 Introduction

Cells are naturally responsive to their environments. Topographic response of cells (for instance, reaction to situation of the surface) to wells, ridges, grooves, and other topographies at the scale of micron and, lately, the nanoscale is now well recognized. Current reports are exposing impacts on behavior of the cells varying from intracellular signaling pathway modulations that control gene expression and transcriptional activity to alterations in cell adhesion. The affiliation between nanoscale topographic properties and cell behavior and protein adhesion is complicated and stays to be clarified in full, changing according to the size and shape of the topographic characteristics, along with the type of cell and protein [1]. Bone homeostasis provided by osteoblasts, osteoclasts and osteocytes (Fig. 3.1). Osteoblasts are initiated from mesenchymal cells, whereas osteocytes are mature osteoblasts and osteoclasts are of hematopoietic source. Between all existing cell resources, embryonic stem cells (ESCs) are extensively investigated for BTE containing differentiation into osteoblasts. Co-culturing of ESCs with fetal fibroblast has exposed improved development of bone nodules. Formation of teratoma, however, restricts medical usages of ESCs [2].

Bone tissue endures an endless cycle of osteoblastic formation of bone and osteoclastic resorption of bone with the purpose of being able to perform appropriately over a lifetime [4–8]. In the context of bone fracture curing, osteoclasts are involved in the resorption of small bone fragments immediately following a crack, as well as the immature matrix of bone during the later stages of curing, while osteoblasts are responsible for the synthesis of bone matrix to facilitate the fixing of bone fragments [9–13]. MSCs are identified to differentiate into mature cells such as chondrocytes, chondroblasts, and osteoblasts on external chemical stimuli. MSCs segregated from adipose tissue, peripheral blood, and bone marrow have been differentiated to chondrocytes, osteoblasts, and cured critical sized bone defects in vivo [14]. Conventionally, the field of material design has mostly focused on the characteristics of bulk tissue, neglecting the comprehensive range of biological length-scale topography that is acknowledged to impact cell function, which spans from 100 μm to 10 nm. Nanophase supports (including carbon nanotube polymer nanocomposites or HA-collagen nanocomposites) are previously causing enhancements in mechanical features and bioactivity like compressive and flexural moduli [1]. Responses of the cells are consistently tailored to distinct binding affinities and ligand surface densities. They are often biphasic such as migratory response, and are influenced by co-localization of ligands that have synergistic effects [1]. Another method to produce a biomimetically boosted situation is to reform the topographical circumstance of natural ECM via engineered 3D nanofibrous matrices [15, 16].

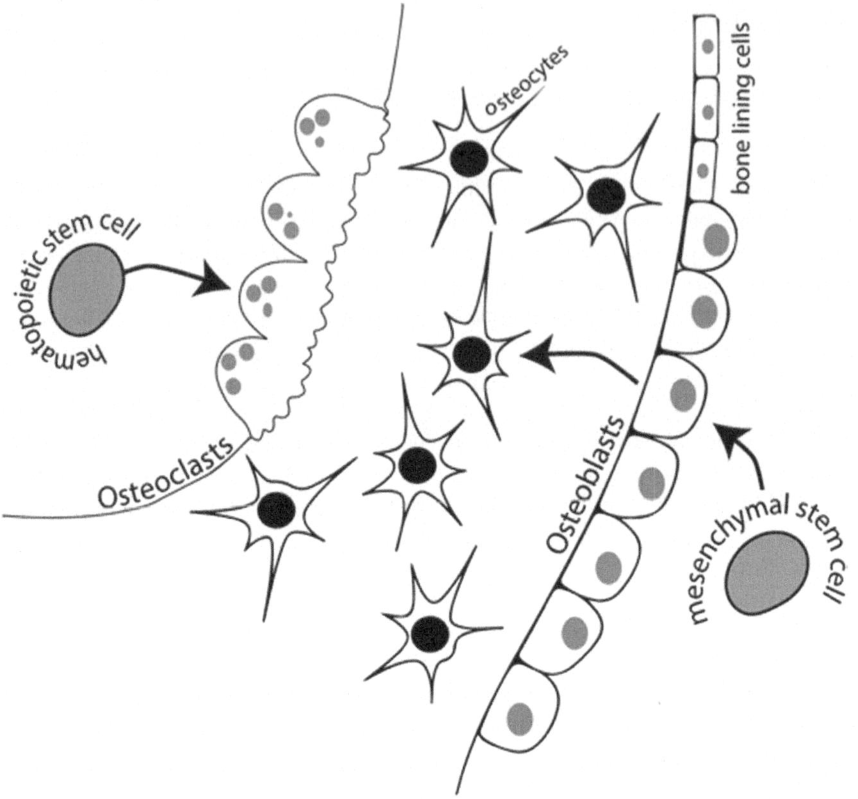

Fig. 3.1 Cell localization within bone tissue. While osteoblasts are responsible for the formation of new bone, osteoclasts are accountable for the resorption of bone. These processes, known as bone modeling and remodeling, are essential for maintaining bone health. Reprinted with permission from Renthal and Ma [3]

3.2 Osteoprogenitors

Extra-skeletal formation of bone is a well-explained cascade: MSCs are attracted to certain locations by chemical signals, proliferate in response to induced signals, differentiate into cartilage, blood vessels grow into the area, and ultimately differentiate into bone [17–19]. Nevertheless, the osteoprogenitor cell sources for the freshly growing bone are hazy [20–26]. Most data indicate that bone marrow stromal cells are the initial source of osteoprogenitors in the body [20, 27–36]. Different sources for osteoprogenitor cells involve muscle cells, pericytes, fibroblasts, endothelial cells, epithelial cells, and blood-borne lymphoid cells [24, 26, 36]. Urist and colleagues assumed that the endomysium was a critical resource of mesenchymal cells that act as osteoprogenitors [37, 38].

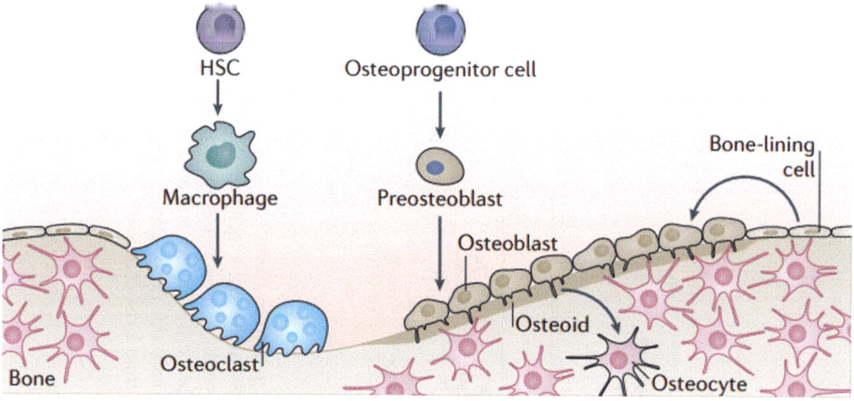

Fig. 3.2 Bone homeostasis. Regulation of bone stability is maintained by the functioning of cells in the osteoblast and osteoclast lineages. Osteoblast lineage cells, like the osteoid, produce calcium and HA to facilitate mineralization of bone and formation of osteocytes. Osteoclast lineage cells are responsible for the resorption of bone. Bone hemostasis is achieved through the equilibrium of the activity of the two cell lineages. HSC: Hematopoietic stem cells. Reprinted with permission from Salhotra et al. [39]

The process of osteoblasts specializing in the skeletal lineage can be categorized into three stages of progressive differentiation: osteoprogenitors, pre-osteoblasts, and osteoblasts (Fig. 3.2). The initial step in the commitment of osteoprogenitor cells is the transcription factor SOX9 expression. Expression of SOX9 also guides differentiation of cells toward a chondrocyte cell fate. The subsequent expression of Runx2 in the osteoprogenitor cells indicates the dedication to pre-osteoblasts [39].

Skeletal stem cells (SSCs) and mesenchymal stem cells (MSCs) have individually demonstrated the ability to differentiate into osteoblasts. The identification and separation of SSCs in humans and mice based on unique immunophenotypic cell-surface markers is a recently developed idea (Fig. 3.3) [39].

Contaminant blood-derived cells, endothelial cells, and pericytes cannot be ignored as the osteoprogenitor cell sources. Eventually, in order to utilize osteoprogenitor cells for new therapies, it is necessary to isolate them directly from tissue samples, particularly from skeletal muscles [36]. The osteoprogenitor stromal and hemopoietic systems are mutually interdependent. Osteoblasts release granulocyte stimulating factors and monocytes, among others, which increase production of blood cells. On the contrary, the blood cell monocyte and osteoclast system release factors from bone matrix that promote osteogenesis and are essential for preserving trabecular and cortical structures [40]. On additional examination, it became apparent that the capability to make osteoprogenitor clusters when marrow is cultured in vitro differs from individual to individual and from species to species [41, 42]. The capability to make osteoblast-like clusters in vitro can be influenced by the existence or lack of factors like platelet derived growth factor and by the marrow preparation methods, and specifically by blood loss [40]. Haynesworth and coworkers [43] and Bruder and coworkers [44] displayed that osteoprogenitors or MSCs can be enlarged

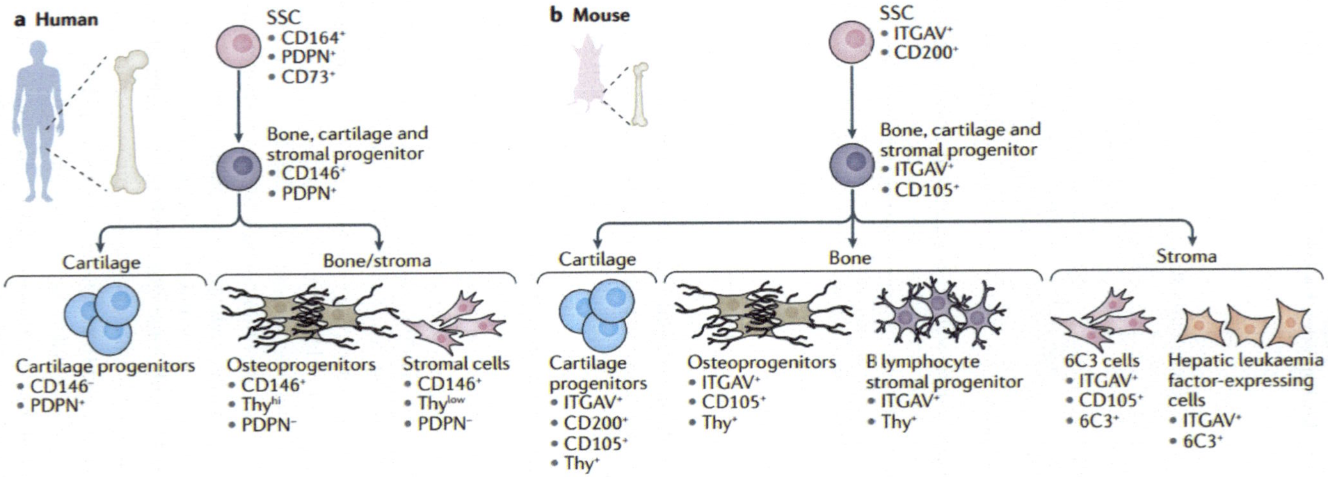

Fig. 3.3 Hierarchical organization of skeletal stem cells. SSCs and their progenitors in humans and mice can be separated by identifying unique immunophenotypic cell-surface markers. **a** The immunophenotypic profile of the human SSC hierarchy starts with a human SSC at the top, which then differentiates into a stromal progenitor, cartilage, and human bone. They then differentiate into their committed progenitors, namely cartilage progenitors, osteoprogenitors, and stromal cells of humans. **b** Immunophenotypic profile of mouse SSC, starts with a mouse SSC at the top, which then differentiates into a stromal progenitor, cartilage, and mouse bone. From there, they further differentiate into their committed progenitors, including hepatic leukemia factor-expressing cells, 6C3 cells, B lymphocyte stromal progenitors, osteoprogenitors, and cartilage progenitors of mice. Reprinted with permission from Urist et al. [39]

drastically by different cell culturing methods. These and other researchers have tried to enhance osteogenic efficiency by merging autologous marrow with carriers like demineralized bone matrix [40, 45–50] and by centrifugation methods [51]. During the last ten years, the impacts of microtopography have been well detailed, and involve gene expression, macrophage activation, apoptosis, cytoskeletal organization, contact guidance, and changes in cell adhesion [52–56]. Significantly, data is currently collected on the value of dimensions in the scale of nano in the design of the next generation of tissue engineering materials. Though, little is acknowledged about the impacts of nanotopography on precursor cell differentiation [57]. Current focus of cell response to accurate nanofabrication is the regulation of progenitor and stem cell differentiation [57]. These MSCs can bring about cells of the fibroblastic, myoblastic, osteoblastic, reticular and adipogenic lineages and produce progenitor cells involved in one or more cell lines with a marked degree of interconversion or plasticity. Therefore, the MSCs bring about a hierarchy of bone cell populations artificially divided into several stages of development, involving: MSCs, regulated osteoprogenitor cells (DOPC), pre-osteoblasts, osteoblasts and finally, osteocytes [57–59].

3.3 Osteoblasts

The phrase “bone formation” is occasionally used to explain differentiation of osteoblast through embryonic advancement, for instance, skeletogenesis. As a matter of transparency, the use of this phrase is limited to the function of osteoblasts, specifically the bone ECM deposition and synthesis. Formation of bone plays a significant role in bone remodeling, bone mineralization, and longitudinal growth of bone, which are complex processes that pose challenges for in vitro studies. The osteoblast lineage cells consist of many types of cells, including osteocytes, bonelining cells, pre-osteoblasts, osteoblasts, and their mesenchymal progenitors (Fig. 3.4).

Osteoblasts play a crucial role in synthesizing ECM components such as type I collagen, alkaline phosphatase, and osteocalcin. They also control the process of ECM mineralization by releasing matrix vesicles. In addition, osteoblasts control osteoclasts by releasing the osteoprotegerin (OPG) and RANKL. Osteoblast differentiation is regulated by the BMP, WNT, and RUNX2 signaling pathways [61].

3.3.1 Osteoblasts Origin

A comprehensive comprehension of the origin of bone-forming osteoblasts and the mechanisms involved in regeneration, remodeling, and development of bone is crucial for the development of innovative therapeutic strategies. This is because

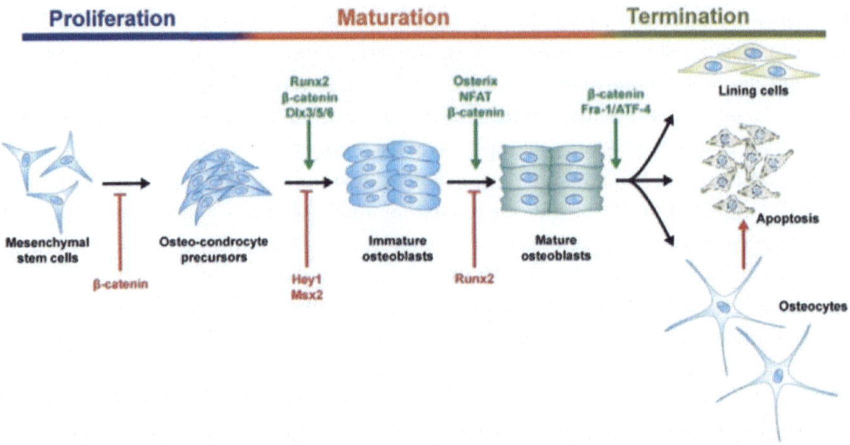

Fig. 3.4 Osteoblastogensis and fate: The diagram shows the process of osteoblastogenesis and the primary transcription factors that control the proliferation and specialization of osteoblast precursor cells. Once bone production is completed, mature osteoblasts have three possible outcomes: they can flatten and cover the inactive bone surface as lining cells, undergo programmed cell death (apoptosis), or encase themselves in bone matrix as they transform into osteocytes. Reprinted with permission from Gasser and Kneissel [60]

any interruption in the regulation of osteoblasts can lead to various bone disorders. Significant advancements in mouse lineage-tracing techniques and other single-cell technologies have provided valuable new knowledge about the development of osteoblasts in recent years. These findings suggest that osteoblasts, which contribute to formation of bone under different therapeutic, pathological, and physiological circumstances, can be produced from several sources. Osteoblasts can originate from various sources, including quiescent bone-lining cells on the bone surface, stromal cells in the bone marrow, and chondrocytes in the growth plate, specialized fibroblasts in the craniofacial structures, like periodontal ligaments and sutures [62].

Osteoblasts originate from mesenchymal cells. In the context of cell culture, osteoblasts and fibroblasts exhibit a high degree of similarity. All the genes that are expressed in fibroblasts are also expressed in osteoblasts. Contrarily, just two transcripts that are special to osteoblasts have been found. One transcript encodes osteocalcin, a secreted protein that suppresses osteoblast role [63], and the other encodes Cbf1, a transcription factor [64]. Osteoblasts also possess a singular morphological characteristic, which is situated externally to the cell, in the shape of a mineralized ECM. Therefore, from a genetic perspective, osteoblasts might be considered as advanced fibroblasts. The process of osteoblast differentiation remains ambiguous; however, several fundamental principles have already been identified. As mentioned above, osteoblasts originate from multipotent MSCs, whereas alternate sources include muscle tissue, fat, and cartilage. Therefore, efforts to differentiate osteoblasts from stem cells should aim to simultaneously promote the osteoblast lineage and suppress these other ones. The role of osteoblasts is balanced by that of osteoclasts, and this equilibrium has a pivotal function in the general bone turnover. Moreover, the availability of pre-osteoblasts is limited, and the exhaustion of this reservoir will result in the development of osteopenia. Furthermore, mechanical stimulation has a crucial function in the functioning and differentiation of osteoblasts. Any alterations in the typical physical environment would result in significant modifications to bone homeostasis [65]. Osteoblasts can be derived from nearby cellular sources, allowing bones to effectively respond to anabolic and regenerative signals. Nevertheless, the inquiry into whether osteoblasts originating from various cellular origins had separate activities has yet to be explored [62]. While there is significant knowledge about the classical pathways involved in determining and differentiating osteoblasts, there are also several new mechanisms that control and adjust this process to meet the ever-changing requirements of the organism, whether it is in a remodeling or growth phase. Figure 3.5 presents a concise description of the key individuals engaged in the process of osteogenesis. Crucially, not only do extracellular vesicles from other cells control osteoblast pathways, but also osteoblast-extracellular vesicles also have the ability to influence other cells. Interestingly, it has been shown that osteoblast-extracellular vesicles present RANKL both inside and on their surface, enabling them to enhance survival and differentiation of osteoclasts [66].

Cbfa1 plays a crucial function not only in the process of osteoblast differentiation but also in the regulation of bone formation by differentiated osteoblasts. It

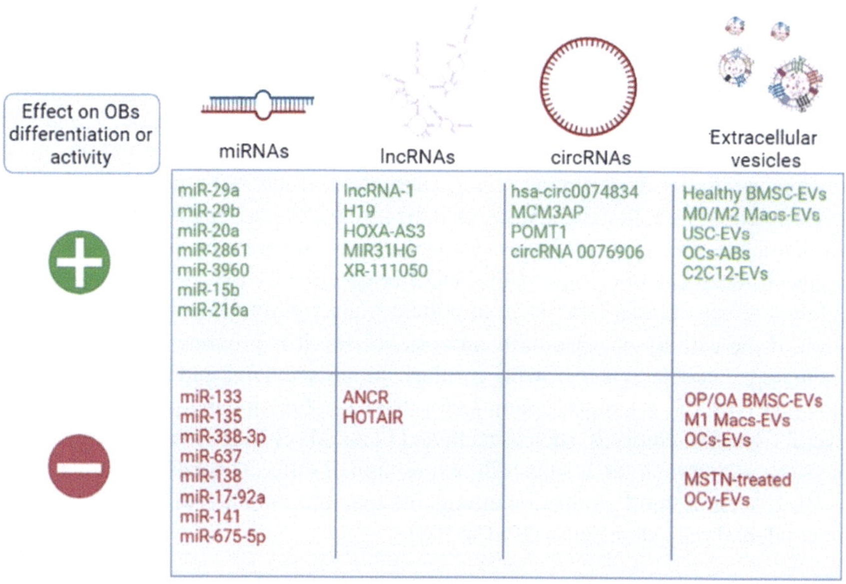

Fig. 3.5 Novel participants in the process of bone formation. The figure illustrates the extracellular vesicles and non-coding RNAs. Abbreviations: osteocytes (OCy), myostatin (MSTN), osteoporotic/osteoarthritic patients (OP/OA), osteoclasts (OCs), macrophages (Macs). Reprinted with permission from Ponzetti and Rucci [66]

has a controlling function in the expression of osteocalcin (Bgp), a gene that is exclusively produced in osteoblasts that have undergone terminal differentiation. The regulatory sequences of the majority of genes involved in the formation of a bone ECM also contain Cbfa1 binding sites [64, 67]. The transcription factor *cbfa/runx2* have an important function in the commitment of mesenchymal progenitors to osteoblast lineage, which has a diverse range of targets and several upstream regulators. Hedgehog signaling, Msx2, Sox9, and Wnt/Notch systems are upstream. Certain endocrine and paracrine factors act as coactivators, specially parathyroid hormone and bone morphogenetic proteins (BMPs). The process undergoes additional refinement through the involvement of histone deacetylases and vitamin D. The differentiation of osteoblasts is regulated by physical stimuli in order to facilitate the bone formation that is sufficient for providing dynamic and structural support to the body [65].

3.3.2 Transcription Factor of Osteoblasts

Runx2, also known as core binding factor α1 (Cbfa1), is involved in every stage of osteoblast differentiation. Runx2 is a dominant transcription factor that is expressed on chondrocytes and osteoblast lineage cells. Cells that are in the early stages of becoming osteoblasts and express the Runx2 transcription factor are called "pre-osteoblasts". Osterix (Osx), also known as Sp7, is a transcription factor that has a role in the initial phases of osteoblast development. Osx is crucial in inhibiting differentiation of chondrocytes. Its overexpression hindered the final phase of osteoblast development. ATF44 has a function in differentiation of osteoblasts from MSCs and homeostasis of bone. It is a member of the ATF/CREB protein family and functions as a leucine-zipper transcription factor. Disruption of the ATF4 gene impeded the development of osteoblasts and decreased the levels of β-catenin inhibited osteoblast differentiation and reduced levels. Conversely, overexpression of ATF4 in vitro enhanced the β-catenin. Importantly, Dlx3 is specifically expressed in osteocytes and mature osteoblasts. Dlx5 is an important transcription factor involved in regulation of the activity of osteoblasts and osteoclasts. It is prominently expressed in less mature osteoblasts and growing cartilage, along with Dlx2 and Dlx6. Msx1, also known as Hox 7.1, is a transcription factor that has a function in the development of several tissues in embryos, such as teeth and bone. Msx1 functions as a regulator of the osteocalcin promoter. It adjusts the expression of different genes, including those involved in cholesterol production throughout the differentiation of osteoblasts from human dental pulp stem cells (DPSCs) [68].

3.3.3 Osteoblasts and Materials Interaction

Material/osteoblast interaction relies on features of the surface of material which might be expressed according to its surface energy, chemistry, and topography. These surface properties control the behavior of cells on interaction. The initial step for cells in contact with a surface involves attachment, adhesion, and spreading. Afterward, the quality of this adhesion will affect their capability for differentiation and proliferation, and their morphology. They also control how biological molecules are adsorbed on the surface and, more specifically, regulate the adsorbed molecule orientation. Primary in vitro cell compatibility investigations concentrated on growth capability, morphological features and the situation of cell differentiation on materials with different chemical structures. The observed variation in cellular reactions to the several tested materials underscores the ability of cells to differentiate between diverse chemical compositions. Nevertheless, it was shown that the sensitivity of in vitro biological studies occasionally fell below the threshold required to differentiate the impacts of small alterations in substratum surface chemistry. For thermodynamic logics, surfaces comprise unsaturated bonds, establishing reactive layers on the surface and adsorb impurity. Preparation method influences including sterilization impacts have been explored by numerous investigators. They have provided evidence of the significant impact that sterilizing techniques on commercially pure titanium (cp–Ti) have on the adherence of cells in vitro. Certain experimental surface treatments have been observed to significantly adjust the chemical properties of titanium-based alloys. Inadequate utilization of glow discharge plasma treatment (a commonly employed technique for biomaterial modification, preparation, and cleaning) may yield undesired and non-reproducible outcomes [69].

SEM studies have shown that particles produced from implant wear exhibit a wider range of shapes compared to commercially available particles, which are usually spherical. Furthermore, these particles are more inflammatory or harmful to osteoblasts. Various in vitro cell models have been employed thus far to investigate the responses of osteoblasts to wear particles [61]. Increasing data suggests that wear debris not only activates bone resorption but also hinders the formation of new bone by blocking the activity of osteoblasts. Osteoblasts engage with particles through both phagocytic and non-phagocytic processes. Osteoblasts have been observed to ingest particles through micropinocytosis, as well as through caveolin- and clathrin-mediated endocytosis. Nevertheless, there has been a limited number of research conducted on this subject thus far [61].

3.4 Osteoclasts

Osteoclasts undergo differentiation from cells of macrophage/monocyte lineage [70]. Osteoclasts are found only in bone while macrophages, monocytes, and osteoclast progenitors, are obtained in most body tissues. Thus, cells particularly presented in bone tissue were thought to maintain osteoclast formation [71]. Osteoclasts are of hematopoietic origin and are derived from macrophage/monocyte lineage cells. Osteoclasts, like all of the hematopoietic lineage cells, initiate from bone marrow hematopoietic stem cells (HSCs), which undergo differentiation and self-renewal, giving rise to many types of hematopoietic cells. They do not undergo direct maturation; rather, they produce various kinds of oligopotent progenitor cells, which subsequently differentiate into lineage-restricted progenitor cells, so establishing a hierarchical differentiation tree (Fig. 3.6). They undergo a loss of their capacity to regenerate themselves and convert into multipotent progenitors (MPPs), capable of maintaining pluripotency.

MPPs differentiate into oligopotent progenitors, CLPs, MEPs, and CMPs [73, 74]. CMPs are the origin of osteoclast progenitor cells. In a primary investigation, it was stated that the $c - Kit^{+}CD11b^{lo/-}$ population in the bone marrow generates osteoclasts [75]. RANKL is present in osteoblasts and immunocytes, however it is more prevalent in bone tissues (mesenchymal cells, osteocytes, and osteoblasts) compared to other tissues. Purified osteocytes have a higher level of RANKL expression compared to osteoblasts and demonstrate superior ability to facilitate the osteoclast formation. RANKL can be classified into free and membrane-bound forms. The membrane-bound form exhibits greater activity compared to the free form. RANKL primarily regulates the differentiation process of osteoclasts. Monocytes that express RANK, a member of the TNF receptor family, merge with RANKL when exposed to it, resulting in the formation of multinucleated osteoclasts. This process involves the activation of many pathways, such as MAPK, nuclear factor Kappa-B (NF-κB), and co-stimulatory pathways (Fig. 3.7). All of these pathways contribute to the development of the primary transcription factor responsible for osteoclastogenesis named NFATc1 [72].

3.4.1 Osteoclastogenesis and Cell Fusion

During the process of osteoclastogenesis, osteoclast progenitor cells undergo fusion, resulting in the formation of a multinucleated large cell. The process of cell expansion facilitates the extensive sealing of the bone surface, hence enhancing the efficiency of bone resorption. Different enzymes, such as MMP-9 and CtsK are responsible for the degradation of organic components. The acidic condition created by HCl facilitates the resorption of inorganic components. Within the sealed area, bone tissue undergoes degradation by catalytic enzymes, H^{+}, and Cl^{-} [73]. Figure 3.8 provides a schematic representation of the historical and current perspective on osteoclastogenesis. Bone organ culture systems are valuable bioassays for investigating factors that manage resorption of bone by osteoclasts and, in certain instances, the formation of osteoclasts. These systems have proven to be valuable in the identification of novel agents that possess the ability to both promote and inhibit resorption of bone [76].

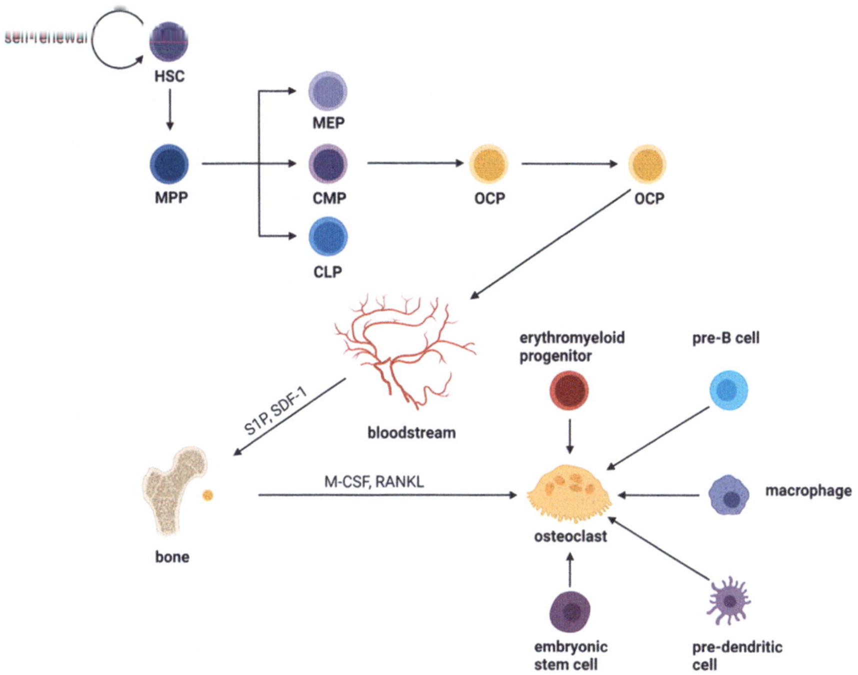

Fig. 3.6 Osteoclast differentiation. Osteoclasts are generated through the fusion of OCPs derived from the monocyte/macrophage lineage of the bone marrow. These precursor cells are derived from HSCs. HSCs do not directly produce fully developed cells. Instead, they generate a range of oligopotent progenitor cells, which then differentiate into lineage-specific progenitor cells, resulting in the formation of a hierarchical branching structure. They lose their capacity for renewal themselves and undergo a transformation into MPPs that have the ability to develop into several cell types. Following a sequence of deviations, MPPs differentiate into oligopotent progenitors, specifically CLPs, MEPs, and CMPs. Pre-monocytes are derived from CMPs and subsequently differentiate into OCPs. Subsequently, OCPs travel through the bloodstream and join BRUs due to the influence of substances such as SDF-1 and S1P. Ultimately, the fully developed osteoclasts were created through the merging of OCPs in the BRUs, prompted by substances such as M-CSF and RANKL. Moreover, within the context of inflammation and the immune system, monocytes or macrophages that are specialized to certain tissues have a vital function in the generation of osteoclasts. Erythromyeloid progenitors in the fetus have an important function in the development of osteoclasts, which are engaged in both hematopoiesis and fracture repair. Prior research has demonstrated that initial cells including embryonic stem cells, pre-B cells, and pre-dendritic cells have the ability to differentiate into osteoclasts. Abbreviations: osteoclast precursor cells (OCPs), hematopoietic stem cells (HSCs), stroma-derived factor 1 (SDF-1), common myeloid progenitors (CMPs), common lymphoid progenitors (CLPs), bone remodeling units (BRUs), sphingosine-1-phosphate (S1P), megakaryocyte–erythrocyte progenitors (MEPs). Reprinted with permission from Sun et al. [72]

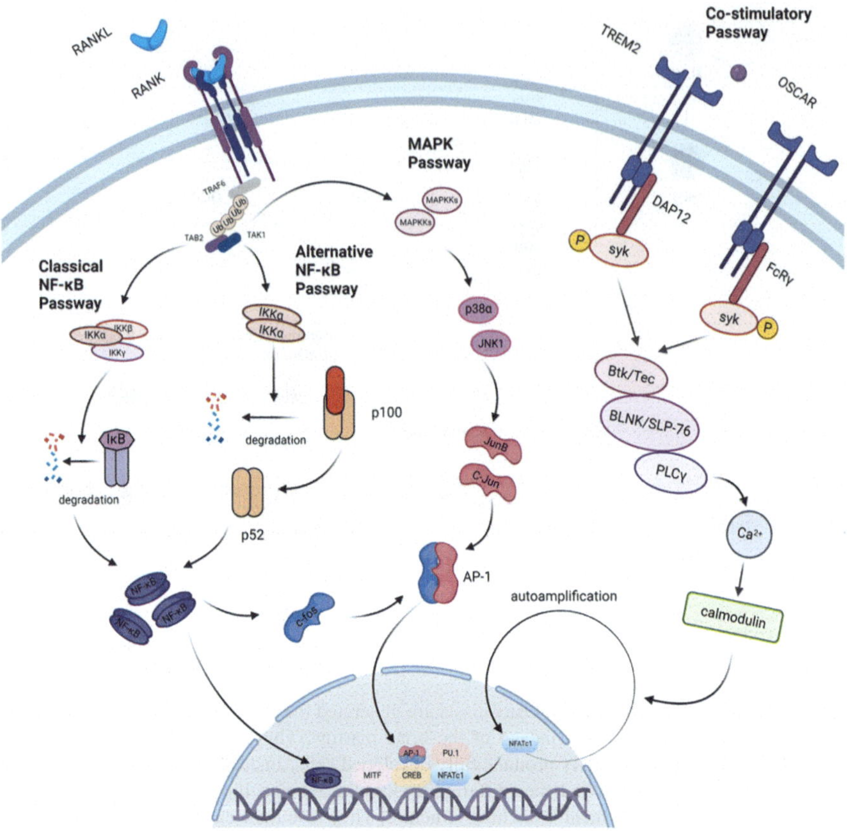

Fig. 3.7 Signal pathways involved in the process of osteoclastogenesis in response to RANKL stimulation. Once RANKL attaches to its receptor RANK, it initiates signaling through the linker TRAF6, which brings in TAB1 and TAB2 to activate the NF-κB and the MAPK pathways. The NF-κB pathway can be categorized into two distinct types: the classic and the alternative pathways. TRAFs initiate the activation of the IKK complex (consisting of IKKα, IKKβ, IKKγ) in the classical pathway. The complex then causes ubiquitination and phosphorylation of the IκB molecule, leading to its degradation and subsequent release of NF-κB. Through the alternative pathway, the homodimer of IKKα performs a cleavage of p100, resulting in the production of p52, which in turn triggers the activation of NF-κB. NF-κB is an essential transcription factor required for the process of osteoclast differentiation. In the aforementioned pathways, the activated NF-κB dimer stimulates the production of factors like AP-1 and c-Fos, which subsequently enhances the expression of NFATc1, hence facilitating the differentiation of osteoclasts. Furthermore, the regulation of NFATc1 expression is also influenced by the co-stimulatory pathway. DAP12 and FcRγ play crucial roles as receptor molecules in the co-stimulatory pathway. Upon binding to the corresponding ligand, they have the ability to initiate the activation of the downstream tyrosine kinase Syk. Syk develops an intricate association with Btk/Tec, BLNK/SLP76, and then stimulates the PLC pathway to initiate differentiation of osteoclasts. Research has demonstrated that TRAF6 has the ability to directly stimulate c-Src and subsequently initiate the PI3K/Akt pathway, which leads to the enhancement of proliferation and differentiation of osteoclasts. Reprinted with permission from Sun et al. [72]

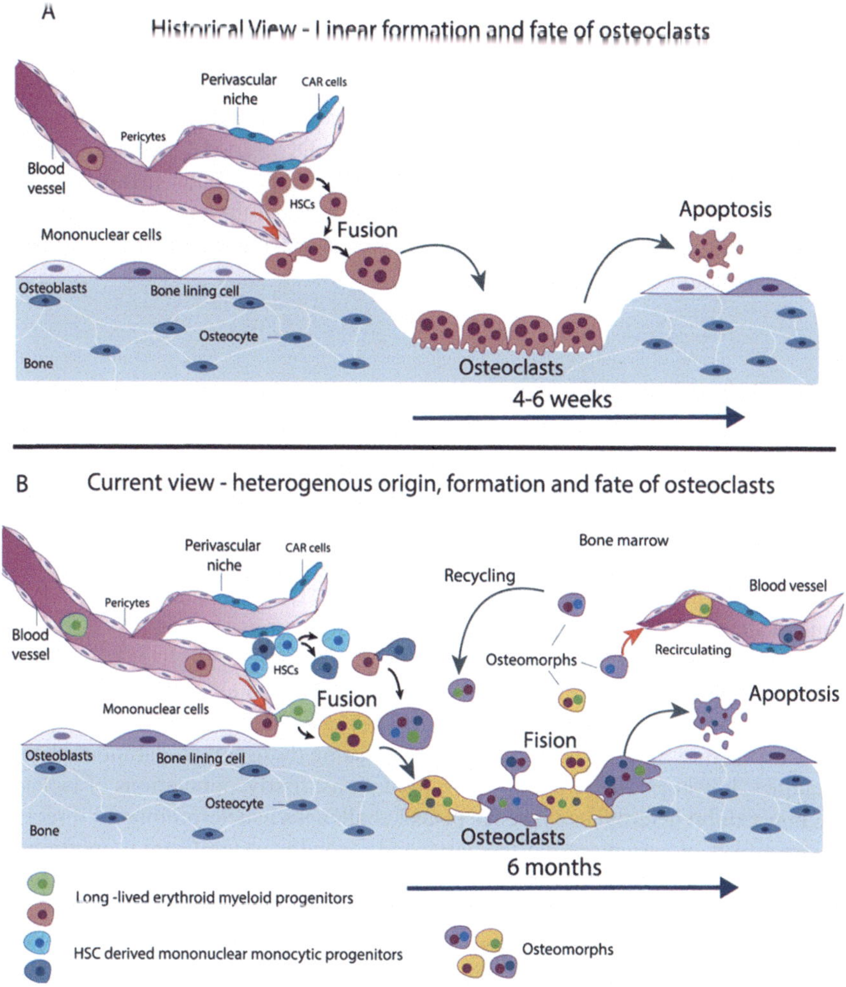

Fig. 3.8 Historic and current view of osteoclastogenesis. **a** Historical perspective: Osteoclasts originate from monocyte precursors produced from hematopoietic stem cells. These precursors fuse together and eventually transform into fully developed osteoclasts that reside on the surface of the bone. Osteoclasts undergo programmed cell death, known as apoptosis, within a period of 4 to 6 weeks following bone resorption. **b** Present perspective: Osteoclasts are created by a repetitive process of merging hematopoietic stem cell-derived monocyte precursors (blue and dark blue) with syncytia that formed from the erythroid-myeloid progenitors throughout embryonic development (red and green). During the process of bone resorption, fully developed osteoclasts divide into osteomorphs, which are characterized by their blue, green, and red colors. These substances either enter the bloodstream or undergo recycling and merge with other groups of osteoclasts to create new osteoclasts or experience programmed cell death. The lifespan of osteoclasts in mice is approximately 6 months. Reprinted with permission from McDonald et al. [77]

The process of osteoclast progenitor cell fusion is believed to have a critical role in optimizing the resorption activity. The process involves the restructuring of phospholipids, which are the constituents of the lipid bilayer of the cellular membrane [73, 78, 79]. Several molecules that are particular to osteoclasts have also been identified to have a function in the process of osteoclast fusion. DC-STAMP, a gene known for its osteoclastogenic properties, has a crucial function in the cellular fusion of osteoclast progenitor cells. The lack of this substance leads to the failure of osteoclast multinucleation, resulting in the formation of numerous $TRAP^+$ mononuclear cells both in vivo and in vitro [73, 80, 81].

3.4.2 Osteoclast Isolation

Advancements in comprehending the molecular processes involved in osteoclastic resorption of bone and differentiation of osteoclasts have been challenging due to the scarcity of osteoclasts, the difficulty in isolating them from a calcified matrix and their fragility when separated from bone. The first to state a cell line that effectively generated a significant quantity of osteoclasts were Hentunen and coworkers [82]. Numerous methods have been advanced to isolate fully grown osteoclasts from long bones. The endosteal surface of chick long bones has been widely utilized as a primary source for osteoclasts [83]. Osteoclasts can also be obtained in significant quantities from long bones of adolescent rabbits, and they can be favorably enriched. Osteoclasts cDNA expression libraries have been generated using these cell preparations. These techniques, however, are plagued by two limitations: firstly, populations of isolated cells may exhibit heterogeneity, hence posing challenges in determining the direct or indirect effects of variables on osteoclasts; and secondly, a significant proportion of these cells may lack viability. Most investigations utilizing isolated osteoclasts have indicated that the majority of osteotropic agents perform indirectly on osteoclasts through stromal cells or osteoblasts. Moreover, there are significant changes in the capability of these cells to undergo bone resorption, requiring a substantial quantity of bone pieces in order to achieve consistent and replicable outcomes [76].

3.4.3 Factors that Increase Activity of Osteoclasts

Interleukin 1: Interleukin-1 (IL-1) is a cytokine that exhibits a broad range of functions. It promotes the production of prostaglandin in bone that might improve its resorptive function since prostaglandins are powerful stimuli for resorption. Impacts of IL-1 on resorption also seem to be both direct on osteoclasts and indirect through its capability to promote the formation of RANKL, the crucial osteoclastogenic factor. IL-1α and IL-1β, which have identical functions, are two gene products that make IL-1. Osteoclast precursor cells are a source of IL-1 derived from bone cells. When these cells interact with the bone matrix, they generate IL-1. IL-1 receptor antagonist that is a natural inhibitor of IL-1 functions as a counterpart of IL-1 by binding the biologically significant type I IL-1 receptors without inducing their activation [84].

Interleukin 6: The involvement of interleukin-6 (IL-6) family cytokines in the skeletal system was initially identified by the observation that IL-6 strongly promotes the development of osteoclasts in culture. This was not an outcome as a result of direct intervention on osteoclasts or their precursor cells, despite their expression of the transmembrane IL-6 receptor (IL-6R). In contrast, IL-6 exerted an indirect influence on resorption of bone by facilitating the RANKL expression in osteoblast precursor cells. This local IL-6 activity inside the microenvironment of bone can be attributed to its production by several cell types, encompassing osteoblasts, osteoclasts, neutrophils, T cells, and macrophages. The expression of RANKL by pathogenic cells is also facilitated by IL-6. When IL-6 family cytokines create a receptor:ligand complex, LIFR, OSMR, and gp130 undergo phosphorylation, creating specific locations where STAT1, STAT3, and STAT5 proteins can attach (Fig. 3.9) [84].

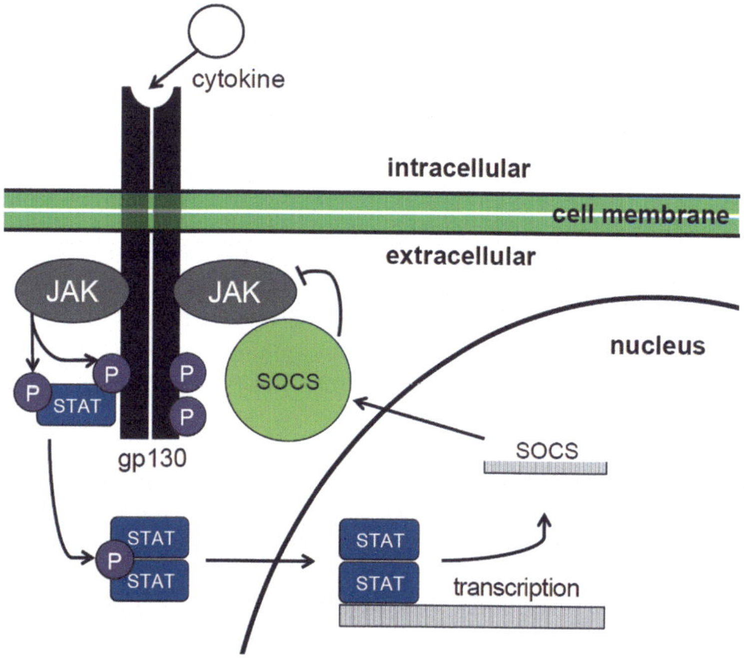

Fig. 3.9 Principle of intracellular signaling through IL-6 family cytokines: Janus kinases (JAK) phosphorylate the receptor complex, when a transmembrane signaling complex is formed, creating docking sites for STAT proteins. After being phosphorylated, the subsequent dimerization (either as homo or heterodimers) and movement to the nucleus triggers the initiation of transcription of certain genes that are targeted by STAT, including the transcription of SOCS proteins. SOCS proteins are transported to the cytoplasm where they deliver negative feedback to the receptor complex, inhibiting further signaling and phosphorylation. Reprinted with permission from Sims and Lorenzo [84]

To promote formation of osteoclasts in vitro, it is necessary to add soluble IL-6 receptors into murine marrow cultures. IL-6 enhances the impact of other hormones, like parathyroid hormone-related protein (PTHrP), on the resorption of bone by osteoclasts and the regulation of calcium levels in the body. Without additional IL-6 receptors, IL-6 can promote the development of osteoclast-like cells in human marrow cultures [85].

Interleukin 11: Stromal cells in the bone marrow produce interleukin 11 (IL-11) [86]. Similar to IL-6, IL-11 acts by attaching to a receptor that is specific to a particular ligand. In co-culture, IL-11 also incites osteoclastogenesis by promoting the production of RANKL in osteoblast lineage cells. While IL-11 has the ability to utilize a soluble receptor known as sIL-11R for trans-signaling, it is not necessary for IL-11 to add soluble receptors to induce osteoclastogenesis [84].

TGF-α: Transforming growth factor alpha (TGF-α) is a growth factor that promotes the proliferation of primary osteoclast precursor cells. Many solid tumors associated with the hypercalcemia of malignancy produce this polypeptide, which shares some homology with epidermal growth factor (EGF). The binding of TGF-α with EGF has been observed to induce osteoclastic resorption of bone in mouse organ cultures [87, 88].

M-CSF: CSF1, alternatively referred to as macrophage colony-stimulating factor (M-CSF), have a critical function as a growth factor that controls the differentiation, proliferation, and survival of hematopoietic lineage cells, and functions as a ligand for colony-stimulating factor 1 receptor (CSF1R, also identified as c-FMS), which is a receptor tyrosine kinase. CSF1 exhibits extensive expression across several cell types, encompassing epithelial cells, fibroblasts, stromal cells, osteoblasts, and some metastatic tumor cells. It has been shown that the differential regulation of osteoclastic resorption of bone is mediated by secreted CSF1 and cell-surface CSF1. CSF1 alone has the ability to rectify deficiency of osteoclasts in CSF1-deficient op/op mice. Additionally, they exhibit delayed resorption of trabecular bone [89]. M-CSF interacts with c-Fms and has an important function in the differentiation of osteoclast lineage by controlling several pathways, such as the MAPK, PI3K-Akt, and Co-Stimulatory pathway (Fig. 3.10) [72].

TNFs: The function of tumor necrosis factors (TNFs) in various inflammatory illnesses is significant. Consequently, there is a strong interest in understanding the mechanisms by which TNFs control the differentiation of cell lineages involved in the remodeling of bone, such as both osteoblasts and osteoclasts, as well as associated coupling factors. However, TNFs alone do not efficiently stimulate differentiation of osteoclasts, they have an essential function in stimulating bone resorption and pathologic osteoclastogenesis, particularly in cooperation with RANKL. TNFs have been observed to potentially enhance the production of osteoclast precursors in vivo [90]. TNFs can also influence the mature osteoclast activities [76]. The growth of osteoclast-like multinucleated cells in human bone marrow cultures is significantly induced by both TNF-α and TNF-β (lymphotoxin) [91]. The impacts of IL-1 on the formation of osteoclasts are enhanced by TNF [92].

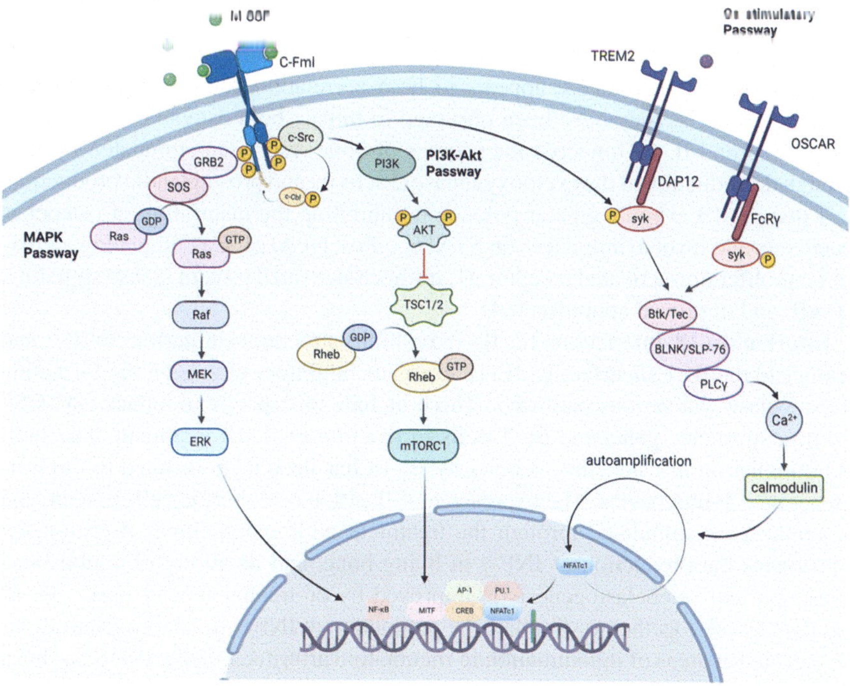

Fig. 3.10 Signaling pathways involved in osteoclastogenesis under the stimulation of M-CSF. MAPK pathway: Upon receptor stimulation, the receptor binds with Grb2/Sos, initiating the MAPK signaling cascade. As a consequence, the ERK undergoes activation and phosphorylation, which facilitates osteoclast formation. PI3K-Akt pathway: M-CSF and c-Fms (its receptor) attach together, causing limited phosphorylation of important sites in the c-Fms tail, and attracting the tyrosine kinase c-Src. The receptor, which has undergone partial phosphorylation, attaches to c-Src creating a complex of c-Fms/c-Src Protein. Next, the complex enlists E3 ubiquitin ligase (c-Cbl) and triggers its tyrosine phosphorylation. In response, the phosphorylated c-Cbl induces receptor ubiquitination, resulting in full receptor phosphorylation. Afterwards, the c-Fms/c-Src protein complex attracts PI3K to initiate the PI3K-Akt pathway, which leads to the proliferation and activation of OCPs. Co-Stimulatory pathway: M-CSF additionally enhances the co-stimulatory pathway of NFATc1. M-CSF has the ability to activate the DAP12 and FcRγ, which in turn triggers the activation of the tyrosine kinase Syk. Syk forms a complex with Btk/Tec, BLNK/SLP76. The complex enhances the activity of PLCγ, triggers the PLC pathway, and removes phosphate groups from the transcription factor NFATc1, facilitating its transportation into the nucleus and stimulating the differentiation of precursor cells into osteoclasts. Abbreviations: extracellular signal-regulated kinase (ERK), dnax activating protein 12 (DAP12), phosphatidylinositol 3-kinase (PI3K), Fc receptor common gamma subunit (FcRγ). Reprinted with permission from Sun et al. [72]

3.4.4 *Inhibitory Factors*

Interleukin 4 and 13: Interleukin-4 (IL-4) and Interleukin-13 (IL-13) belong to a class of locally functioning factors known as “inhibitory cytokines” [84]. The data from investigations propose that elevated levels of IL-4 have the potential to impede both resorption and formation of bones. IL-4 production by activated T cells has been observed to exert influence on both hematopoietic and immunologic procedures [76]. There appears to be a correlation between the actions of IL-4 and IL-13, as they have been observed to impact both osteoblasts and osteoclasts. IL-4 and IL-13 hindered the process of bone resorption stimulated by IL-1 through the reduction of the cyclooxygenase-2 activity and prostaglandin production. IL-4 than IL-13 exhibits greater potency in inhibiting the maturation of osteoclast precursors and involve influences on STAT6, c-Fos, NFATc1, Ca^{++} signaling, peroxisome proliferator-activated receptor g1, mitogen-activated protein kinase signaling, NF-κB, and improved apoptosis [84].

Interleukin 18: Interleukin-18 (IL-18) exhibits structural similarities to IL-1 and belongs to the IL-1 superfamily. IL-18 exerts its inhibitory effects on the formation of osteoclasts via various pathways. These include its capacity to induce GM-CSF (100), a substance generated by T cells in reaction to IL-18 treatment. The indirect stimulation of osteoclastogenesis by IL-18 has been demonstrated through its impact on T lymphocytes. The expression of IL-18 in osteoblastic cells is seen, and its synthesis is stimulated through the treatment with endothelin-1. Additionally, it promotes the production of INF-γ in living bone, and its ability to inhibit bone resorption and osteoclastogenesis is improved by co-treatment with IL-12. IL-18 and IL-12 work together to stimulate the production of INF-γ, and its concentrations are elevated at areas of inflammation in rheumatoid arthritis. Finally, IL-18 has been found to enhance OPG production [84].

TGF-β: De Larco and Todaro made the discovery of transforming growth factor beta (TGF-β) [93]. As of the present time, the TGF-β superfamily consists of around 30 members. TGF-β superfamily members are secreted growth factors that function as versatile regulatory proteins in bone. They play a role in several activities, such as the function, differentiation, and proliferation of bone cells. The members of TGF-β family can directly or indirectly (through substances released by osteoblasts or osteocytes) influence the bone resorption activity, osteoclastogenesis, proliferation of osteoclast progenitors, and the mature osteoclast survival. The coordination of communication between osteoclasts and osteoblasts is also facilitated by them in order to ensure a suitable response [94]. The release of TGF-β occurs in an inactive precursor state, wherein it remains attached to a latency related peptide, which holds it in an activated state. Active TGF-β is released during bone resorption, as a result of proteolytic cleavage from latent peptides and pH fluctuations in the surrounding environment [95].

Nitric Oxide: Nitric oxide (NO) is a global signaling molecule that has a critical function in the regulation of bone homeostasis. NO is generated by NO synthase (NOS) by the conversion of L-arginine (L-arg) to L-citrulline in the presence of nicotinamide adenine dinucleotide phosphate and oxygen. Inducible NOS is a calcium-independent enzyme, while endothelial NOS and neuronal NOS are constitutively expressed calcium-dependent enzymes. Collectively, all these isoforms are expressed in bone cells. Inducible NOS has a critical role in regulating differentiation of osteoclasts in the presence of bacterial infection-induced inflammatory environments. Several investigations have shown that NO has biphasic impacts on the function and formation of osteoclasts. NO has been found to promote bone resorption and osteoclastogenesis in several cases. NO also facilitates the fusion of pre-osteoclasts by increasing the activity of actin cytoskeleton remodeling [96]. Some significant cytokines, including TNF and IL-1, exhibit strong stimulatory effects on the production of NO. At a certain concentration range, NO has the ability to stimulate growth of osteoblasts and prevent bone resorption mediated by osteoclasts. In bone, NO has the ability to stimulate osteoblast differentiation, survival and proliferation at low concentrations (pico-nano molar range). At low concentrations, NO enhances osteoclastic bone resorption induced by IL-1. However, NO can hinder the process of bone formation/resorption at high concentrations (micromolar range). At high concentrations, NO hinders osteoclastic activity and formation, leading to osteoclastic progenitor apoptosis. The exogenous delivery of NO may have significant biomedical applications because of its possible positive impacts on bone tissue regeneration [97].

Osteoprotegerin: Osteoprotegerin (OPG) is a secretory glycoprotein of the tumor necrosis factor (TNF) receptor and has a crucial function in regulating bone resorption. It is generated by most types of the cells and has been observed to inhibit the differentiation/fusion phase of osteoclast differentiation, as opposed to the proliferation stage. The identification of OPG has been conducted by Simonet et al. [98]. The identification of OPG as an inhibitor of osteoclast maturation and activity has prompted a novel avenue of investigation into the possible therapeutic application of OPG in bone disorder treatments. It possesses a ligand known as RANKL. Mature osteoclasts also exhibit the expression of RANK and RANKL, which both promote bone-resorbing activity of osteoclasts and support survival of them. The bond between RANK and RANKL promotes differentiation of monocyte/macrophage progenitors into matured and active osteoclasts through several signaling pathways. The decoy receptor OPG has the ability to inhibit the activity of RANK ligands. Prevention of osteoclastic resorption of bone can result in an increase in bone mass through the inhibition of RANKL–RANK signaling in bone. It inhibits the maturation of osteoclasts and protects the physiological mechanisms of them in ovariectomy-associated bone resorption and bone remodeling. The differentiation of osteoclasts from progenitor cells is dose-dependently inhibited via recombinant osteoprotegerin both in vivo and in vitro [76, 99, 100].

Interferons: Interferons (IFNs) are a distinct class of cytokines that exert a significant influence on the immune response. IFNs are categorized into three primary classes based on their biological functions, structural characteristics, and the type of receptor they use to transduce signals: type I, type II, and type III. Among the type I interferons, IFN-β hinders the process of osteoclastogenesis through either autoregulatory or exogenous regulatory means, whereas IFN-α has recently been determined to have a function in regulating the formation of osteoclasts. IFN-γ, which belongs to the category of type II interferons, exhibits biphasic influences on the process of osteoclastogenesis. IFN-γ stimulates the generation of superoxide in osteoclasts, leading to the suppression of osteoclast function and induction of apoptosis in osteoclast progenitor cells. Nevertheless, IFN-γ serves as the physiologic inducer for the expression of MHC class II and, consequently, presentation of antigens. Consequently, IFN-γ triggers the activation of T cells and T cell secretion of the osteoclastogenic factors TNF-α and RANKL. Collectively, these data indicate that IFN-γ hinders the formation of osteoclasts by directly targeting mature osteoclasts, whereas it stimulates the activation of T cells, hence indirectly stimulates osteoclastogenesis. Previous studies have also demonstrated the involvement of type III interferons in osteoclast bone resorption [101, 102].

Sex Steroids: Sex steroid deficiency, along with old age and glucocorticoid excess, do not induce loss of bone mass through the activation of an entirely novel mechanism. Conversely, they induce a disruption in the typical mechanism of bone renewal. Bone marrow stromal cells, chondrocytes, osteoblasts, and osteoclasts and their progenitors include classical receptors for androgens (AR) or estrogens (ERα and ERβ). This indicates that the impact of sex steroids on bone is partially mediated by direct mechanisms. Estrogen is recognized as a significant suppressor of osteoclast formation. Androgens and estrogens decelerate the process of bone remodeling and provide protection against bone loss. In contrast, the lack of estrogen leads to a higher rate of remodeling and favors the process of bone resorption over the formation of bone. The presence of androgens and estrogens decreases the number of remodeling cycles by decreasing the proliferation of osteoblasts and osteoclasts from their particular progenitor cells [76, 103].

Sex steroids exhibited an increase in apoptosis of osteoclasts in vivo after treatment with either testosterone or estradiol in orchiectomized or ovariectomized mice, respectively. The addition of testosterone, tamoxifen or estradiol to bone marrow cultures in murine resulted in a two- to threefold increase in apoptosis of osteoclasts in comparison with control groups [76].

3.4.5 Osteoclast Apoptosis

Apoptosis (type I cell death) is a unique kind of regulated programmed cell death that is specified by genes and involves specific biochemical and morphological alterations. Other kinds of programmed cell death include oncosis, pyroptosis, and autophagy. Apoptosis leads to the elimination of cells, with minimal tissue disturbance. It serves as a supplementary, although contrasting function to mitosis in controlling the population of animal cells and takes place at a specific time point in cellular growth and demonstrates exceptional precision [104, 105]. Osteoclasts can undergo apoptosis by either the internal (mitochondria) or external (death receptor) mechanisms. Both routes have the capability to initiate caspase activation, hence triggering apoptosis by the cleavage of particular substrates. In addition to the aforementioned two traditional apoptotic routes, recent years have yielded further insights into the apoptotic mechanisms of osteoclasts (Fig. 3.11) [72].

Osteomorphs are a distinct form of cells that is different from macrophage progenitors and osteoclasts. They are generated through the process of osteoclast fission, resulting in the formation of smaller, highly mobile daughter cells that have the

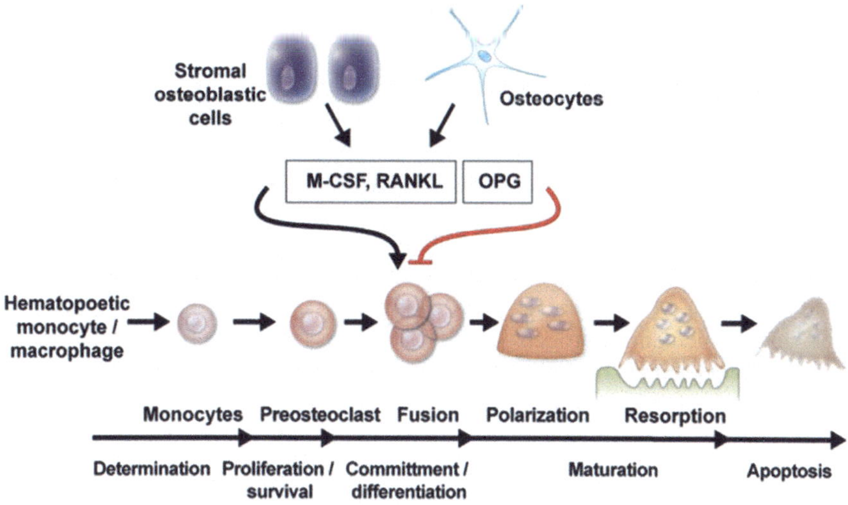

Fig. 3.11 Osteoclastogenesis and apoptosis: The process of osteoclast differentiation is regulated by two key factors, M-CSF and RANKL, along with other cytokines released by osteoblasts and osteocytes that have a role in controlling several stages of osteoclastogenesis, such as maturation, differentiation, commitment, and precursor proliferation. OPG, which is also produced by osteoblasts and osteocytes, functions as a decoy receptor for RANKL and decreases the osteoclast differentiation. Reprinted with permission from Gasser and Kneissel [60]

capacity to merge and give rise to new osteoclasts. Osteomorphs are located in the blood and bone marrow, while osteoclasts are connected to bone. Recent findings indicate that osteoclasts undergo recycling via osteomorphs during resorption of bone triggered by RANKL. Mature osteoclasts undergo a process of differentiation into smaller osteomorphs, enabling them to survive and persist for prolonged durations until their reutilization becomes necessary. Accordingly, under certain circumstances, osteomorphs have the ability to undergo re-fusion into multinucleated osteoblasts. This methodology circumvents the conventional apoptotic routes and offers a novel perspective on the destination of fully developed osteoclasts [72]. Osteoclast apoptosis is believed to be influenced by several endogenous factors, such as osteoprotegerin (OPG), semaphorin-3A (sema3A), TGF-β1, estrogen, and elevated extracellular calcium. Several of these factors function by inhibiting RANKL signaling, suggesting that they may primarily serve as opposing pro-osteoclastic factors rather than directly inducing apoptosis. Furthermore, there is contradictory data about some of these purported pro-apoptotic variables. Studies have shown that TGF-β signaling stimulates the expression of pro-osteoclast survival factors, including SOCS3 and LIF. Additionally, previous studies have demonstrated that OPG exhibits the ability to inhibit osteoclast apoptosis by impeding TNF-related apoptosis-inducing ligand (TRAIL). Moreover, estrogen therapy has been demonstrated to reduce the quantity of osteoclasts on the surfaces of bones, while increasing osteoclasts in the marrow area away from bone. Instead of osteoclast apoptosis being considered as a mechanism, it can be seen as accumulation of precursor cells and a suppression in formation of osteoclasts or alterations in osteoclast morphology [77].

Significantly, alterations in the control of osteoclast apoptosis are observed in various pathological bone disorders, including Paget's disease of bone (PDB), osteoarthrosis, hyperparathyroidism, rheumatoid arthritis (RA), and postmenopausal osteoporosis. The occurrence of apoptosis in osteoclasts can perhaps be attributed to an imbalance between survival factors and their opposing factors. While the survival of cells relies on the signaling through the TNF-α, the receptor activator of NF-κB ligand (RANKL), M-CSF, and IL-1 via NF-κB and Akt, osteoprotegerin (OPG) and other factors, including estrogen, induce osteoclast apoptosis [105]. While osteoclast stimulatory factors, including PTH and 1.25−*dihydroxyvitamin* D_3, have been shown to hinder in vitro stimulation of osteoclast apoptosis, it can be induced by bisphosphonates, which hinder resorption of bone [106]. The data from investigations recommend that regulation of osteoclast lifespan is a significant factor in the typical process of bone remodeling, as it can either promote or prevent osteoclastic resorption of bone. Cytokines that improve activity of osteoclasts partially achieve this by extending the lifespan of osteoclasts, and factors that hinder activity of osteoclasts seem to prevent formation of osteoclasts and resorption of bone while provoking apoptosis of osteoclasts [76].

3.5 Osteocytes

The osteocytes serve as a reservoir of factors which are soluble that have the ability to target many distant organs, including muscle, kidney, and other tissues, as well as cells on the surface of the bone. They constitute the majority of bone cells, ranging from 90 to 95%, in adult bones. These cells exhibit the longest lifespan, up to decades inside their mineralized environment. The osteocytes are exclusive cells that exhibit the expression of genes from both the hematopoietic and mesenchymal lineages. Contrary to their conventional role as a "passive placeholder in bone", these cells have been discovered to possess several roles. These functions include serving as a regulator of bone remodeling by adjusting the activities of both osteoblasts and osteoclasts, as well as performing as endocrine cells. Osteocytes possess the ability to remodel their perilacunar matrix and are involved in both calcium availability and phosphate metabolism [107]. They are situated between lamellae, inside small cavities recognized as lacuna, interconnected by a series of channels named canaliculi [108]. More than a hundred years ago, osteocyte was thought to be able to remodel its ECM; more than 40 years ago, osteocyte was believed to express tartrate-specific acid phosphatase, remodel bone and respond to parathyroid hormone; and more than 20 years ago, osteocyte was described as mechanosensory cell [107]. The first person who isolated avian osteocytes was Peter Nijweide [109]. Marotti and Palumbo illustrated notable figures to explain their concepts concerning communication and function of osteocytes [110]. Histology was the main tool used by these primary innovators to make their ideas, Kumegawa et al. recorded some of the first videos of bone cells containing osteocytes [111]. In the past two decades, there has been a significant surge in knowledge about osteocyte biology due to advancements in technology, including advanced instrumentation, systems biology, cell lines, imaging, transgenic and molecular approaches, and others. This has resulted in the confirmation of existing theories and the development of new ones [107].

The process of osteocyte formation has been considered to be a passive process in which a certain subset of osteoblasts converts passively encased in osteoid, which subsequently undergoes passive mineralization. It was conceived that the entrapment of an osteoblast-producing matrix/osteoid occurs when neighboring osteoblasts deposit the osteoid onto the embedding cells [112]. Nevertheless, there exist multiple disagreements opposing the notion that osteocyte formation is a passive process [107]. The primary roles of the osteoid osteocyte are to form connective dendritic processes and regulate mineralization at the same time. When the osteoblasts change to osteocytes, there is an increase in casein kinase II and osteocalcin levels and a decrease in alkaline phosphatase levels. Unexpectedly, osteocytes have the ability to express osteoclast markers, including cathepsin K and acid phosphatase with the purpose of remodeling their perilacunar matrix under specific situations like lactation [107].

3.5.1 Cell Differentiation and Origin

Osteocytes, which are the predominant cells in mature bone, originate from osteoprogenitor cells, some of which undergo differentiation into active osteoblasts. They have an assumed function in mechanotransduction, which is supposed to regulate activities related to bone turnover and remodeling. The body of osteocytes and their processes are enclosed within spaces named lacunae and channels named canaliculi, respectively, in mature bone [113]. A functional syncytium is formed by the expanded osteocytic network, which consists of cells interconnected by several cellular processes and bound at gap junctions [114–116]. Therefore, apart from intercellular communication via gap junctions, the cells comprising the syncytial network maintain contact through their shared environment, which is represented by a space filled with a contiguous fluid-saturated proteoglycan network [114, 116]. It is now understood that osteocytes are derived from matrices producing mature osteoblasts. The precise mechanisms underlying the determination of cell fate in mature osteoblasts remain unknown. Embedding cells are enveloped by collagen, which is synthesized either by their own matrix-producing apparatus or by adjacent cells. During the process of cellular embedding, these cells produce cellular extensions, which serve as the future dendritic processes of mature osteocytes that undergo constant extension and contraction until they establish a suitable connection with previously embedded cells. Subsequently, the dendrites seem to attach to the existing cell, initiating the mineralization process as the cell encloses itself within a hydroxyapatite cavity of lacunae.

Upon the completion of the process, the embedded cells become functionally part of the lacunocanalicular network of osteocytes [117].

Through osteoblast transformation to the newborn osteocytes, cellular processes first radiate toward the mineralizing matrix. These processes are believed to have an essential function in the calcifying matrix vesicle extrusion [118]. Highly selective or osteocyte-specific proteins have been displayed to act in mineral metabolism. Sclerostin, which was explained formerly, is one of them. Osteocytes also seem to control biomineralization and phosphate through molecules including FGF-23, MEPE, DMP-1, and PHEX [119].

3.5.2 Perilacunar Matrix Remodeling

Elimination of a small amount of mineral per osteocyte would have a considerable impact on the levels of circulating, systemic ions due to the substantial difference in surface area between the area of the bone surface and the lacunocanalicular system of osteocytes in bone [107, 120]. Contrary to the healthy and young osteocytes within a hypomineralized matrix, the aging osteocytes may experience hyper-mineralization of their perilacunar matrix. Additionally, studies have reported the occurrence of micropetrosis in aging bone, where the lacunae get filled with minerals. The hyper-mineralization can result in age-related death of cells, therefore causing the filling of the lacunae with minerals. This can significantly alter the bone fluid flow dynamics throughout the osteocyte lacunocanalicular network, possibly influencing viability and function of osteocytes [107].

3.5.3 Osteocytes as Mechanosensory Cells

Another primary proposed role for osteocytes is their role as mechanosensory cells due to their complicated dendritic network and their positioning in bone. Timing, intensity, and the optimal frequency of loading in vivo have been determined to effectively enhance bone mass. The lining cells are also postulated as significant mechanosensory cells in the adult skeleton, although there is limited knowledge regarding this particular bone cell type [121–123]. A significant challenge has been the identification of in vitro studies that can accurately reproduce the outcomes observed in vivo while factors of in vivo loading have been extensively studied. The observed alteration in the expression of osteocyte genes in vivo as a result of mechanical loading and unloading indicates the significant impact of load on function of osteocytes [124–127]. The osteocytes can detect load by various mechanisms, including cilia bending, the dendritic processes, or through the cell body. There exists a proposition suggesting that the osteocytes sense load exclusively via their processes or via both the processes and the cell body [107].

3.5.4 Osteocyte Organelles

Life expectancy of osteocytes might be very flexible [110], but their estimated mean half-life is approximately 25 years [128]. The cellular mechanism of a certain osteocyte indicates both its functional state within the tissue and its stage in the life cycle. In the nascent osteocytes, while the numbers and size of organelles are reduced [113, 118], the organelles structure closely resembles that of osteoblasts [129].

3.5.5 The Osteocyte Apoptosis

Osteocyte viability obviously has a critical function in the preservation of bone integrity and homeostasis. Though, osteocyte apoptosis might be critical for normal skeletal replacement and damage repair whereas blocking osteocyte apoptosis might progress diseases like bone loss because of aging or to glucocorticoid therapy. Any agents that prevent this procedure might intensify situations in which repair is needed [107]. A significant number of apoptotic osteocytes participate in various important pathological conditions, such as inflammation, androgens (As) or estrogens (Es) deficiency, excess glucocorticoids (GCs), unloading/disuse, fatigue/microdamage, and aging. These conditions are linked to a decrease in bone mineral density and an increase in bone loss. Presently, there is a prevailing consensus that the pulsatile fluid flow (PFF) surrounding osteocytes has the potential to induce fluid shear stress, hence impacting osteocytic vitality and metabolism. The regulation is achieved by modulating the pulsatile fluid flow inside the lacunar-canalicular network. It is worth mentioning that transportation of oxygen and nutrient by pulsatile fluid flow within lacunar-canalicular network plays a crucial role in preserving the vitality of osteocytes, which is particularly significant in the maintenance of bone mechanotransduction and mechanosensation [130]. Figure 3.12 provides a concise overview of the functions that pathological conditions play in the osteocyte apoptosis process.

Death of osteocyte cells is associated with pathological situations such as osteoarthritis and osteoporosis, directing to increased skeletal brittleness. Such brittleness is thought to be caused by the loss of capability to detect microdamages and/or healing signals. Similar to estrogen withdrawal and glucocorticoid treatment, oxygen deficiency has been displayed to stimulate apoptosis in osteocytes (such as happens through immobilization) [107, 131, 132]. Osteocyte death can be inhibited by many substances such as monocyte chemotactic proteins 1 and 3 (MCP1 and MCP3), calbindin-D28k, CD40 ligand, calcitonin, bisphosphonates, and estrogen and selective estrogen receptor modulators [107].

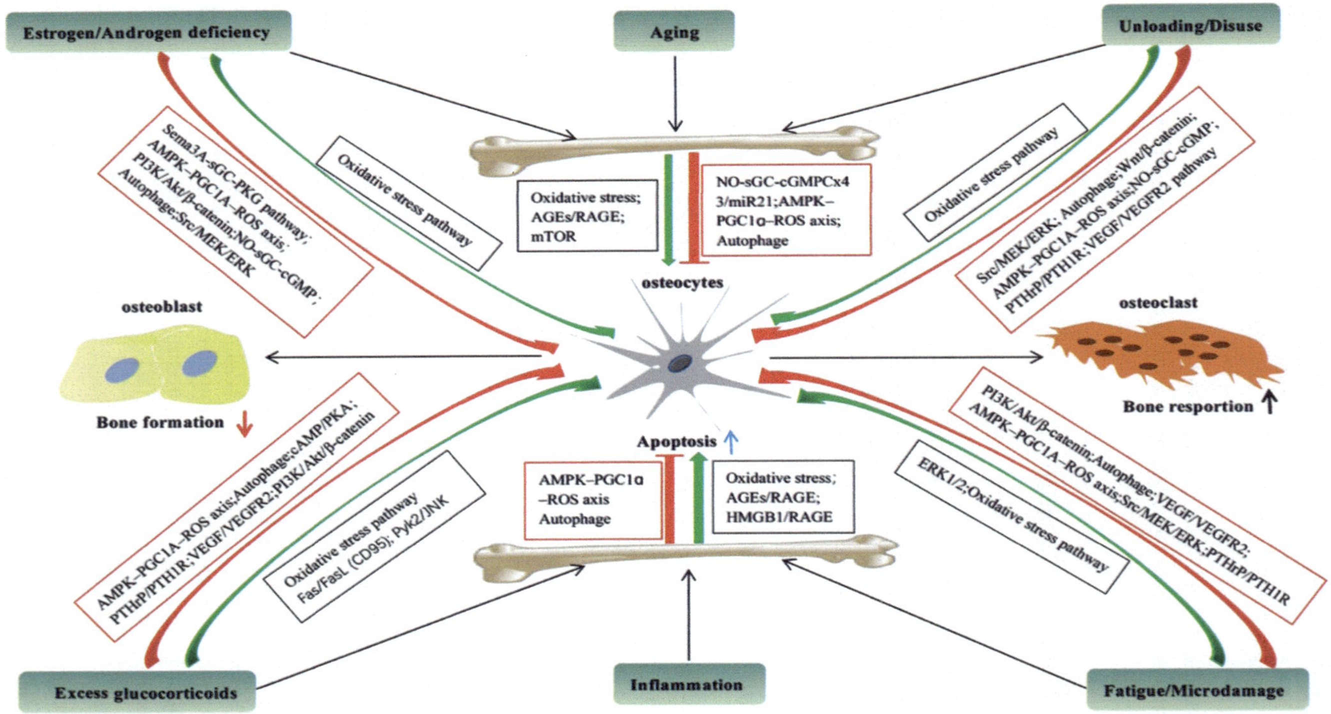

Fig. 3.12 Crucial anti-apoptotic and pro-apoptotic paths in osteocytes throughout various pathological circumstances. These signaling paths, comprising anti-apoptotic pathways (shown by green arrows) and pro-apoptotic pathways (shown by red arrows), have a strong correlation with reduced formation and elevated resorption of bone. Reprinted with permission from Ru and Wang [130]

3.6 Conclusion

The regulation of bone homeostasis is attained through the synchronized functions of osteoblasts, osteoclasts, and osteocytes. Osteoblasts originate from mesenchymal cells, while osteocytes are fully developed osteoblasts, and osteoclasts are derived from hematopoietic cells. Osteoblasts originate from MSCs. Osteoblasts and fibroblasts demonstrate a significant level of resemblance within the context of cell culture. The set of genes that are active in fibroblasts is likewise active in osteoblasts. Osteoclasts are derived from cells in the macrophage/monocyte lineage. Osteocytes function as a storage site for soluble substances that can effectively target various distant organs, such as muscle, kidney, and other tissues, as well as cells located on the bone's surface. An osteocyte is a cellular component located within the bone matrix. They arise from mesenchymal stem cells during the process of differentiating into osteoblasts. Osteocytes that have been purified have a greater degree of RANKL expression in comparison to osteoblasts, and they display a stronger capacity to promote the production of osteoclasts. The origin of osteoprogenitor cells responsible for formation of bone is unclear. The majority of available data suggests that bone marrow stromal cells are the main origin of osteoprogenitors within the body. The osteoprogenitor stromal and hemopoietic systems have a mutually reliant relationship. The interaction between material and osteoblasts depends on the characteristics of the material's surface, such as its surface energy, chemistry, and topography. Cell behavior is affected by these surface features during contact. Traditionally, the discipline of material design has mostly concentrated on the properties of large amounts of tissue, disregarding the whole spectrum of biological surface features that have been shown to affect cell activity. These surface features can range in size from 100 μm to 10 nm. Nanophase supports have been found to enhance mechanical properties and bioactivity, such as compressive and flexural moduli.

References

1. Stevens, M. M. (2008). Biomaterials for bone tissue engineering. *Materials Today, 11*(5), 18–25.
2. Levi, B., et al. (2012). In vivo directed differentiation of pluripotent stem cells for skeletal regeneration. *Proceedings of the National Academy of Sciences, 109*(50), 20379–20384.
3. Renthal, N. E., & Ma, N. S. (2018). Normal bone physiology 101. In: *A practical approach to adolescent bone health: A guide for the primary care provider* (p. 11–25).
4. Kalantari, E., et al. (2018). Nanostructured monticellite: An emerging player in tissue engineering. *Materials Today: Proceedings, 5*(7, Part 3), 15744–15753.
5. Kalantari, E., et al. (2019). Biocomposites based on hydroxyapatite matrix reinforced with nanostructured monticellite ($CaMgSiO_4$) for biomedical application: Synthesis, characterization, and biological studies. *Materials Science and Engineering: C, 105*, 109912.
6. Kalantari, E., & Naghib, S. M. (2019). A comparative study on biological properties of novel nanostructured monticellite-based composites with hydroxyapatite bioceramic. *Materials Science and Engineering: C, 98*, 1087–1096.

7. Kalantari, E., et al. (2018). Nanostructured monticellite for tissue engineering applications—Part I: Microstructural and physicochemical characteristics. *Ceramics International, 44*(11), 12731–12738.
8. Kalantari, E., et al. (2018). Nanostructured monticellite for tissue engineering applications—Part II: Molecular and biological characteristics. *Ceramics International, 44*(12), 14704–14711.
9. Einhorn, T. A., & Gerstenfeld, L. C. (2015). Fracture healing: Mechanisms and interventions. *Nature Reviews Rheumatology, 11*(1), 45.
10. Ono, T., & Takayanagi, H. (2017). Osteoimmunology in bone fracture healing. *Current Osteoporosis Reports, 15*(4), 367–375.
11. Rahmanian, M., et al. (2019). Multifunctional gelatin–tricalcium phosphate porous nanocomposite scaffolds for tissue engineering and local drug delivery: In vitro and in vivo studies. *Journal of the Taiwan Institute of Chemical Engineers, 101*, 214–220.
12. Seyfoori, A., et al. (2019). Multifunctional magnetic $ZnFe_2O_4$-hydroxyapatite nanocomposite particles for local anti-cancer drug delivery and bacterial infection inhibition: An in vitro study. *Journal of the Taiwan Institute of Chemical Engineers, 96*, 503–508.
13. Askari, E., et al. (2022). Highly bioactive Akermanite-Monticellite nanocomposites for bone tissue engineering: A tunable three-dimensional biological study. *Journal of Materials Research and Technology, 20*, 4066–4076.
14. Kashte, S., Jaiswal, A. K., & Kadam, S. (2017). Artificial bone via bone tissue engineering: Current scenario and challenges. *Tissue Engineering Regenerative Medicine, 14*(1), 1–14.
15. Li, W.-J., et al. (2005). Multilineage differentiation of human mesenchymal stem cells in a three-dimensional nanofibrous scaffold. *Biomaterials, 26*(25), 5158–5166.
16. Li, W. J., et al. (2002). Electrospun nanofibrous structure: A novel scaffold for tissue engineering. *Journal of Biomedical Materials Research: An Official Journal of The Society for Biomaterials, The Japanese Society for Biomaterials, The Australian Society for Biomaterials, the Korean Society for Biomaterials, 60*(4), 613–621.
17. O'Connor, J. P. (1998). Animal models of heterotopic ossification. *Clinical Orthopaedics Related Research, 346*, 71–80.
18. Reddi, A. H., & Huggins, C. B. (1975). Formation of bone marrow in fibroblast-transformation ossicles. *Proceedings of the National Academy of Sciences, 72*(6), 2212–2216.
19. Sampath, T., & Reddi, A. H. (1981). Dissociative extraction and reconstitution of extracellular matrix components involved in local bone differentiation. *Proceedings of the National Academy of Sciences, 78*(12), 7599–7603.
20. Friedenstein, A. (1968). Induction of bone tissue by transitional epithelium. *Clinical Orthopaedics Related Research, 59*, 21–38.
21. Hamdy, F. C., et al. (1997). Immunolocalization and messenger RNA expression of bone morphogenetic protein-6 in human benign and malignant prostatic tissue. *Cancer Research, 57*(19), 4427–4431.
22. Hirano, H., & Urist, M. R. (1991). Induced bone development in transplants of fresh human pseudomalignant heterotopic ossification tissue in athymic nude mice. *Clinical Orthopaedics Related Research, 263*, 113–120.
23. McKibbin, B. (1978). The biology of fracture healing in long bones. *The Journal of Bone Joint Surgery. British Volume, 60*(2), 150–162
24. Włodarski, K. (1978). Failure of heterotopic osteogenesis by epithelial mesenchymal cell interactions in xenogeneic transplants in the kidney. *Calcified Tissue Research, 25*(1), 7–11.
25. Yoo, J. U., & Johnstone, B. (1998). The role of osteochondral progenitor cells in fracture repair. *Clinical Orthopaedics Related Research, 355*, S73–S81.
26. Włodarski, K. (1990). Properties and origin of osteoblasts. *Clinical Orthopaedics Related Research, 252*, 276–293.
27. Benayahu, D., et al. (1989). Bone marrow-derived stromal cell line expressing osteoblastic phenotype in vitro and osteogenic capacity in vivo. *Journal of Cellular Physiology, 140*(1), 1–7.

28. Bruder, S. P., et al. (1998). Mesenchymal stem cells in osteobiology and applied bone regeneration. *Clinical Orthopaedics Related Research®, 355*, S247–S256.
29. Caplan, A. I. (1991). Mesenchymal stem cells. *Journal of Orthopaedic Research, 9*(5), 641–650.
30. Friedenstein, A., Chailakhyan, R., & Gerasimov, U. (1987). Bone marrow osteogenic stem cells: In vitro cultivation and transplantation in diffusion chambers. *Cell Proliferation, 20*(3), 263–272.
31. Owen, M. (1980). The origin of bone cells in the postnatal organism. *Arthritis Rheumatism: Official Journal of the American College of Rheumatology, 23*(10), 1073–1080.
32. Prockop, D. J. (1997). Marrow stromal cells as stem cells for nonhematopoietic tissues. *Science, 276*(5309), 71–74.
33. Pereira, R., et al. (1995). Cultured adherent cells from marrow can serve as long-lasting precursor cells for bone, cartilage, and lung in irradiated mice. *Proceedings of the National Academy of Sciences, 92*(11), 4857–4861.
34. Reddi, A. H. (1995). Bone morphogenetic proteins, bone marrow stromal cells, and mesenchymal stem cells. Maureen Owen revisited. *Clinical Orthopaedics Related Research, 1995*(313), 115–119.
35. Rickard, D. J., et al. (1996). Isolation and characterization of osteoblast precursor cells from human bone marrow. *Journal of Bone Mineral Research, 11*(3), 312–324.
36. Bosch, P., et al. (2000). Osteoprogenitor cells within skeletal muscle. *Journal of Orthopaedic Research, 18*(6), 933–944.
37. Urist, M. R., et al. (1969). 23 osteogenetic competence. *Clinical Orthopaedics Related Research®, 64*, 194–220.
38. Urist, M. R., et al. (1968). Inductive substrates for bone formation. *Clinical Orthopaedics Related Research®, 59*, 59–96.
39. Salhotra, A., et al. (2020). Mechanisms of bone development and repair. *Nature Reviews Molecular Cell Biology, 21*(11), 696–711.
40. Connolly, J. F. (1998). Clinical use of marrow osteoprogenitor cells to stimulate osteogenesis. *Clinical Orthopaedics Related Research®, 355*, S257–S266.
41. Strates, B. S., Stock, A. J., & Connolly, J. F. (1988). Skeletal repair in the aged: A preliminary study in rabbits. *The American Journal of the Medical Sciences, 296*(4), 266–269.
42. Majors, A. K., et al. (1997). Characterization of human bone marrow stromal cells with respect to osteoblastic differentiation. *Journal of Orthopaedic Research, 15*(4), 546–557.
43. Haynesworth, S., et al. (1992). Characterization of cells with osteogenic potential from human marrow. *Bone, 13*(1), 81–88.
44. Bruder, S. P., Jaiswal, N., & Haynesworth, S. E. (1997). Growth kinetics, self-renewal, and the osteogenic potential of purified human mesenchymal stem cells during extensive subcultivation and following cryopreservation. *Journal of Cellular Biochemistry, 64*(2), 278–294.
45. Bolander, M. E., & Balian, G. (1986). The use of demineralized bone matrix in the repair of segmental defects. Augmentation with extracted matrix proteins and a comparison with autologous grafts. *The Journal of Bone Joint Surgery. American Volume, 68*(8), 1264–1274.
46. Glowacki, J., & Mulliken, J. (1985). Demineralized bone implants. *Clinics in Plastic Surgery, 12*(2), 233–241.
47. Green, E., Hinton, C., & Triffitt, J. T. (1986). The effect of decalcified bone matrix on the osteogenic potential of bone marrow. *Clinical Orthopaedics Related Research, 205*, 292–298.
48. Lindholm, T. S., & Urist, M. R. (1980). A quantitative analysis of new bone formation by induction in compositive grafts of bone marrow and bone matrix. *Clinical Orthopaedics Related Research, 150*, 288–300.
49. Nade, S., et al. (1983). Osteogenesis after bone and bone marrow transplantation. The ability of ceramic materials to sustain osteogenesis from transplanted bone marrow cells: preliminary studies. *Clinical Orthopaedics Related Research*, (181), 255–263.
50. Salama, R., & Weissman, S. (1978). The clinical use of combined xenografts of bone and autologous red marrow. A preliminary report. *The Journal of Bone Joint Surgery. British Volume, 60*(1), 111–115.

51. Connolly, J., et al. (1989). Development of an osteogenic bone-marrow preparation. *The Journal of Bone & Joint Surgery, 71*(5), 684–691.
52. Clark, P., et al. (1987). Topographical control of cell behaviour. I. Simple step cues. *Developmental Dynamics: An Official Publication of the American Association of Anatomists, 99*(3), 439–448.
53. Dalby, M. J., et al. (2003). Nucleus alignment and cell signaling in fibroblasts: Response to a micro-grooved topography. *Experimental Cell Research, 284*(2), 272–280.
54. Wójciak-Stothard, B., et al. (1995). Activation of macrophage-like cells by multiple grooved substrata. Topographical control of cell behaviour. *Cell Biology International, 19*(6), 485–490.
55. Chen, C. S., et al. (1997). Geometric control of cell life and death. *Science, 276*(5317), 1425–1428.
56. Britland, S., et al. (1996). Synergistic and hierarchical adhesive and topographic guidance of BHK cells. *Experimental Cell Research, 228*(2), 313–325.
57. Dalby, M. J., et al. (2006). Osteoprogenitor response to defined topographies with nanoscale depths. *Biomaterials, 27*(8), 1306–1315.
58. Bianco, P., et al. (2001). Bone marrow stromal stem cells: Nature, biology, and potential applications. *Stem Cells, 19*(3), 180–192.
59. Friedenstein, A. J. (1976). Precursor cells of mechanocytes. *International Review of Cytology, 47*(327), 59.
60. Gasser, J. A., & Kneissel, M. (2017). Bone physiology and biology. *Bone Toxicology,* 27–94.
61. Zhang, L., et al. (2020). The effects of biomaterial implant wear debris on osteoblasts. *Frontiers in Cell and Developmental Biology, 8*, 352.
62. Mizoguchi, T., & Ono, N. (2020). The diverse origin of bone-forming osteoblasts. *Journal of Bone and Mineral Research, 36*(8), 1432–1447.
63. Ducy, P., et al. (1996). Increased bone formation in osteocalcin-deficient mice. *Nature Materials, 382*(6590), 448–452.
64. Ducy, P., et al. (1997). Osf2/Cbfa1: A transcriptional activator of osteoblast differentiation. cell proliferation. *89*(5), 747–754.
65. Rutkovskiy, A., Stensløkken, K.-O., & Vaage, I. J. (2016). Osteoblast differentiation at a glance. *Medical Science Monitor Basic Research, 22*, 95.
66. Ponzetti, M., & Rucci, N. (2021). Osteoblast differentiation and signaling: Established concepts and emerging topics. *International Journal of Molecular Sciences, 22*(13), 6651.
67. Ducy, P., Schinke, T., & Karsenty, G. (2000). The osteoblast: A sophisticated fibroblast under central surveillance. *Science, 289*(5484), 1501–1504.
68. Khotib, J., et al. (2021). Signaling pathway and transcriptional regulation in osteoblasts during bone healing: Direct involvement of hydroxyapatite as a biomaterial. *Pharmaceuticals, 14*(7), 615.
69. Anselme, K. (2000). Osteoblast adhesion on biomaterials. *Biomaterials, 21*(7), 667–681.
70. Edwards, J. R., & Mundy, G. R. (2011). Advances in osteoclast biology: Old findings and new insights from mouse models. *Nature Reviews Rheumatology, 7*(4), 235.
71. Rodan, G. A., & Martin, T. J. (1981). *Role of osteoblasts in hormonal control of bone resorption—A hypothesis.* Springer.
72. Sun, Y., et al. (2021). Recent advances in osteoclast biological behavior. *Frontiers in Cell and Developmental Biology, 9*, 788680.
73. Ono, T., & Nakashima, T. (2018). Recent advances in osteoclast biology. *Histochemistry Cell Biology, 149*(4), 325–341.
74. Seita, J., & Weissman, I. L. (2010). Hematopoietic stem cell: Self-renewal versus differentiation. *Wiley Interdisciplinary Reviews: Systems Biology Medicine, 2*(6), 640–653.
75. Arai, F., et al. (1999). Commitment and differentiation of osteoclast precursor cells by the sequential expression of c-Fms and receptor activator of nuclear factor κB (RANK) receptors. *The Journal of Experimental Medicine, 190*(12), 1741–1754.
76. Roodman, G. D. (1999). Cell biology of the osteoclast. *Experimental Hematology, 27*(8), 1229–1241.

77. McDonald, M. M., et al. (2021). New insights into osteoclast biology. *Journal of Bone and Mineral Research Plus, 5*(9), e10539.
78. Zachowski, A. (1993). Phospholipids in animal eukaryotic membranes: Transverse asymmetry and movement. *Biochemical Journal, 294*(Pt 1), 1.
79. Irie, A., et al. (2017). Phosphatidylethanolamine dynamics are required for osteoclast fusion. *Scientific Reports, 7*, 46715.
80. Kukita, T., et al. (2004). RANKL-induced DC-STAMP is essential for osteoclastogenesis. *The Journal of Experimental Medicine, 200*(7), 941–946.
81. Yagi, M., et al. (2005). DC-STAMP is essential for cell–cell fusion in osteoclasts and foreign body giant cells. *The Journal of Experimental Medicine, 202*(3), 345–351.
82. Hentunen, T. A., et al. (1998). Immortalization of osteoclast precursors by targeting Bcl-XL and Simian virus 40 large T antigen to the osteoclast lineage in transgenic mice. *The Journal of Clinical Investigation, 102*(1), 88–97.
83. Osdoby, P., Martini, M. C., & Caplan, A. I. (1982). Isolated osteoclasts and their presumed progenitor cells, the monocyte, in culture. *Journal of Experimental Zoology, 224*(3), 331–344.
84. Sims, N. A., & Lorenzo, J. A. (2020). Local regulators of bone: Interleukin-1, tumor necrosis factor, interferons, the IL-6 family, and additional cytokines. *Principles of bone biology* (pp. 1205–1246). Elsevier.
85. Kurihara, N., et al. (1990). IL-6 stimulates osteoclast-like multinucleated cell formation in long term human marrow cultures by inducing IL-1 release. *The Journal of Immunology, 144*(11), 4226–4230.
86. Paul, S., et al. (1990). Molecular cloning of a cDNA encoding interleukin 11, a stromal cell-derived lymphopoietic and hematopoietic cytokine. *Proceedings of the National Academy of Sciences, 87*(19), 7512–7516.
87. Yates, A., et al. (1992). Expression of human transforming growth factor α by Chinese hamster ovarian tumors in nude mice causes hypercalcemia and increased osteoclastic bone resorption. *Journal of Bone Mineral Research, 7*(7), 847–853.
88. Takahashi, N., et al. (1986). Recombinant human transforming growth factor-alpha stimulates the formation of osteoclast-like cells in long-term human marrow cultures. *The Journal of Clinical Investigation, 78*(4), 894–898.
89. Mun, S. H., Park, P. S. U., & Park-Min, K.-H. (2020). The M-CSF receptor in osteoclasts and beyond. *Experimental & Molecular Medicine, 52*(8), 1239–1254.
90. Zhao, B. (2017). TNF and bone remodeling. *Current Osteoporosis Reports, 15*, 126–134.
91. Pfeilschifter, J., et al. (1989). Interleukin-1 and tumor necrosis factor stimulate the formation of human osteoclastlike cells in vitro. *Journal of Bone Mineral Research, 4*(1), 113–118.
92. Fried, R., et al. (1989). Two squamous cell carcinomas not associated with humoral hypercalcemia produce a potent bone resorption-stimulating factor which is interleukin-1α. *Endocrinology, 125*(2), 742–751.
93. De Larco, J. E., & Todaro, G. J. (1978). Growth factors from murine sarcoma virus-transformed cells. *Proceedings of the National Academy of Sciences, 75*(8), 4001–4005.
94. Jann, J., et al. (2020). Influence of the TGF-β superfamily on osteoclasts/osteoblasts balance in physiological and pathological bone conditions. *International Journal of Molecular Sciences, 21*(20), 7597.
95. Trivedi, T., et al. (2021). The role of TGF-β in bone metastases. *Biomolecules, 11*(11), 1643.
96. Yan, T., et al. (2021). Role of nitric oxide in orthodontic tooth movement. *International Journal of Molecular Medicine, 48*(3), 1–14.
97. Nascimento, M. H., et al. (2020). How can nitric oxide help osteogenesis? *AIMS Molecular Science, 7*(1), 29–48.
98. Simonet, W. S., et al. (1997). Osteoprotegerin: A novel secreted protein involved in the regulation of bone density. *Cell, 89*(2), 309–319.
99. Jayash, S. N., Al-Namnam, N. M., Shaghayegh, G. (2021). Osteoprotegerin (OPG) pathways in bone diseases and its application in therapeutic perspectives. *Biointerface Research in Applied Chemistry*.

100. Udagawa, N., et al. (2021). Osteoclast differentiation by RANKL and OPG signaling pathways. *Journal of Bone and Mineral Metabolism, 39*, 19–26.
101. Xiong, Q., et al. (2016). The roles of interferons in osteoclasts and osteoclastogenesis. *Joint, Bone, Spine, 83*(3), 276–281.
102. Gao, Y., et al. (2007). IFN-γ stimulates osteoclast formation and bone loss in vivo via antigen-driven T cell activation. *The Journal of Clinical Investigation, 117*(1), 122–132.
103. Manolagas, S., Kousteni, S., & Jilka, R. (2002). Sex steroids and bone. *Recent Progress in Hormone Research, 57*, 385–410.
104. Elmore, S. (2007). Apoptosis: A review of programmed cell death. *Toxicologic Pathology, 35*(4), 495–516.
105. Soysa, N. S., & Alles, N. (2019). Positive and negative regulators of osteoclast apoptosis. *Bone Reports, 11*, 100225.
106. Hughes, D. E., et al. (1995). Bisphosphonates promote apoptosis in murine osteoclasts in vitro and in vivo. *Journal of Bone Mineral Research, 10*(10), 1478–1487.
107. Bonewald, L. F. (2011). The amazing osteocyte. *Journal of Bone Mineral Research, 26*(2), 229–238.
108. Lopes, D., et al. (2018). Bone physiology as inspiration for tissue regenerative therapies. *Biomaterials, 185*, 240–275.
109. Van der Plas, A., & Nijweide, P. (1992). Isolation and purification of osteocytes. *Journal of Bone Mineral Research, 7*(4), 389–396.
110. Marotti, G., et al. (1990). Structure-function relationships in the osteocyte.
111. Tanaka, K., et al. (1995). Time-lapse microcinematography of osteocytes. *Mineral Electrolyte Metabolism, 21*(1–3), 189–192.
112. Franz-Odendaal, T. A., Hall, B. K., & Witten, P. E. (2006). Buried alive: How osteoblasts become osteocytes. *Developmental Dynamics: An official Publication of the American Association of Anatomists, 235*(1), 176–190.
113. Tate, M. L. K., et al. (2004). The osteocyte. *The International Journal of Biochemistry Cell Biology, 36*(1), 1–8.
114. Doty, S. B. (1981). Morphological evidence of gap junctions between bone cells. *Calcified Tissue International, 33*(1), 509–512.
115. Knothe, M. T. (2003). Whither flows the fluid in bone? An osteocyte's perspective. *Journal of Biomechanics, 36*(10), 1409–1424.
116. Aarden, E. M., Nijweide, P. J., & Burger, E. H. (1994). Function of osteocytes in bone. *Journal of Cellular Biochemistry, 55*(3), 287–299.
117. Robling, A. G., & Bonewald, L. F. (2020). The osteocyte: New insights. *Annual Review of Physiology, 82*, 485–506.
118. Palumbo, C. (1986). A three-dimensional ultrastructural study of osteoid-osteocytes in the tibia of chick embryos. *Cell Tissue Research, 246*(1), 125–131.
119. Bonewald, L. F. (2007). Osteocytes as dynamic multifunctional cells. *Annals of the New York Academy of Sciences, 1116*(1), 281–290.
120. Marotti, G., et al. (1995). Quantitative evaluation on osteocyte canalicular density in human secondary osteons. *Bone, 16*(1), 125–128.
121. Robling, A. G., et al. (2002). Shorter, more frequent mechanical loading sessions enhance bone mass. *Medicine Science in Sports Exercise, 34*(2), 196–202.
122. Turner, C. H., Forwood, M. R., & Otter, M. W. (1994). Mechanotransduction in bone: Do bone cells act as sensors of fluid flow? *The FASEB Journal, 8*(11), 875–878.
123. Rubin, C. T. (1984). Skeletal strain and the functional significance of bone architecture. *Calcified Tissue International, 36*(1), S11–S18.
124. Robling, A. G., et al. (2008). Mechanical stimulation of bone in vivo reduces osteocyte expression of Sost/sclerostin. *Journal of Biological Chemistry, 283*(9), 5866–5875.
125. Gluhak-Heinrich, J., et al. (2003). Mechanical loading stimulates dentin matrix protein 1 (DMP1) expression in osteocytes in vivo. *Journal of Bone Mineral Research, 18*(5), 807–817.
126. Gluhak-Heinrich, J., et al. (2007). MEPE expression in osteocytes during orthodontic tooth movement. *Archives of Oral Biology, 52*(7), 684–690.

127. Skerry, T. M., et al. (1989). Early strain-related changes in enzyme activity in osteocytes following bone loading in vivo. *Journal of Bone Mineral Research, 4*(5), 783–788.
128. Frost, H. M. (1963). Bone remodelling dynamics. Thomas.
129. Dudley, H. R., & Spiro, D. (1961). The fine structure of bone cells. *The Journal of Cell Biology, 11*(3), 627–649.
130. Ru, J.-Y., & Wang, Y.-F. (2020). Osteocyte apoptosis: The roles and key molecular mechanisms in resorption-related bone diseases. *Cell Death & Disease, 11*(10), 846.
131. Weinstein, R. S., Nicholas, R. W., & Manolagas, S. C. (2000). Apoptosis of osteocytes in glucocorticoid-induced osteonecrosis of the hip. *J Clin Endocrinol Metab, 85*(8), 2907–2912.
132. Tomkinson, A., et al. (1997). The death of osteocytes via apoptosis accompanies estrogen withdrawal in human bone. *J Clin Endocrinol Metab, 82*(9), 3128–3135.

Chapter 4
Bone Extracellular Matrix

Abstract In order for regenerative medicine strategies to achieve desired outcomes, the materials employed, mostly consisting of combinations of stem cells, growth factors, and scaffolds, must possess the ability to augment the regeneration of the original tissue or serve as a replacement for damaged tissue, functioning in a manner similar to the natural tissue. Due to variations in regeneration capabilities among different tissues, certain tissues may just require biologics and biomaterials, while others with partial regenerative capabilities necessitate the presence of cells, biomolecules, and biomaterials for regeneration. The ECM is accountable for preserving the flexibility and structural integrity of tissues. The structure undergoes constant remodeling in response to changes in the local pH levels, growth factors, and receptor abundance. The extracellular matrix (ECM) of bone and cartilage is unique, as it specifically caters to the distinct functional needs of these tissues. Collagen is the primary constituent of all extracellular matrices (ECMs). The bone extracellular matrix is predominantly composed of collagenous proteins, accounting for 90% of its makeup. The remaining 10% consists of non-collagenous proteins. Osteoblasts are accountable for the production of these proteins, primarily involved in cellular adhesion. Hydroxyapatite is the primary inorganic constituent of hard tissues. Collagen provides a structural base that aids in the deposition of hydroxyapatite during the mineralization of tissues. The extracellular matrix (ECM) of bone can be categorized into two components: organic and mineral. Each entity has distinct and individual attributes. It is important to consider that materials intended for insertion into bone tissue should replicate these characteristics. This chapter presents a thorough overview of the extracellular matrix of bones.

Abbreviations

BMSCs	Bone marrow stromal cells
BTE	Bone tissue engineering
ClnD1	Cyclin D1
COL	Collagen

R. Gholami and S. M. Naghib, *Nanobioceramics for Bone Tissue Engineering and Regenerative Biomedicine*, Biological and Medical Physics, Biomedical Engineering,
https://doi.org/10.1007/978-981-96-0041-0_4

DCB	Decellularized bone matrix
DEB	Dystrophic epidermolysis bullosa
DMB	Demineralized bone matrix
ECM	Extracellular matrix
ECMRs	ECM receptors
FACIT	Fibril-associated collagens with interrupted triple helix
FGF	Fibroblast growth factor
FRAP	FK506-binding protein/rapamycin-associated protein
HAP	Hydroxyapatite
IVD	Intervertebral disk
MF	Microfilament
MSCs	Mesenchymal stem cells
MSS	Mechanical stress sensor
MTs	Microtubules
NCPs	Non-collagenous proteins
p70S6K	P70 ribosomal S6 protein kinase
TGF-β	Transforming growth factor-β

4.1 Introduction

The fields of regenerative medicine and/or tissue engineering encompass the application of biological and engineering concepts in order to facilitate the regeneration of diseased or damaged organs and tissues, as well as generate new organs and tissues [1]. With the help of regenerative medicine, injuries can be repaired that the body is unable to repair [2]. Inspired by the demand for development of reliable and safe suppliers of organs and tissues, clinicians and scientists are developing technologies and therapies that can regenerate tissues and, in some cases, generate completely new tissues [1, 3–5]. Connective tissue is classified as one of the four basic forms of human tissue predominantly consisting of fibrous extracellular matrix constituents [6]. Intervertebral disk (IVD), cartilage, bone, adipose tissue, ligaments, and tendons are connective tissues that play several physiological roles such as structural integrity, endocrine function, and nutrient storage [7, 8]. The role of connective tissues is contingent upon the biochemical and physical characteristics of their ECM, which are influenced by the composition of ECM proteins [7]. Bone tissue possesses the unique ability to undergo self-repair without the formation of scar tissue. Nevertheless, there exists a crucial threshold of 2.5 cm for the majority of bones [9], beyond which no regeneration will happen. In such instances, the support and induction of osteogenesis is essential for healing of the defect [2]. Bone tissue, like other body tissues, has a hierarchical configuration on longitudinal scales that covers many important arrangements from macro (cm) to nanostructured (ECM) constituents (Fig. 4.1). The nanocomposite structure (supple and hard collagen fibers supported by HAP crystals) is essential for the high fracture toughness and compressive strength required by

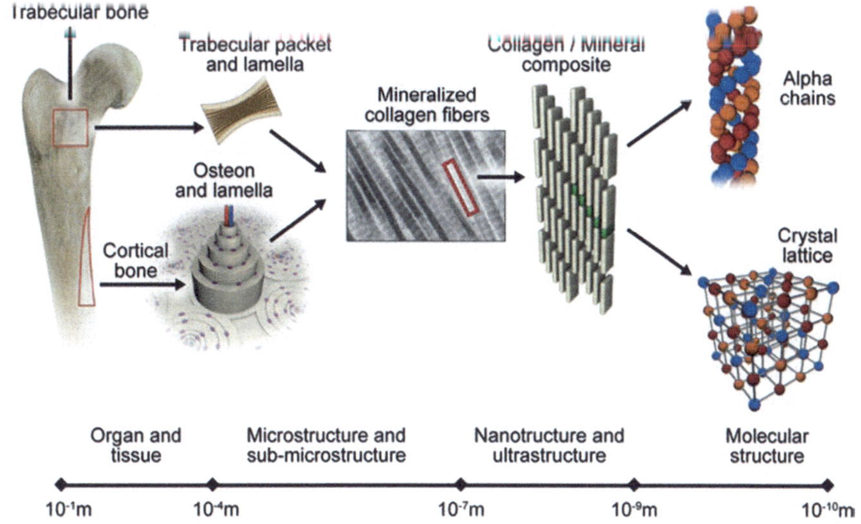

Fig. 4.1 Hierarchical structure of bone: The bone at the organ level consists of cancellous (trabecular) bone found in the marrow cavity, and dense cortical bone that forms the outer shell. Cortical bone consists of several Haversian systems, also known as osteons, at the microscopic level. Osteons consist of a core canal that contains lymphatics, nerves, and a blood vessel surrounded by concentric lamellae layers. The structure of trabecular bone, while also lamellar, is composed of a mixture of lamellae that are mostly aligned with the trabecular surface and the fragments of earlier bone that have been remodeled and may resemble osteons in certain regions. At the microscopic and nanoscopic levels, bone consists of collagen fibers that are accompanied by mineral plates, which are found both between and within the collagen fibrils. The collagen fibrils consist of molecules that assemble into a triple helix structure, comprising two α_1 and one single α_2 chain. Reprinted with permission from Gasser and Kneissel [11]

bone. There is an expanding acknowledgment that a nanoscale inorganic substance is probably more biologically active than a micro-sized substance. For example, collagen-HAP nanocomposite systems in tissue engineering are showing promise and coming to light quickly [10].

The advancement of tissue engineering technology has facilitated the utilization of biomaterials made by 3D printing, nanotechnology, and materials engineering in the development of new implants for the purpose of bone regeneration. Nevertheless, many of these new materials have certain limitations, including high immunogenicity, low osteoinductivity, and inadequate biocompatibility [12].

4.2 Tissue Engineering and Regenerative Medicine

Tissue engineering and regenerative medicine terms are used by clinicians and scientists with considerable overlap. Regenerative medicine has exhibited promising results for the replacement and regeneration of various organs and tissues including liver, kidneys, heart, and skin, and the potential to even correct some congenital defects [1, 13, 14]. The potential of regenerative medicine is based on the ability and potential to replace and regenerate damaged organs and tissues [13, 15–19]. Many innovations and advances are taking place in the areas of tissue engineering and regenerative medicine, and it has a great effect on the 3D bioprinting of organs and tissues [20–22]. 3D bioprinting holds great promise for bioprinting of organs and artificial tissues, thus creating a revolution in the field of regenerative medicine [1]. For regenerative medicine approaches to be effective, the material used, mainly combinations of stem cells, growth factors, and scaffolds, should be able to enhance regeneration of the original tissue or substitute the damaged tissue and act as the original tissue [23, 24]. Cells used in tissue engineering and regenerative medicine can be obtained from another patient (allogeneic) or from the same patient (autologous). Moreover, xenogeneic cells like those from animals can also be used in regenerative medicine approaches. Keratinocytes, chondrocytes, fibroblasts, and stem cells are the cells that have been used so far [1, 25, 26].

Stem cells have received a great deal of attention in regenerative medicine, such as bone regeneration, due to their suitable capability for osteogenic differentiation and proliferation ability. Furthermore, they are able to synthesize an ECM that can accelerate repair and calcification, thus returning damaged bone function [12, 27–29]. Depending on the patient's age, some regenerative medicine strategies can use the natural treatment process of the body and accelerate it [30, 31]. These approaches are performed with the aim of modifying the environment of the tissue by introducing biological factors and exogenous materials to improve and accelerate the healing process of the body. ECM biomimetics and materials have been used for many years and do more than just providing the physical structure [1, 32–34].

Because of the different regeneration capabilities of different tissues, some tissues may not need cells but only the biologics and biomaterials, while other tissues have partial regenerative capabilities and need the cells, biomolecules, and biomaterial to regenerate. Organs and tissues that are not able to regenerate at all or are limited include the cornea and cartilage, while those that have a high regenerative capacity include the lungs and liver [1, 16, 35, 36]. Various factors, such as the cellular source and the biomaterial used, should be considered through the manufacture of graft or tissue [37, 38]. Such considerations allow for suitable interactions between cells and matrices (biomaterials) and enhance the scaffold function. Regenerated tissue for transplantation should recapitulate natural tissue in having a particular cell type, with a particular function [1, 39, 40].

The field of regenerative medicine uses the inherent capabilities of the human body with the help of engineered biomaterials to effectively regenerate and repair injured tissues [41–44]. The modulation of cellular response to direct curing of tissues can be achieved through the design of responsive biomaterials that possess the proper biochemical and biophysical properties. Inorganic biomaterials have recently demonstrated the ability to regulate cellular responses, such as interactions between matrices and cells. Furthermore, these mineral-based biomaterials release ions that have an important impact on the facilitation of tissue-specific functions and determination of cellar identity. The inherent characteristics of inorganic biomaterials, including the release of bioactive ions (for instance lanthanum, europium, vanadium, silver, gold, manganese, molybdenum, cobalt, chromium, zinc, copper, iron, boron, silicon, strontium, magnesium, and calcium), can be used to modulate the immune microenvironment or provoke phenotypic alterations in cells in order to guide regeneration and healing of tissues. It has been shown that biophysical properties of biomaterials, including stiffness, interactions, electrostatic, size, charge, and topography can be modulated by the incorporation of inorganic micro- and nanoparticles into polymeric networks [45].

Biological molecules, including minerals, metabolites, and proteins, are employed as chemical stimulants to direct and control cellular processes. Furthermore, metabolic products possessing the capacity to guide cellular processes as nutrients undergo degradation into smaller components that serve as essential components in several significant biochemical pathways [46]. Similarly, the body uses ions released from minerals as signaling molecules to regulate several physiological activities including oxygen transport, nerve response, and heartbeat [47]. Although minerals play a significant role in the human body, their significance in regenerative medicine is often overlooked. Minerals play critical roles in homeostasis, and also perform as fundamental constituents for crucial tissue systems, including bone, by imparting structural integrity for the skeletal system [45].

Inorganic biomaterials have several applications in regenerative medicine due to their adjustable characteristics. Some of these applications are shown in Fig. 4.2a. Their characteristics can be categorized as biochemical or biophysical signs that can lead tissue regeneration [48]. Cellular functions can be directed by biochemical characteristics, including the release of therapeutics biomolecules or ionic dissolution products, through intracellular signaling. For instance, studies have demonstrated that ions like lithium $\left(Li^{+}\right)$ can stimulate osteogenic differentiation and enhance the wingless INT-1 (WNT) signaling pathway in stem cells. Furthermore, mineral ions have important functions in biochemical processes including cofactor activation in the case of magnesium ions $\left(Mg^{2+}\right)$ [49], or apoptosis regulation in the case of calcium ions $\left(Ca^{2+}\right)$ [50]. Mineral-based biomaterials can store and release biomolecules to direct and control cellular functions [45]. The addition of inorganic biomaterials can modify the biophysical characteristics of biomaterials used in regenerative medicine, including their charge, stiffness, topography, surface-to-volume ratio, size, and shape [51, 52]. For instance, while soft biomaterials can enable chondrogenic development, stiff biomaterials have been exhibited to enhance osteogenic differentiation of stem cells and ease adhesion of cells [53]. Therefore, the interactions between inorganic-based biomaterials and biological systems are influenced by their biochemical and biophysical properties, which are crucial in the field of regenerative medicine [45]. Micro- and nanoscale mineral particles can easily penetrate into cells, allowing them to produce ionic dissolution products that can offer bioactive signals to cells. Also, the combination of these micro- and nanoscale particles with different polymers is used to design scaffolds such as hydrogels, microporous structures, and fibrous scaffolds (Fig. 4.2b) [45].

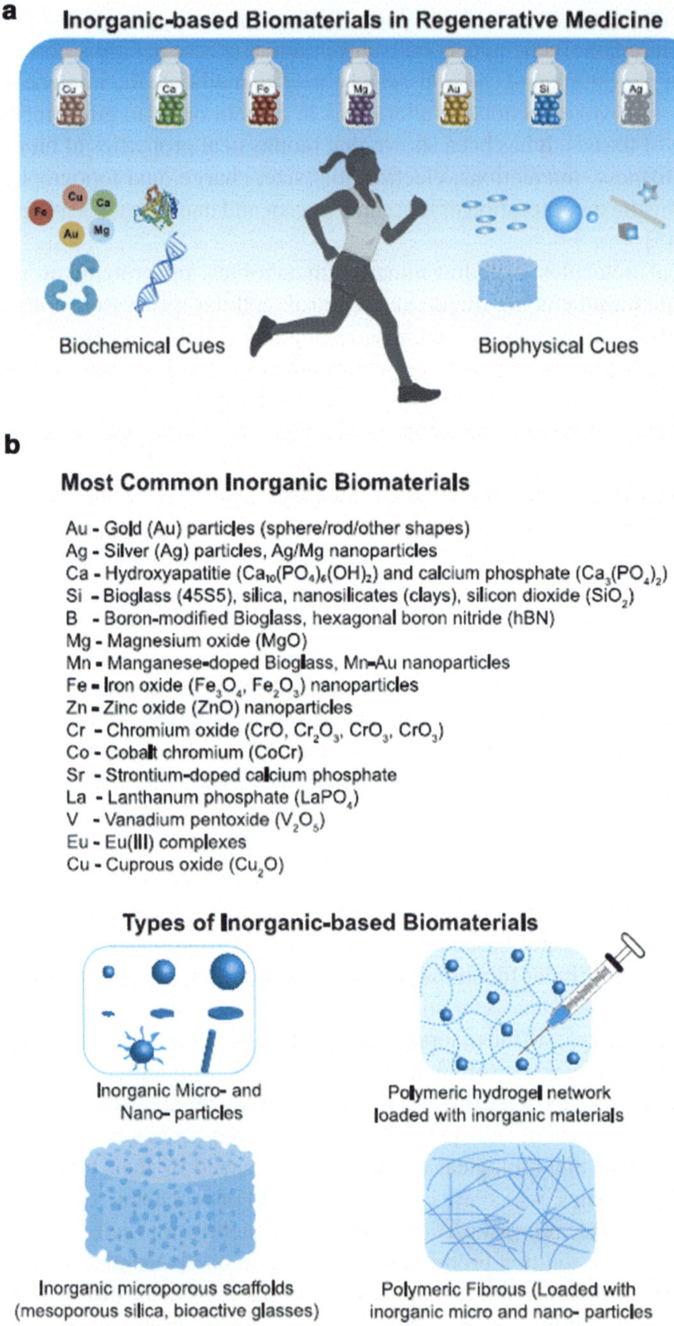

Fig. 4.2 Utilization of inorganic biomaterials in the field of regenerative medicine. **a** Inorganic biomaterials offer both biophysical and biochemical cues that promote regeneration and healing of tissues. **b** Some of the most often studied inorganic nanomaterials that have been studied in the field of regenerative medicine. Reprinted with permission from Brokesh and Gaharwar [45]

Biophysical signs refer to the physical properties of biomaterials, including material stiffness, electrostatic interactions, size, charge, and topography, which direct biological activities (Fig. 4.3a). Biochemical signs refer to signaling molecules, cytokines, cofactors, and ions that are utilized to activate particular pathways or genes in order to guide cellular responses. For instance, cellular absorption of mineral-based nanomaterials can release bioactive ions into the cytosol, often because of lower pH (i.e., 5.5 in the endosome) (Fig. 4.3b). The simultaneous utilization of biochemical and biophysical signs of biomaterials can effectively guide cellular responses and promote the healing of different tissues, including cartilage, bone, vascular, and neural tissues (Fig. 4.3c) [45].

In general, inorganic biomaterials have the ability to affect cellular activities through direct or indirect ways. Considering the significance of macrominerals in growth and homeostasis of tissue, the use of these essential nutrients is imperative for regenerative medicine [45]. Table 4.1 provides a list of important ions and mineral-based biomaterials that possess the capability to enable regeneration and healing of tissue.

Because of several disadvantages of organ and tissue replacement, including inadequate success rate (transplant rejection), the need to suppress the immune system and shortage of available donors, there is a growing demand in tissue engineering and regenerative medicine solutions. The increasing need for precise control over nanoscale systems and the progress made in the production of nanoparticles have greatly influenced the fields of tissue engineering and regenerative medicine. It has integrated the fields of engineering, materials science and biology to manufacture and develop artificial structures that look like the natural tissues. These structures can be used as implantable systems or as miniature models of organs [54, 55].

Nanotechnology has shifted simple and traditional attitudes in tissue engineering and regenerative medicine to more effective and sophisticated techniques. In addition to nanoparticles, other nanotechnology products including nanopatterned surfaces and nanofibers have been operated to direct cellular behavior in these fields. Increasing biocompatibility and reducing toxicity through targeted delivery to particular tissues, modifying mechanical strength of scaffolds for applications involving hard tissues, controlling the release of various bioactive substances particularly growth factors to direct morphogenesis and stem cells fate, incorporating new biomaterials with precise spatial and temporal control in scaffolds, and using both imaging and therapeutic systems simultaneously are some of the numerous applications of nanoparticles in these fields [56–58]. Nanoparticles with precise control over behavior through external stimuli including thermal approaches, stimuli-responsive and targeted delivery potential, tunable properties, contrasting agent properties, and low toxicity have made it possible to operate them to improve engineered tissues and overcome problems in tissue engineering and regenerative medicine [56].

To achieve satisfactory results in imitation of tissue ECM composition by fabricating a 3D scaffold for cells with suitable mechanical strength, facilitating the delivery of bioactive agents and monitoring cellular functions, a nanoscale approach is necessary instead of a macroscopic. Nanoparticles offer precise manipulation of scaffold features, including the ability to provide controlled release of bioactive

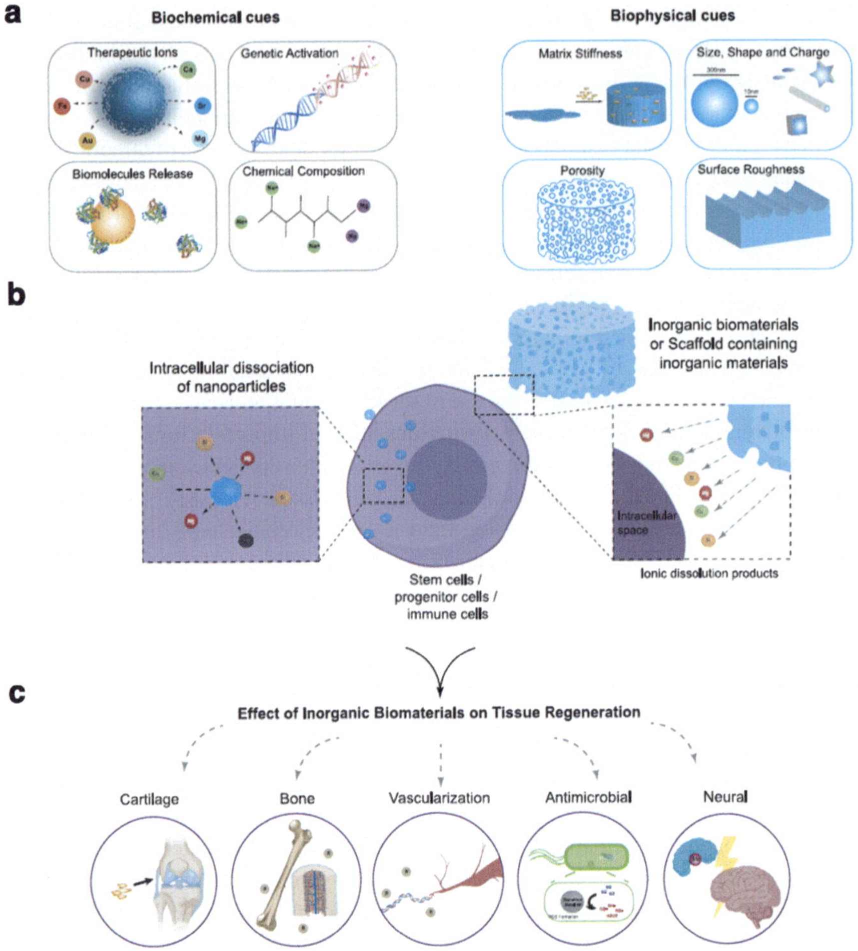

Fig. 4.3 Properties of inorganic biomaterials. **a** Biophysical signs are related to the physicochemical properties of a biomaterial including the charge, size, or shape. On the other hand, biochemical signs refer to the release of biomolecules and therapeutic ions from inorganic biomaterials. **b** Extracellular and intracellular release of ions from bulk biomaterials and nanoparticles respectively can regulate and guide cellular activity to promote regeneration and healing of tissue. Cells absorb inorganic nanoparticles, and the acidic conditions in the endosome lead to the degradation of nanoparticles into individual ions within the cell. Comparably, extracellular releases of therapeutics ions from biomaterials might affect the function of cells. It has been shown that both of these ion release mechanisms can direct and control regeneration and healing of tissue. **c** When inorganic biomaterials are degraded, the released ions can lead particular cellular processes including adipogenesis, chondrogenesis, and osteogenesis. Mineral-based biomaterials also release ions that play a role in the homeostasis of tissues, including neural, vascular, bone, and cartilage. Several ions also possess antibacterial properties. Reprinted with permission from Brokesh and Gaharwar [45]

Table 4.1 Effects of different elements/biomaterials on tissue healing and regeneration [45]

Element	Ions	Effects
Magnesium	Mg^{2+}	Enzyme cofactor: DNA polymerase, glucose-6-phosphatase, hexokinase
Calcium	Ca^{2+}	Bone/lattice structure, calcium signaling pathways
Silicon	Si^{4+}, Si^{4-}	Material component which increase biodegradation Collagen fibril alignment
Strontium	Sr^{2+}	Causes apoptosis of osteoclasts, upregulates formation of bone
Zinc	Zn^{2+}	Component of neural function and signaling Stent material component enhancing biodegradation Zinc-finger-binding domains, transcription cofactor, enzymatic cofactor
Copper	Cu^{2+}	Enhance neural differentiation as a part of standard neural function Increase vascularization, upregulate HIF-1a Mitochondrial enzymatic cofactor
Iron	Fe^{2+}, Fe^{3+}	Stent structural component, hemoglobin component, enzymatic cofactor
Boron	B^{3-}	Increase vascularization, upregulate FAK and ERK Upregulate osteoblast specific genes (Col1)
Silver	Ag^{1+}	ROS generation, antimicrobial (antiadhesion)
Gold	Au^{1+}, Au^{3+}	Osteogenesis activated by RUNX2 stimulation via p38/MAPK cascade, ROS generation
Molybdenum	Mo^{3+}, Mo^{6+}	Imaging capacities because of SPR Structural component in implants Enzymatic cofactor
Manganese	Mn^{2+}, Mn^{3+}, Mn^{4+}, Mn^{6+}, Mn^{7+}	Neuronal differentiation Important to BMD, influence glycosylation of proteoglycans (part of cell adhesion)
Cobalt	Co^{2+}, Co^{3+}	Stimulate VEGF to increase proliferation of osteoblasts Imitates hypoxia and rises ROS, which upregulate FGF and VEGF
Chromium	Cr^{3+}, Cr^{6+}	Play a role in pathways of vascularization but also potentially damages at high oxidative state (+6) or high concentrations, increases ROS which upregulates (HIF)-1a

(continued)

Table 4.1 (continued)

Element	Ions	Effects
Lanthanum	La^{3+}	Osteogenesis activation, although the exact process is unknown
Europium	Eu^{3+}	ROS production increases vascularization and possibly helps in osteogenesis
Vanadium	V^{5+}, V^{6+}	Antifouling films Potentially enhances angiogenesis, which is believed to be because of the ROS production Accumulates in bone increasing BMD and possibly substitutes P in HAP

Some of these elements may have valuable applications in the field of treatments and the capacity to be used in nanomaterials to adjust cellular fate. It could be related to observation of various types of minerals in the composition, or the use of their potential cytotoxic influences as a therapeutic agent for treatment of cancer. However, for designing new nanomaterials, it is worth considering all possibilities, since even the smallest piece of knowledge could potentially serve as the key to addressing challenging biological issues.

agents and regulate their mechanical strength. Furthermore, limiting factors and disadvantages including unstable bioactivity, low solubility and short circulating half-life of contrast agents and bioactive molecules (drugs, genes, inhibitors, cytokines, growth factors, etc.) have turned the nanoparticles as highly appropriate options for delivering and monitoring bioactive agents in various applications [56, 59–63].

4.3 Extracellular Matrix

The ECM is a non-cellular 3D structure that is produced by cells and released into the extracellular space. It consists of particular polysaccharides and proteins. During development, each tissue type has an exclusive topology and composition of the ECM [64]. The ECM is responsible for maintaining the elasticity and integrity of tissues. It undergoes continuous remodeling in response to variations in the local pH levels, growth factors, and receptor abundance. This process regulates the homeostasis, function, and development of organs and tissues [65, 66]. The ECM is regarded as the fourth component in the BTE advancement [67]. In addition, the precise composition of the ECM varies depending on health conditions, age, and gender. Trace elements and calcium-deficient apatite are the major inorganic constituents of the ECM. On the contrary, the organic ECM is considerably more complex, consisting generally of 90% COL-I, and 10% NCPs. Osteoblastic cells mostly produce it prior to the mineralization [68]. Bone ECM interacts dynamically with osteoclasts and osteoblasts to control new bone formation through regeneration [12].

ECM of Cartilage consists mainly of a collagenous network [69], with different forms of collagen depending on the cartilage types. Proteoglycans are another key constituent of these networks. While fibrous cartilage is a combination of both COL-I and COL-II, hyaline cartilage is mostly COL-II [2, 70, 71]. Bone ECM consists of a mineralized inorganic constituent (consisting of plate-like carbonated apatite minerals 4 nm thick) and a non-mineralized organic constituent (mainly COL-I). Furthermore, more than 200 distinct kinds of non-collagenous matrix proteins (sialoproteins, proteoglycans, and glycoproteins) participate in the signal profusion in the direct extracellular conditions [72].

The significance of the ECM in skeletal advancement has been ignored till the recent advances in recombinant DNA technology and advanced extraction processes. Earlier than these methodological developments, the ECM was defined as a structural scaffold consisting of an amorphous ground substance (proteoglycan) and a proteinaceous fiber component (collagen). This description indicated that the ECM was a static structure with restricted capability to affect gene expression, development, function, or tissue structure. Currently, it is identified that the ECM is not a static structure but a dynamic network of molecules secreted by cells. Various types of cells secrete specific matrices that alter with growing age. In addition to forming a complicated construction around the cell, ECM molecules adjust behavior of cells by modulating their differentiation and proliferation, signal transduction pathways, and response to growth factors [73]. ECM signaling is influenced by alterations in cytoarchitecture. The extent of cell spreading, as determined by the characteristics of the ECM substrate, regulates the growth of cells that require attachment for proliferation. Similarly, cell spreading is necessary for the ECM-mediated inhibition of apoptosis and also influences gene expression and differentiation. Several research on anchorage-dependent cell-cycle progression have revealed various insights, which are described in Fig. 4.4 [74].

Growth factors also manage differentiation of osteoblasts. Members of all major growth factor families have been involved in regulating osteoblast differentiation through embryonic advancement. Numerous BMPs can provoke expression of Cbfa1 in vitro. Transforming growth factor-β (TGF-β) prevents the expression of Cbfa1 in cultured osteoblasts and manages the steady-state level of differentiation of osteoblasts in vivo. When limb modeling happens the fibroblast growth factor (FGF) group members look to perform formerly in skeletogenesis. At a subsequent step, they play serious roles mainly during differentiation of chondrocytes. FGF2, however, might organize mesenchymal progenitor cell differentiation into mature osteoblasts [75]. Bone and cartilage ECM is exclusive and reflects the individual functional requirements of the tissues. Both have a main collagen matrix that forms fibrils while their matrices are unique. In contrast to bone, cartilage comprises a large proteoglycan called aggrecan, which permits for the cartilage flexibility and distributes load. The bone matrix comprises a high degree of mineralization that is essential to form a functional structure which is rigid [73].

Collagen is the main component of all ECMs. They participate in most of the structural and extracellular functions of various vertebrates and invertebrates. To date, 28 distinctive vertebrate collagens, type I to XXVIII, have been identified. These collagens can be divided into the following categories based on size or function: long chain, short chain, filamentous, network-forming, fibril associated, and fibrillary. They have

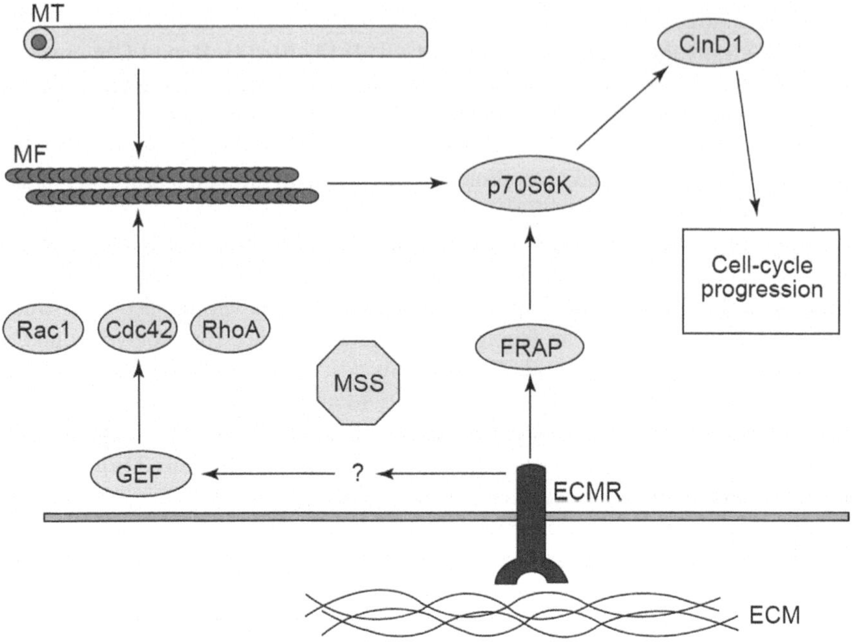

Fig. 4.4 Cell shape-dependent regulation of cell-cycle progression by the ECM. This diagram illustrates a hypothetical signal-transduction mechanism by which the ECM can control the expression of ClnD1 and the function of p70S6K. Cell adhesion to the ECM, facilitated by the ECMRs, triggers activation of p70S6 via two mechanisms: activation of the FRAP and regulation of MF arrangement by activating Rho-family small GTPases (Rac1, Cdc42, and RhoA). Based on the suggested model, the regulation of MFs may be controlled by a MSS that affects the activity of GDP–GTP exchange factors, which in turn govern the activation state of the small Rho-like GTPases. The regulatory mechanism described in this scheme is probably influenced by the MTs through the regulation of the organization and assembly of MFs and MF-associated focal adhesions. Abbreviations: cyclin D1 (ClnD1), p70 ribosomal S6 protein kinase (p70S6K), ECM receptors (ECMRs), FK506-binding protein/rapamycin-associated protein (FRAP), microfilament (MF), mechanical stress sensor (MSS), microtubules (MTs). Reprinted with permission from Lukashev and Werb [74]

distinctive functional characteristics, specific spatial and temporal distributions, and tissue-specific distributions [73, 76, 77]. Collagen fibrils are acknowledged for their rare stability rising from the cross-linking and aggregation of the molecules. Collagens are structural proteins that aid tissues maintain their shape, organization and mechanical properties. They are found in the ECM, where they establish supramolecular constructions for the most portion. In connective tissue, cornea, ligaments, and tendons, the collagen is the major protein and forms the matrix of teeth and bones. They even control differentiation, migration, and proliferation of cells by interacting with cells throughout various receptor families [77].

Another classification for collagen superfamily presented at the following sentences: multiplexins, transmembrane, fibril-associated collagens with interrupted triple helix [FACIT], hexagonal network-forming, anchoring fibrils, microfibrillar, basement membrane, and the fibril-forming [78]. Collagen molecules have a wide range of applications. They are employed as a stabilizing agent in glass and beads for cell culture, as well as for the synthesis of orally administered hydrolysates [79], and for additional purposes outlined in Table 4.2. In addition, they are utilized as raw materials for cosmetics, adhesives, and gelatin [80]. Furthermore, they are employed for the development of tissue scaffolds [81], as well as for subcutaneous injection in the form of microparticles [81], and as a biomaterial for vascular prosthesis [82].

While bone is composed of fibrillar COL-I, cartilage comprises fibrillar COL-II. The fibrillar collagens, including COL-I and COL-II, are located in tissues exposed to tensile stress and compression. These collagens create a fibrillar network in the ECM and strengthen the tissue to endure tensile and compressive stresses [73]. The cartilage ECM is generally composed of COL-II and proteoglycan aggrecan. Though, other ECM macromolecules, including the cartilage matrix protein, fibronectin, fibromodulin, and leucine-rich proteoglycans, are found in cartilage. The proteoglycan matrix is mechanically vulnerable but has a high density of negative charges that directs to an ionic interaction with H_2O, causing an osmotic swelling of the tissue. Swelling of the proteoglycan constituent is limited by the COL-II fiber network. The COL-II fibers counteract the proteoglycan-induced swelling that causes intensified collagen fiber tension and allows the cartilage to reinforce load-bearing stress. Bone needs a mineralized rigid matrix that can sustain weight, in contrast to the cartilage matrix, which mostly reduces compressive force. Thus, a progressive modification from a cartilage matrix to one particular for function of bone should befall. Through this modification, the chondrocytes in the central part of the growing bone increase and start to generate an exclusive non-cartilaginous ECM. The cells are now called hypertrophic chondrocytes and secrete a matrix consisting of alkaline phosphatase, osteopontin, fibronectin, and type X collagen [83]. Through hypertrophy, the formation of vascular channels and mineral deposition will increase. Finally the hypertrophic chondrocytes and cartilage matrix are totally reduced and substituted by bone [73].

Porous biomaterials that mimic the composition and structure of natural ECM have exhibited remarkable outcomes in tissue renewal [84–86]. Electrospinning nanofibrous scaffolds might imitate the nanoscale structure of ECM and show effective advantages in bone tissue engineering (BTE) [87]. Regenerative medicine has advanced quickly, leading to the recognition of the extracellular matrix as a crucial component in the progress of bone tissue creation (Fig. 4.5). The extracellular matrix serves as a structural framework and foundation for cell adhesion, providing both pharmacological and biomechanical cues that trigger cell migration, differentiation, morphogenesis, and homeostasis [12].

Mature bone ECM comprises collagen nanofibers and inorganic apatite nanocrystals, and has exposed convincing angiogenesis capability through effective mediation of growth factor secretion and cellular differentiation [85, 86, 88]. Native ECM is a potential therapeutic material, which is the non-cellular constituent of tissues that supplies biochemical signals and structural support to cells. Each type of tissue has a particular ECM composition and structure that modulates cellular responses and is

Table 4.2 Collagen types and their roles in the human body or applications in the industries [78]

Family	Type	Function/Application	Tissue Distribution
Fibril-forming	I	Membranes for directed tissue regeneration	Cornea, bone, ligaments and tendon
	II	Arthritis treatment and cartilage repair	Skin, vitreous body, bone, nucleus pulposus, vessel wall, cartilage, reticular fibers, cornea, and lung
	III	Tissue sealants and hemostats	
	V	Raw Materials for biomaterial in treatment of cornea	
	XI	mAbs development for osteoarthritis	Cartilage and Vitreous body
Basement membrane	IV	Diabetic nephropathy indicator and cell culture attachment enhancer (mouse neuroblastoma)	Basement membranes
Microfibrillar	VI	Hemostat	Placenta, intervertebral disk, cartilage, lungs, and dermis
Anchoring	VII	Subdermal injection as treatment for dystrophic epidermolysis bullosa (DEB)	Skin, dermal and epidermal junctions, cervix, and oral mucous
FACIT	IX	It codistributes with COL-II in the vitreous body and cartilage	Vitreous humor, cartilage, and cornea
	XII	Linked to fibrils carrying COL-I. COL-XIV plays a role in controlling the diameter of fibrils during the initial stages of fibril formation	Ligaments, perichondrium, and tendon
	XIV		Lungs, dermis, liver, placenta, and vessel wall
	XIX	Expose antitumor and anti-angiogenic features, and found in the basement membrane area	Human rhabdomyosarcoma
	XX	Similar to COL-XII and COL-XIV. The role of this collagen is almost unspecified	Sternal cartilage, corneal epithelium, tendon, and embryonic skin
	XXI	Strongly associated with COL-XII, COL-XIV, and COL-XX. The expression of this collagen is activated by platelet-derived growth factor, indicating its involvement in the assembly of the vascular network by forming blood vessels	Blood vessel wall
Transmembrane	XIII	Involved in vasculogenesis and inflammation. Influences formation of bone and might have a role in coupling the bone mass regulation to mechanical application	Lungs, epidermis, dermal and epidermal junctions, liver, intestine, and hair follicle

(continued)

Table 4.2 (continued)

Family	Type	Function/Application	Tissue Distribution
	XVII	It has significant functions in the formation of teeth. Supposed function of COL-XVII is to stabilize adhesion of epithelial cells to the adjacent ECM	Epidermal and dermal junctions
Multiplexins	XV	It makes a bridge that connects large, banded fibrils, likely fibrils including COL-I and COL-III	Pancreas, fibroblasts, smooth muscle cells, and kidney
	XVI	It could be probably applied as a biomarker or drug target, since it has been involved in various diseases	Fibroblasts and keratinocytes
	XVIII	It has key function in closure of the neural tube and in determining retinal structure	Lungs and liver

beneficial for cell survival in that tissue [2, 64]. The native ECM in bone has a well-described arrangement that contains oriented COL-I fibers with interfering carbonated hydroxyapatite (HAP) nanocrystals, a kind of mineral calcium phosphate. COL-I, which makes up more than 90% of the bone organic phase, has both cell signaling and structural purposes. Interactions between cell surface receptors like integrins and collagen type I adjust cell adhesion to the ECM and also stimulate mesenchymal stem cells (MSCs) differentiation along the osteoblastic lineage [89]. The residual 10% of the organic part consists of more than one hundred special proteins [90, 91] containing several proteoglycans. Though less rich than COL-I, they play significant roles in the bone homeostasis and organization. Non-collagenous bone matrix proteins participate in cell signaling, matrix mineralization, collagen fibrillogenesis, and release and sequestration of morphogens and growth factors [92]. Several efforts have been focused on advancing scaffolds that reproduce either mechanical properties or the biochemistry of bone matrix, but less attention has been given to addressing both of these factors. It has been recommended that a scaffold composed of molecules located in temporary wound-healing kinds of matrices might better reflect the natural procedure of bone healing. For these last kinds of scaffolds, the bone-inductive molecule delivery from a provisional scaffold is more prior than the mechanical strength of the scaffold. One of the most promising ways to investigate places is to use hyaluronan, fibrin, or heparin-derived scaffolds or hydrogels with embedded cytokines or growth factors [92].

4.4 Bone ECM Proteins

The bone extracellular matrix (ECM) consists primarily of collagenous proteins, making up 90% of its composition, with 97% being type I collagen (COL-I) and 3% type V collagen (COL-V). The remaining 10% is formed by non-collagenous proteins

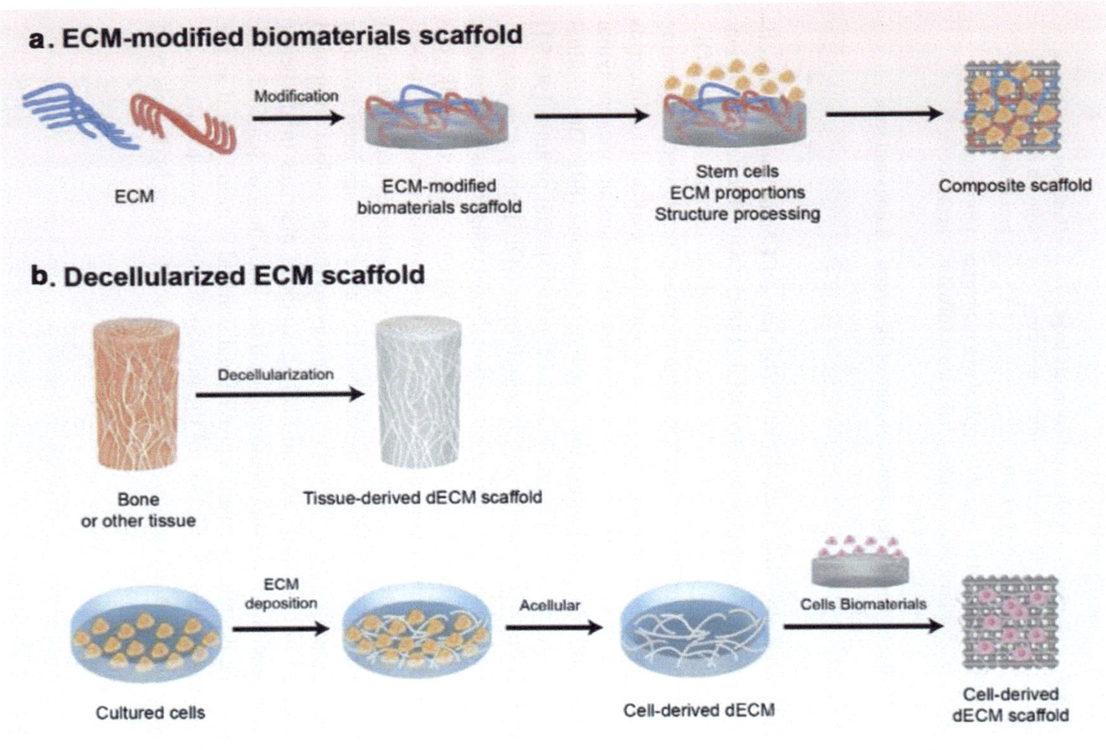

Fig. 4.5 Schematic preparation of ECM-based scaffold in bone regeneration. **a** ECM-modified biomaterials scaffold. Different constituents of ECM altered with biomaterial-based scaffold, and further improved with the addition of stem cells and undergoing structural processing to imitate natural biomaterials. **b** Decellularized ECM scaffold can be derived from either cultured cells in vitro or tissue in vivo through the process of decellularization, which has a great potential in promoting bone regeneration and has demonstrated positive outcomes in clinical practice. Reprinted with permission from Seyfoori et al. [12]

(NCPs), which include osteocalcin and osteonectin at 20% each, bone sialoproteins at 12%, proteoglycans at 10%, along with bone morphogenetic proteins, growth factors, fibronectin, osteopontin, and others. Osteoblasts are responsible for the synthesis of these proteins, which predominantly play a role in cellular adhesion. Proteins such as vitronectin and fibronectin have been identified as key players in the in vitro adhesion process of osteoblasts. Additionally, certain proteins derived from plasma, like transferrin, immunoglobulin, albumin, and glycoprotein, are found within the mineralized bone matrix, although their specific functions in osteoblast adhesion have yet to be fully elucidated [93].

Human osteoblasts have a stronger affinity for fibronectin compared to vitronectin, COL-I, and COL-IV. However, their adhesion to COL-V and laminin is very poor. However, they ignore COL-III [94]. Both bone sialoprotein and osteopontin, which are two types of non-collagenous proteins, exhibit acidic properties and have a robust affinity for hydroxyapatite [95, 96].

The mechanical characteristics of collagen depend on the intra- and interchain crosslinks that sustain the polypeptide chains in a highly ordered fibrillar structure. The significant contribution of collagen in influencing the strength of bones. Changes in collagen structure or COL-I deficiency considerably enhance the risk of fractures by affecting the extracellular matrix [12, 97].

4.5 Cytoskeleton Proteins

Adhesion plaques, commonly referred to as focal contacts, are the areas where tissue-cultured cells connect to substrate surfaces. Formation of these structures in culture is facilitated by extracellular matrix proteins, such as vitronectin and fibronectin, and is particularly evident in cells with limited mobility. Actin filaments and membrane receptor proteins (integrins) may form stronger interactions in adhesion plaques due to the facilitation provided by proteins such as talin, tensin, vinculin, and paxillin (Fig. 4.6). Adhesion plaques possess distinct receptor proteins, such as integrins, on their outer surfaces. Talin and vinculin are found together at focal contacts together with several other proteins, including as cytoskeletal proteins, kinases, phosphatases, signaling molecules, and proteases. Their presence is crucial for transmitting signals [93].

The maintenance of cell adhesion and shape relies on the integrity of the actin cytoskeleton. Lamellipodia are flat protrusions that are supported by F-actin when arranged in a mesh-like pattern. Filopodia, on the other hand, are finger-like protrusions that are supported when arranged in elongated bundles. Adhesion plaques serve the purpose of grouping and connecting actin "stress fibers" which subsequently transmit mechanical forces to the substrate [98]. Interconnected porosity is an essential element in scaffold design within the realm of regenerative medicine as it facilitates vascularization, cell migration, and the circulation of nutrients and regulatory chemicals inside the scaffold. Due to its precise control over the interconnectivity, porosity, and pore size of scaffolds, 3D printing is an effective way

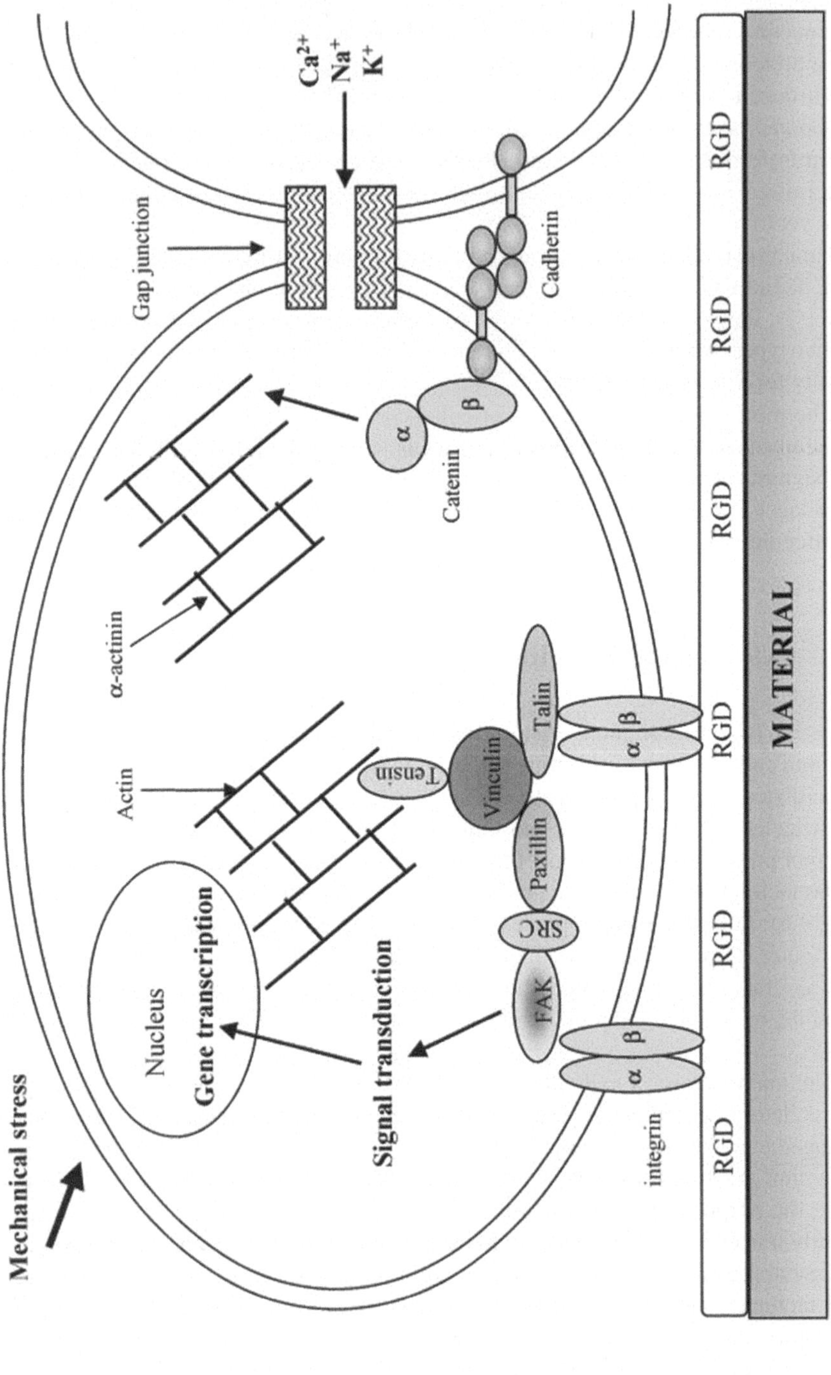

Fig. 4.6 Representation of the cell proteins involved in cell adhesion on biomaterial. Reprinted with permission from De Larco and Todaro [93]

for achieving these goals [99]. The capacity of the scaffold to support the affected area throughout the process of regeneration is an essential need for scaffolds used in bone tissue engineering. To facilitate the transportation of nutrients, cells, and regulatory molecules, as well as to encourage the growth of blood vessels after implantation, it is necessary for the scaffolds to have interconnected holes of the appropriate dimensions [100–102]. In addition, it is necessary for them to provide the necessary biophysical and biochemical signals that stimulate the migration of stem cells or seeded cells from nearby tissues or bone marrow and facilitate their transformation into osteoblasts [99].

The bioactivity of scaffolds may be enhanced by including ECM components. Demineralized bone matrix (DMB) and decellularized bone matrix (DCB) are widely employed in clinical settings as bone extracellular matrix-based products [103]. One example of a kind of decellularized bone matrix often used in dental surgery is Bio-Oss, produced by Geistlich Biomaterials. Bio-Oss is composed of millimeter-sized granules of bovine trabecular bone, which are mostly devoid of organic components [104]. DMB is an allogenic bone graft material that is generated from xenogeneic or allogenic sources. It is utilized in orthopedic treatments to repair bone deficiencies following tumor removal or fractures [103, 105]. The first stages of the preparation process typically include lipid, blood, and soft tissue extraction, followed by acid demineralization, antimicrobial treatment, and many rounds of freeze-drying [106]. The resulting material has osteoinductive properties due to its composition of collagen types I, IV, and X, as well as growth factors such as TGF-β1, TGF-β2, and TGF-β3 [103, 106]. The mechanical properties of DMB remain a disadvantage, particularly for load-bearing bone abnormalities. Moreover, the process of converting biomaterials derived from bone matrix into porous scaffolds with precise shapes suitable for blood vessel and bone development remains challenging [99].

The extracellular matrix in bone is a precisely controlled and constantly changing biological environment with intricate biochemical and mechanical characteristics. Modulating cellular responses to growth factors is vital for regulating the differentiation, proliferation, and adhesion of cells, which are all necessary for the proper functioning of mature bone. The extracellular matrix of bones stimulates the activation of mesenchymal stem cells (MSCs), osteocytes, and osteoblasts, hence facilitating the formation of new bone tissue. Furthermore, it enhances osteoclast activity, which leads to the degradation of bone. The primary appeal of ECM-based scaffolds lies in their osteogenic, osteoconductive, and osteoinductive characteristics, which have become more important due to rapid breakthroughs in the area of bone regeneration treatment. There are two types of scaffolds that may be used in bone tissue engineering: decellularized extracellular matrix scaffolds and biomaterial scaffolds with modified ECM. Functional extracellular matrix has shown to be very beneficial in tissue engineering as it effectively guides the development of certain tissues at the sites of implantation. Bone tissue production relies on several processes associated with bone regeneration, such as the maturation of highly specialized osteocytes, the extracellular matrix, the mineralization of osteoid, the generation of osteoblasts, and the recruitment of mesenchymal stem cells [12, 107].

4.6 Inorganic ECM

Hydroxyapatite ($Ca_5(PO_4)_3OH$) is the major inorganic component of hard tissues, including dentine and bone [108]. HAP deposition occurs during a process termed biomineralization. HAP crystallization is regulated by the interplay of amino acids in non-collagenous proteins (NCPs) with minerals in teeth and bones. Collagen offers a structural foundation that facilitates HAP deposition during tissue mineralization. HAP is commonly utilized for covering metallic implants, bone fillers, and injectable bone replacements because of its osteoconductive and biocompatible qualities, which stem from its physical and chemical closeness to the minerals found in human bone and teeth [12, 108]. The extracellular matrix sends mechanical and biochemical cues to osteoblasts, mesenchymal stem cells, and osteocytes that control their respective activity [12, 109]. In addition to generating significant extracellular matrix proteins such fibronectin, TGF-βI, connective tissue growth factor, integrin chains (α5 and β5), COL-IIIα1, and COL-Vα1, bone marrow stromal cells (BMSCs) have the ability to move, develop, proliferate, and interact with other cells [110]. For MSC adhesion, spread, and proliferation, integrin receptors are essential [12].

Mature and immature osteoblasts function as transitional cells during osteogenesis from MSCs, encouraging differentiation, osteoid mineralization, and ECM secretion. Collagen is necessary for the appropriate architecture and mechanical integrity of bone tissue because osteoblasts cannot manufacture new matrices without a collagen surface, which limits their ability to arrange matrices across a finite region [111]. Different types of type I collagen (COL-I) have different effects on osteoblast activity. In contrast to fibrillar and soluble forms, denatured COL-I stimulates osteoblast growth but inhibits proliferation. Collagen III (COL-III) is also found in bone collagen fibrils. ALP activity, mineralization, osteoblast formation, and osteogenic marker levels are all hampered by COL-III deficiency. By providing a framework for cell attachment, collagen offers structural support, controlling osteoblast growth and bone formation. ECM proteins influence osteoblast growth and mineralization in addition to controlling the synthesis of collagen fibers. In addition to providing a structural framework for cell attachment, the extracellular matrix transmits chemical and mechanical signals that control homeostasis, shape tissues and organs, cellular specialization, and movement. In tissue engineering, ECM-based scaffolds are used to imitate natural biomaterials, fostering cell interactions, tissue growth and remodeling, and bone integration. Collagen, HAP, and fibronectin are examples of common natural biomaterials for scaffolding. ECM is applied to absorbable polymers as a surface coating and is also used to make biodegradable scaffolds for bone regrowth. Synergistic effects can be achieved by combining multiple materials, since the complex osteogenic environment cannot be fully replicated by a single bone ECM component. In the development of materials for bone repair and regeneration, such as injectable bone substitutes and bone fillers, HAP, a crucial component of the natural bone extracellular matrix, is employed. Cellular activity is influenced by ECM components during bone repair. Through replication of the biochemical signals and matrix material composition and distribution, ECM scaffolds effectively guide bone regeneration and tissue repair by imitating the natural bone microenvironment. Conventionally, scaffold designs modified by extracellular matrix employ biomaterial coatings or mixes of ECM elements. Lab-cultivated cells or decellularized allogeneic or autologous tissues preserve structural integrity while eliminating cell-bound antigens to reduce immunological response. In tissue engineering and regenerative medicine, decellularized extracellular matrix and scaffolds modified by ECM hold great potential for augmenting bone regrowth. Nevertheless, there are still difficulties in applying decellularized extracellular matrix or ECM-modified scaffolds for clinical bone healing. During processing, design, ECM modification, decellularization, mass production, and standardization required for research, these include maintaining biochemical signals and growth factors. The goal of decellularization is to preserve ECM features and functions, but thorough analysis of released ECM ingredients has been hampered by the dynamic and complex nature of ECM components. Whether decellularized ECM can accurately replicate the biochemical processes inherent in bone ECM is still unknown. Regulating ECM components released by cells for large-scale production is also challenging. Genetic manipulation of cells to produce desired substances, combined with appropriate bioreactors for monitoring substance release and cell growth, can set standards for controlled ECM release, enhancing graft quality. Adding bioactive compounds and growth factors to ECM scaffolds during production can improve bone defect repair efficacy. Additives should promote specific bone regeneration at the defect site without affecting surrounding tissue growth or causing inflammation. Integrating ECM scaffolds with organ-specific progenitor cells or autologous pluripotent stem cells can enhance therapeutic outcomes. Improved ECM scaffold processing and design can allow them to accurately fill damaged areas, resulting in better functional and mechanical properties [12].

3D printing technology enables processing ECM through biological printing, creating scaffolds with diverse topologies and morphologies that match defect areas. This enhances implants' bone regeneration capacity and biomechanical properties. Despite challenges, ECM utilization in bone regeneration offers significant prospects [12]. Decellularized ECM has emerged as a biomaterial providing signals for cell differentiation, boosting cell proliferation, and maintaining the original tissue environment. While its therapeutic use is growing, constraints hinder its widespread adoption. Tissue-specific ECM is gaining popularity in tissue regeneration, especially in cartilage and bone engineering, due to its resemblance to natural tissue and modulatory properties facilitating tissue differentiation and growth [2]. Decellularization involves using chemical/enzymatic agents and physical stress to remove cellular components from tissue, preserving the non-cellular ECM for therapeutic use. The decellularization method varies by tissue type [2]. Decellularized ECM retains the biological environment's components [64]. Proper decellularization preserves ECM's intricate physical and biomolecular signals, supporting cell viability and growth. Decellularized ECM is used in various tissue engineering applications and has a lower immune response risk than transplanted tissues due to the absence of cellular DNA [112]. However, decellularization faces challenges, primarily removing cellular components while minimizing ECM damage [2, 112].

4.7 Conclusion

Regenerative medicine and tissue engineering hold immense promise for overcoming the limitations of conventional medicine by leveraging biological and engineering principles to repair damaged organs and tissues, as well as to generate new ones. The need for reliable and safe organ suppliers drives ongoing research and innovation. Connective tissues, such as intervertebral disks, cartilage, bone, adipose tissue, ligaments, and tendons, are crucial to our bodies, with their extracellular matrix composition determining their function and regenerative potential. Understanding the biochemical and physical characteristics of ECM is essential for effective tissue engineering. While bone tissue has remarkable self-repair abilities, it fails to regenerate beyond a critical defect size of approximately 2.5 cm, necessitating the induction of osteogenesis. Nano-bioceramics, with their hierarchical structure of collagen fibers supported by hydroxyapatite crystals, offer promise for enhancing bone strength and toughness. The nanocomposite nature of bone, particularly the collagen-HAP system, highlights the significance of nanoscale interactions, with inorganic substances at this scale exhibiting enhanced biological activity. Researchers are exploring nano-bioceramics to optimize bone regeneration. Advances in tissue engineering technology, including 3D printing and nanotechnology, have enabled the development of new implants. However, challenges such as high immunogenicity, low osteoinductivity, and inadequate biocompatibility of some materials remain. Addressing these limitations is a priority. In summary, the convergence of nano-bioceramics, tissue engineering, and regenerative medicine holds immense potential for transforming healthcare. As scientists delve deeper into the intricacies of biological characteristics, we move closer to a future where damaged bones can be effectively repaired, significantly improving lives.

References

1. Dzobo, K., et al. (2018). Advances in regenerative medicine and tissue engineering: innovation and transformation of medicine. *Stem Cells International, 2018.*
2. Kim, Y. S., et al. (2019). Applications of decellularized extracellular matrix in bone and cartilage tissue engineering. *Bioengineering and Translational Medicine, 4*(1), 83–95.
3. Kalantari, E., et al. (2019). Biocomposites based on hydroxyapatite matrix reinforced with nanostructured monticellite ($CaMgSiO_4$) for biomedical application: Synthesis, characterization, and biological studies. *Materials Science and Engineering: C, 105*, 109912.
4. Kalantari, E., et al. (2018). Nanostructured monticellite for tissue engineering applications—Part I: Microstructural and physicochemical characteristics. *Ceramics International, 44*(11), 12731–12738.
5. Kalantari, E., et al. (2018). Nanostructured monticellite for tissue engineering applications—Part II: Molecular and biological characteristics. *Ceramics International, 44*(12), 14704–14711.
6. Bhagavan, N. V., & Ha, C.-E. (2011). Essentials of medical biochemistry: With clinical cases. Academic Press.
7. McKee, T. J., et al. (2019). Extracellular matrix composition of connective tissues: A systematic review and meta-analysis. *Scientific Reports, 9*(1), 1–15.

8. Halper, J., & Kjaer, M. (2014). Basic components of connective tissues and extra-cellular matrix: elastin, fibrillin, fibulins, fibrinogen, fibronectin, laminin, tenascins and thrombospondins. In *Progress in heritable soft connective tissue diseases* (pp. 31–47).
9. Schemitsch, E. H. (2017). Size matters: Defining critical in bone defect size! *Journal of Orthopaedic Trauma, 31*, S20–S22.
10. Liao, S., Cui, F., & Zhu, Y. (2004). Osteoblasts adherence and migration through three-dimensional porous mineralized collagen based composite: NHAC/PLA. *Journal of Bioactive Compatible Polymers, 19*(2), 117–130.
11. Gasser, J. A., & Kneissel, M. (2017). Bone physiology and biology. *Bone Toxicology*, 27–94.
12. Lin, X., et al. (2020). The bone extracellular matrix in bone formation and regeneration. *Frontiers in Pharmacology, 11*, 757.
13. Rahmanian, M., et al. (2019). Multifunctional gelatin–tricalcium phosphate porous nanocomposite scaffolds for tissue engineering and local drug delivery: In vitro and in vivo studies. *Journal of the Taiwan Institute of Chemical Engineers, 101*, 214–220.
14. Askari, E., et al. (2021). Local delivery of chemotherapeutic agent in tissue engineering based on gelatin/graphene hydrogel. *Journal of Materials Research and Technology, 12*, 412–422.
15. Mendelson, A., & Frenette, P. S. (2014). Hematopoietic stem cell niche maintenance during homeostasis and regeneration. *Nature Medicine, 20*(8), 833–846.
16. Atala, A. (2008). Advances in tissue and organ replacement. *Current Stem Cell Research & Therapy, 3*(1), 21–31.
17. Naghib, S. M., et al. (2012). Bioactivation of 304 stainless steel surface through 45S5 bioglass coating for biomedical applications. *International Journal of Electrochemical Science, 7*(4), 2890–2903.
18. Dehghani, N., et al. (2023). Chitosan/silk fibroin/nitrogen-doped carbon quantum dot/α-tricalcium phosphate nanocomposite electrospinned as a scaffold for wound healing application: In vitro and in vivo studies. *International Journal of Biological Macromolecules, 238*, 124078.
19. Soleymani, S., & Naghib, S. M. (2023). 3D and 4D printing hydroxyapatite-based scaffolds for bone tissue engineering and regeneration. *Heliyon, 9*(9), e19363.
20. Moheb Afzali, A., Kheradmand, M. A., & Naghib, S. M. (2024). Bioreactor design-assisted bioprinting of stimuli-responsive materials for tissue engineering and drug delivery applications. *Bioprinting, 37*, e00325.
21. Kangarshahi, B. M., et al. (2024). Bioprinting of self-healing materials and nanostructures for biomedical applications: Recent advances and progresses on fabrication and characterization techniques. *Bioprinting, 38*, e00335.
22. Gonçalves, I. M., et al. (2022). Recent trends of biomaterials and biosensors for organ-on-chip platforms. *Bioprinting, 26*, e00202.
23. Guan, X., et al. (2017). Development of hydrogels for regenerative engineering. *Biotechnology Journal, 12*(5), 1600394.
24. Gao, G., & Cui, X. (2016). Three-dimensional bioprinting in tissue engineering and regenerative medicine. *Biotechnology Letters, 38*(2), 203–211.
25. Mistry, H., et al. (2017). Autologous chondrocyte implantation in the knee: Systematic review and economic evaluation. *Health Technology Assessment, 21*(6), 1–294.
26. Kraeutler, M. J., et al. (2018). Microfracture versus autologous chondrocyte implantation for articular cartilage lesions in the knee: A systematic review of 5-year outcomes. *The American Journal of Sports Medicine, 46*(4), 995–999.
27. Gao, C., et al. (2017). Bone biomaterials and interactions with stem cells. *Bone Research, 5*(1), 1–33.
28. Clough, B. H., et al. (2015). Bone regeneration with osteogenically enhanced mesenchymal stem cells and their extracellular matrix proteins. *Journal of Bone and Mineral Research, 30*(1), 83–94.
29. Darghiasi, S. F., Askari, E., & Naghib, S. M. (2021). The effect of a novel in situ akermanite-monticilite nanocomposite on osteogenic differentiation of mouse mesenchymal stem cells. *Progress in Canadian Mechanical Engineering, 4*, 10–32393.

30. Mao, A. S., & Mooney, D. J. (2015). Regenerative medicine: Current therapies and future directions. *Proceedings of the National Academy of Sciences, 112*(47), 14452–14459.
31. Guex, A., et al. (2012). Fine-tuning of substrate architecture and surface chemistry promotes muscle tissue development. *Acta Biomaterialia, 8*(4), 1481–1489.
32. Pina, S., et al. (2017). Biofunctional ionic-doped calcium phosphates: Silk fibroin composites for bone tissue engineering scaffolding. *Cells, Tissues, Organs, 204*(3–4), 150–163.
33. Gonçalves, A. I., Rodrigues, M. T., & Gomes, M. E. (2017). Tissue-engineered magnetic cell sheet patches for advanced strategies in tendon regeneration. *Acta Biomaterialia, 63*, 110–122.
34. Alves da Silva, M., et al. (2017). Electrospun nanofibrous meshes cultured with Wharton's jelly stem cell: An alternative for cartilage regeneration, without the need of growth factors. *Biotechnology Journal, 12*(12), 1700073.
35. Kotton, D. N., & Morrisey, E. E. (2014). Lung regeneration: Mechanisms, applications and emerging stem cell populations. *Nature Medicine, 20*(8), 822–832.
36. Atala, A. (2012). Regenerative medicine strategies. *Journal of Pediatric Surgery, 47*(1), 17–28.
37. Murphy, S. V., & Atala, A. (2014). 3D bioprinting of tissues and organs. *Nature Biotechnology, 32*(8), 773–785.
38. Witten, C. M., McFarland, R. D., & Simek, S. L. (2015). Concise review: The US Food and Drug Administration and regenerative medicine. *Stem Cells Translational Medicine, 4*(12), 1495–1499.
39. Do, A. V., et al. (2015). 3D printing of scaffolds for tissue regeneration applications. *Advanced Healthcare Materials, 4*(12), 1742–1762.
40. Chhaya, M. P., et al. (2015). Additive manufacturing in biomedical sciences and the need for definitions and norms. *Expert Review of Medical Devices, 12*(5), 537–543.
41. Askari, E., et al. (2021). Reduced graphene oxide-grafted bovine serum albumin/bredigite nanocomposites with high mechanical properties and excellent osteogenic bioactivity for bone tissue engineering. *Bio-Design and Manufacturing, 4*(2), 243–257.
42. Askari, E., et al. (2021). A hybrid approach for in-situ synthesis of bioceramic nanocomposites to adjust the physicochemical and biological characteristics. *Journal of Materials Research and Technology, 14*, 464–474.
43. Kalantari, E., & Naghib, S. M. (2019). A comparative study on biological properties of novel nanostructured monticellite-based composites with hydroxyapatite bioceramic. *Materials Science and Engineering: C, 98*, 1087–1096.
44. Kalantari, E., et al. (2018). Nanostructured monticellite: An emerging player in tissue engineering. *Materials Today: Proceedings, 5*(7, Part 3), 15744–15753.
45. Brokesh, A. M., & Gaharwar, A. K. (2020). Inorganic biomaterials for regenerative medicine. *ACS Applied Materials & Interfaces, 12*(5), 5319–5344.
46. DeBerardinis, R. J., & Thompson, C. B. (2012). Cellular metabolism and disease: What do metabolic outliers teach us? *Cell, 148*(6), 1132–1144.
47. Gharibzahedi, S. M. T., & Jafari, S. M. (2017). The importance of minerals in human nutrition: Bioavailability, food fortification, processing effects and nanoencapsulation. *Trends in Food Science & Technology, 62*, 119–132.
48. Gaharwar, A. K., Singh, I., & Khademhosseini, A. (2020). Engineered biomaterials for in situ tissue regeneration. *Nature Reviews Materials, 5*(9), 686–705.
49. Jahnen-Dechent, W., & Ketteler, M. (2012). Magnesium basics. *Clinical Kidney Journal, 5*(Suppl_1), i3–i14.
50. Pinton, P., et al. (2008). Calcium and apoptosis: ER-mitochondria Ca^{2+} transfer in the control of apoptosis. *Oncogene, 27*(50), 6407–6418.
51. Keratívitayanan, P., Carrow, J. K., & Gaharwar, A. K. (2015). Nanomaterials for engineering stem cell responses. *Advanced Healthcare Materials, 4*(11), 1600–1627.
52. Cross, L. M., et al. (2019). Sustained and prolonged delivery of protein therapeutics from two-dimensional nanosilicates. *ACS Applied Materials & Interfaces, 11*(7), 6741–6750.
53. Gaharwar, A. K., et al. (2011). Highly extensible, tough, and elastomeric nanocomposite hydrogels from poly (ethylene glycol) and hydroxyapatite nanoparticles. *Biomacromolecules, 12*(5), 1641–1650.

54. Dvir, T., et al. (2011). Nanotechnological strategies for engineering complex tissues. *Nature Nanotechnology, 6*(1), 13–22.
55. Zorlutuna, P., Vrana, N. E., & Khademhosseini, A. (2012). The expanding world of tissue engineering: The building blocks and new applications of tissue engineered constructs. *IEEE Reviews in Biomedical Engineering, 6*, 47–62.
56. Fathi-Achachelouei, M., et al. (2019). Use of nanoparticles in tissue engineering and regenerative medicine. *Frontiers in Bioengineering and Biotechnology, 7*, 113.
57. Colson, Y. L., & Grinstaff, M. W. (2012). Biologically responsive polymeric nanoparticles for drug delivery. *Advanced Materials, 24*(28), 3878–3886.
58. Shi, J., et al. (2010). Nanotechnology in drug delivery and tissue engineering: From discovery to applications. *Nano Letters, 10*(9), 3223–3230.
59. Mi, P., et al. (2016). A pH-activatable nanoparticle with signal-amplification capabilities for non-invasive imaging of tumour malignancy. *Nature Nanotechnology, 11*(8), 724–730.
60. Bahal, R., et al. (2016). In vivo correction of anaemia in β-thalassemic mice by γPNA-mediated gene editing with nanoparticle delivery. *Nature Communications, 7*(1), 1–14.
61. Cheng, C. J., et al. (2015). A holistic approach to targeting disease with polymeric nanoparticles. *Nature Reviews Drug Discovery, 14*(4), 239–247.
62. Perez, R. A., et al. (2013). Naturally and synthetic smart composite biomaterials for tissue regeneration. *Advanced Drug Delivery Reviews, 65*(4), 471–496.
63. Park, J., et al. (2012). Combination delivery of TGF-β inhibitor and IL-2 by nanoscale liposomal polymeric gels enhances tumour immunotherapy. *Nature Materials, 11*(10), 895–905.
64. Frantz, C., Stewart, K. M., & Weaver, V. M. (2010). The extracellular matrix at a glance. *Journal of Cell Science, 123*(24), 4195–4200.
65. Mouw, J. K., Ou, G., & Weaver, V. M. (2014). Extracellular matrix assembly: A multiscale deconstruction. *Nature Reviews Molecular Cell Biology, 15*(12), 771–785.
66. Bonnans, C., Chou, J., & Werb, Z. (2014). Remodelling the extracellular matrix in development and disease. *Nature Reviews Molecular Cell Biology, 15*(12), 786–801.
67. Ravindran, S., et al. (2012). Biomimetic extracellular matrix-incorporated scaffold induces osteogenic gene expression in human marrow stromal cells. *Tissue Engineering Part A, 18*(3–4), 295–309.
68. Mansour, A., et al. (2017). Extracellular matrices for bone regeneration: A literature review. *Tissue Engineering Part A, 23*(23–24), 1436–1451.
69. Cancedda, R. (2009). Cartilage and bone extracellular matrix. *Current Pharmaceutical Design, 15*(12), 1334–1348.
70. Chen, J.-L., et al. (2014). Extracellular matrix production in vitro in cartilage tissue engineering. *Journal of Translational Medicine, 12*(1), 1–9.
71. Sophia Fox, A. J., Bedi, A., & Rodeo, S. A. (2009). The basic science of articular cartilage: structure, composition, and function. *Sports Health, 1*(6), 461–468.
72. Stevens, M. M. (2008). Biomaterials for bone tissue engineering. *Materials Today, 11*(5), 18–25.
73. Velleman, S. (2000). The role of the extracellular matrix in skeletal development. *Poultry Science, 79*(7), 985–989.
74. Lukashev, M. E., & Werb, Z. (1998). ECM signalling: Orchestrating cell behaviour and misbehaviour. *Trends in Cell Biology, 8*(11), 437–441.
75. Montero, A., et al. (2000). Disruption of the fibroblast growth factor-2 gene results in decreased bone mass and bone formation. *The Journal of Clinical Investigation, 105*(8), 1085–1093.
76. Van der Rest, M., & Garrone, R. (1991). Collagen family of proteins. *The FASEB Journal, 5*(13), 2814–2823.
77. KN, A., Bhatia, P., & Singh, S. (2021). Review on collagen type 1 & type 2.
78. Avila Rodríguez, M. I., Rodriguez Barroso, L. G., & Sánchez, M. L. (2018). Collagen: A review on its sources and potential cosmetic applications. *Journal of Cosmetic Dermatology, 17*(1), 20–26.

79. Ganceviciene, R., et al. (2012). Skin anti-aging strategies. *Dermato-endocrinology, 4*(3), 308–319.
80. Yang, L. (2008). *Mechanical properties of collagen fibrils and elastic fibers explored by AFM*. University of Twente.
81. Silvipriya, K., et al. (2015). Collagen: Animal sources and biomedical application. *Journal of Applied Pharmceutical Science, 5*(3), 123–127.
82. Ramshaw, J. A., Vaughan, P. R., & Werkmeister, J. A. (2001). Applications of collagen in medical devices. *Biomedical Engineering: Applications, Basis and Communications, 13*(01), 14–26.
83. Schmid, T. M., & Linsenmayer, T. F. (1985). Immunohistochemical localization of short chain cartilage collagen (type X) in avian tissues. *The Journal of Cell Biology, 100*(2), 598–605.
84. Lei, B., et al. (2012). Nanofibrous gelatin–silica hybrid scaffolds mimicking the native extracellular matrix (ECM) using thermally induced phase separation. *Journal of Materials Chemistry, 22*(28), 14133–14140.
85. Sridhar, R., et al. (2013). Electrospun inorganic and polymer composite nanofibers for biomedical applications. *Journal of Biomaterials Science, Polymer Edition, 24*(4), 365–385.
86. Gao, W., et al. (2017). A highly bioactive bone extracellular matrix-biomimetic nanofibrous system with rapid angiogenesis promotes diabetic wound healing. *Journal of Materials Chemistry B, 5*(35), 7285–7296.
87. Gupta, K. C., et al. (2014). Nanofibrous scaffolds in biomedical applications. *Biomaterials Research, 18*(1), 5.
88. Lei, B., et al. (2013). Biomimetic and molecular level-based silicate bioactive glass–gelatin hybrid implants for loading-bearing bone fixation and repair. *Journal of Materials Chemistry B, 1*(38), 5153–5162.
89. Alford, A. I., Kozloff, K. M., & Hankenson, K. D. (2015). Extracellular matrix networks in bone remodeling. *The International Journal of Biochemistry Cell Biology, 65*, 20–31.
90. Jiang, X., et al. (2007). Method development of efficient protein extraction in bone tissue for proteome analysis. *Journal of Proteome Research, 6*(6), 2287–2294.
91. Salmon, C. R., et al. (2013). Proteomic analysis of human dental cementum and alveolar bone. *Journal of Proteomics, 91*, 544–555.
92. Curry, A. S., et al. (2016). Taking cues from the extracellular matrix to design bone-mimetic regenerative scaffolds. *Matrix Biology, 52*, 397–412.
93. Anselme, K. (2000). Osteoblast adhesion on biomaterials. *Biomaterials, 21*(7), 667–681.
94. Gronthos, S., et al. (1997). Integrin expression and function on human osteoblast-like cells. *Journal of Bone Mineral Research, 12*(8), 1189–1197.
95. Franzen, A., & Heinegård, D. (1985). Isolation and characterization of two sialoproteins present only in bone calcified matrix. *Biochemical Journal, 232*(3), 715–724.
96. Reinholt, F. P., et al. (1990). Osteopontin–A possible anchor of osteoclasts to bone. *Proceedings of the National Academy of Sciences, 87*(12), 4473–4475.
97. Fonseca, H., et al. (2014). Bone quality: The determinants of bone strength and fragility. *Sports Medicine, 44*(1), 37–53.
98. Zigmond, S. H. (1996). Signal transduction and actin filament organization. *Current Opinion in Cell Biology, 8*(1), 66–73.
99. Freeman, F., et al. (2019). Biofabrication of multiscale bone extracellular matrix scaffolds for bone tissue engineering. *European Cells & Materials, 38*, 168–187.
100. Habibovic, P., et al. (2008). Osteoconduction and osteoinduction of low-temperature 3D printed bioceramic implants. *Biomaterials, 29*(7), 944–953.
101. Müller, B., et al. (2009). Bio-mimetic hollow scaffolds for long bone replacement. In *Biomimetics and bioinspiration*. International Society for Optics and Photonics.
102. Jones, A. C., et al. (2007). Assessment of bone ingrowth into porous biomaterials using MICRO-CT. *Biomaterials, 28*(15), 2491–2504.
103. Cheng, C. W., Solorio, L. D., & Alsberg, E. (2014). Decellularized tissue and cell-derived extracellular matrices as scaffolds for orthopaedic tissue engineering. *Biotechnology Advances, 32*(2), 462–484.

104. Nyberg, E., et al. (2017). Comparison of 3D printed poly-ε-caprolactone scaffolds functionalized with tricalcium phosphate, hydroxyapatite, bio-oss, or decellularized bone matrix. *Tissue Engineering Part A, 23*(11–12), 503–514.
105. Drosos, G. I., et al. (2007). Safety and efficacy of commercially available demineralised bone matrix preparations: A critical review of clinical studies. *Injury, 38*, S13–S21.
106. Gruskin, E., et al. (2012). Demineralized bone matrix in bone repair: History and use. *Advanced Drug Delivery Reviews, 64*(12), 1063–1077.
107. Wang, X., et al. (2013). Role of mesenchymal stem cells in bone regeneration and fracture repair: A review. *International Orthopaedics, 37*(12), 2491–2498.
108. Ramesh, N., Moratti, S. C., & Dias, G. J. (2018). Hydroxyapatite–polymer biocomposites for bone regeneration: A review of current trends. *Journal of Biomedical Materials Research Part B: Applied Biomaterials, 106*(5), 2046–2057.
109. Assis-Ribas, T., et al. (2018). Extracellular matrix dynamics during mesenchymal stem cells differentiation. *Developmental Biology, 437*(2), 63–74.
110. Ren, J., et al. (2011). Global transcriptome analysis of human bone marrow stromal cells (BMSC) reveals proliferative, mobile and interactive cells that produce abundant extracellular matrix proteins, some of which may affect BMSC potency. *Cytotherapy, 13*(6), 661–674.
111. Kerschnitzki, M., et al. (2011). The organization of the osteocyte network mirrors the extracellular matrix orientation in bone. *Journal of Structural Biology, 173*(2), 303–311.
112. Crapo, P. M., Gilbert, T. W., & Badylak, S. F. (2011). An overview of tissue and whole organ decellularization processes. *Biomaterials, 32*(12), 3233–3243.

Chapter 5
Bone ECM Proteins—Part I

Abstract Bone extracellular matrix (ECM) is comprised of organic and mineral components. Proteins, which are the primary constituents in the body that regulate all our actions, comprise the organic part of ECM. Collagen, a crucial and potent protein present in the body, mostly exists in supporting and connective tissues. Type I collagen is extensively used in the field of medicine because of its formation into fibrils consisting of triple helical polypeptide chains, which contribute to its remarkable mechanical properties. COL-V regulates the process of fibril production. There is conjecture that non-collagenous proteins may have a substantial impact on the regulation of hydroxyapatite mineral production, as well as the differentiation and attachment of cells. However, they possess several drawbacks such as restricted accessibility, high water solubility, and a lack of thorough understanding regarding their mechanism of action. Recently, multiple researchers have focused on striding each of these proteins individually, investigating their distinct characteristics and functions. Understanding the characteristics and dynamics of these proteins and their interactions with each other is crucial for the development of bone implants. ECM proteins can be employed either in their complete form or fragmented into peptides with specific sequences that will trigger the required reaction. For instance, collagen is commonly found in a variety of physical forms such as composites, sponges, membranes, films, solutions, gels, fibers, particles, powder, and so on. ECM proteins can also be utilized to enhance the biocompatibility of implants. This chapter will focus on the collagenous and non-collagenous proteins of the bone extracellular matrix.

Abbreviations

AGEs	Advanced glycation end
ALP	Alkaline phosphatase
Asp	Aspartic acid
BD	Binding domains
BMPs	Bone morphogenetic proteins

R. Gholami and S. M. Naghib, *Nanobioceramics for Bone Tissue Engineering and Regenerative Biomedicine*, Biological and Medical Physics, Biomedical Engineering,
https://doi.org/10.1007/978-981-96-0041-0_5

BSE	Bovine spongiform encephalopathy
BSP	Bone sialoprotein
COL	Collagen
CS	Chondroitin sulfate
DS	Dermatan sulfate
ECM	Extracellular matrix
EGF	Epidermal growth factor
FGFs	Fibroblast growth factors
GAG	Glycosaminoglycan
GalN	D-galactosamine
GDF-8	Growth/differentiation factor-8
GlcA	D-glucuronic acid
GlcN	D-glucosamine
Glu	Glutamic acid
Gly	Glycine
HA and HAP	Hydroxyapatite
HGF	Hepatocyte growth factor
HGF/SF	Hepatocyte growth factor/scatter factor
HS	Heparan sulfate
HSCs	Hematopoietic stem cells
HSPGs	Heparan sulfate proteoglycans
IdoA	L-iduronic acid
IGFs	Insulin-like growth factors
Kd	Dissociation constant
KS	Keratan sulfate
LAP	Latency-associated p
LRR	Leucine-rich repeat
mABs	Monoclonal antibodies
MAPKs	Mitogen-activated protein kinases
MGP	Matrix Gla protein
MMPs	Matrix metalloproteinases
NCPs	Non-collagenous proteins
OPN	Osteopontin
PDGFs	Platelet-derived growth factors
PGs	Proteoglycans
RGD	Arginine-glycine-aspartate or Arg-Gly-Asp
rhBMP-2	Recombinant human BMP-2
SAXS	Small-angle X-ray scattering
SLRPs	Small leucine-rich proteoglycans
SPARC	Secreted protein acidic and rich in cysteine
SVVYGLR	Serine-Valine-Valine-Tyrosine-Glutamate-Leucine-Arginine
TGF-β	Transforming growth factor-beta

5.1 Introduction

Peptides and proteins are fundamental areas of study in the fields in Nanomedicine. Currently, professionals in diverse disciplines such as immunology, toxicology, pharmacy, nanobiotechnology, nanomedicine, and other medical sciences are investigating the different aspects of these crucial biomolecules. This includes comprehending how nanostructures interact with the body and their use in treatment and diagnostic subjects [1]. The extracellular matrix (ECM) of bone consists of two main components: the organic section and the mineral section. The mineral section consists of hydroxyapatite, which makes up 70–90% of the composition. The organic section (10–30%) is predominantly composed of collagen accounting for ~90% of the organic matrix. The remaining 10% of the organic section is made up of non-collagenous proteins (NCPs) [2]. Approximately 90% of the organic component of bone consists of type I collagen, whereas minor quantities of collagen type III and type V are also present in the region surrounding the bone cells. NCPs have an essential function in controlling microcrack resistance, cell attachment, mineralization, and fibril size and production of collagen. Out of the limited quantity of NCPs, approximately 85% is located outside of cells, while the remaining portion is present within bone cells [3]. ECM proteins can be utilized in their entirety or broken down into peptides with precise sequences that will initiate the necessary response. Consequently, these novel osteoinductive peptides possess the advantages of being cost-effective and easily producible, less prone to triggering immunological reactions due to their compact size, and exhibiting stability under physiological conditions [2].

5.2 Collagenous Proteins

Collagen, an essential and potent protein found in mammals, mostly resides in supportive and connective tissues like joints, bones, and skin. It plays a vital role in connecting biological components across a wide variety of species [4]. The collagen superfamily consists of around 28 varieties of collagens composed of at least 46 distinct polypeptide chains. These collagens exhibit unique characteristics in terms of functionality, structural integrity, form, and their association with specific disorders [5]. Collagen fibrils are linked together by several types of cross-links, which can significantly impact the material characteristics of the tissue and, consequently, the mechanical properties of the entire bone (Fig. 5.1). These can be categorized into two main groups: those made by enzymatic procedures and those made through non-enzymatic glycation procedures, resulting in the creation of advanced glycation end products (AGEs) [3].

Collagen is frequently present in various physical forms, including composites, sponges, membranes, films, solutions, gels, fibers, particles, powder, etc. The medical applications encompass the use of hemostatic agents, sutures, tissue engineering for

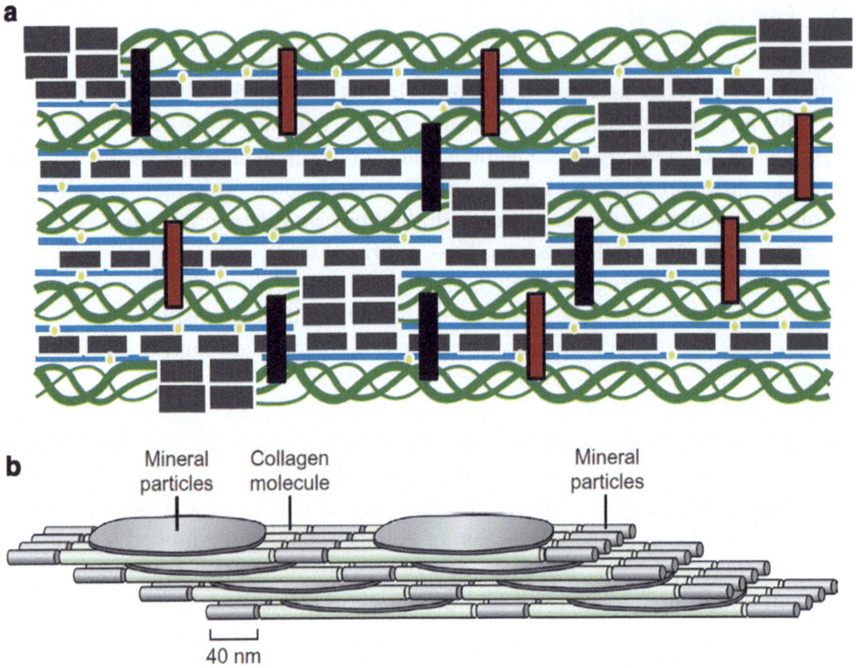

Fig. 5.1 **a** Collagen molecules (*green helix*) are cross-linked within the fibril by enzymatic and non-enzymatic mechanisms. Enzymatically generated cross-links (*black bars*), such as deoxypyridinoline or pyridinoline, are created in close proximity to the C- and N-termini of the molecules. Bonds created non-enzymatically (*red bars*), like pentosidine, are distributed randomly among the molecules. The mineral (*gray blocks*) is deposited within the pore and hole regions located between the collagen fibrils. The linking of mineral and collagen within the fibril is facilitated by the presence of hydrogen bonds and water (*blue lines*). **b** The intermolecular spaces in bone consist of layers of hydroxyapatite, forming pore and hole zones. The presence of water in these spaces creates a connection with collagen, which in turn affects how the load is distributed between the collagen and the mineralized bone in this area. Reprinted with permission from [3]

various body parts such as blood vessels, tendons, skin, cartilages, and bones. Additionally, they include drug delivery systems, contact lenses, implants, membranes for blood oxygenation, skin replacement and supplementation, dental applications, and cosmetic surgery procedures for the skin and lips, among numerous other uses. Biomaterials, which are made based on collagens, are in high demand for the replacement and treatment of human tissues like bones, vascular applications, and tendons since they are highly versatile and biocompatible [5].

5.2.1 Type I Collagen

Type I collagen is widely prevalent in mammals, primarily constituting the organic component of the ECM in non-cartilaginous tissues, including bones, cornea, tendon, and dermis. These tissues consist of intricate fibrils that have unique functionalities and morphologies. The compact arrangement and elevated level of organization of the fibrous structure have a direct impact on cellular functions and the mechanical characteristics, including differentiation, migration, and attachment. Type I collagen is widely employed in medicine due to its fibrillar elaboration in triple helical polypeptide chains, which give it exceptional mechanical qualities. Furthermore, the structure of this substance consists of over 107 molecules organized in three-dimensional arrays, with a relatively small diameter in the nanometer range when dissolved in liquids [5]. Collagen type I (COL-I) selectively binds to α2β1 integrin receptors on cell membranes, which directly affects the development of bone cell structures [6, 7]. The study conducted by Green et al. emphasizes the substantial influence of COL-I on the structure, growth, and communication pathways of osteoblastic UMR-106 cells and recently developed osteoblasts in the organic ECM of the bone [8, 9]. These cells are the preferred target of the DGEA peptide found in COL-I. Mizuno et al. have shown that the osteoblastic characteristics of bone marrow cells are diminished as a consequence of the peptide's loss of affinity for the cells. Additionally, the anti-α2 integrin antibody, which can inhibit the binding of integrin and collagen and interact with the α integrin subunit, effectively inhibited the production of the osteoblastic phenotype. The results suggest that the interaction between collagen and α2β1 integrin has an important function in signaling the osteoblastic development of bone marrow cells [10].

5.2.2 Type V Collagen

The fibril formation process is regulated by COL-V. Although its function in connective tissues that are abundant in Type I collagen (COL-I) may be restricted, it still has a substantial impact. At the very least, the molecular isoforms of COL-V are $\alpha1(V)_2$ α2(V), $\alpha1(V)_3$, and α1(V) α2(V) α3(V). Each isoform is composed of three polypeptide α chains: α1(V), α2(V), and α3(V). The α3(V) chain in COL-V significantly improves the function of pancreatic islet cells [11]. Svelte collagen, or COL-V, is a component of the ECM and is present in the majority of connective tissue matrices. It impacts the overall health and several diseases, such as fibrosis. In COL-V, heterotypic fibril development is frequently observed in conjunction with microfilaments or fibrils that are up to 12 nm in thickness. The strong correlation between the distribution of COL-I and COL-V in the cornea and other tissues is evident from immunohistochemical studies, which demonstrate that they copolymerize. The genetic expression of COL-V, which is produced by a myriad of mesenchymal cells, is controlled by growth factors such as TGF-β. Gelatinases and metalloproteinases can hydrolyze both native and denatured COL-V, but interstitial collagenases are unable to do so. This results in alterations to the ECM. Through interactions with other matrix collagens and structural proteins, COL-V alters cell behavior and activity and improves the structural stability of tissue scaffolds when it binds to matrix macromolecules. Heterotypic fibrils are necessary for collagen fibrillation, and they are produced by COL-V and COL-I in the dermis and cornea of the skin, respectively. Ehlers–Danlos syndrome and corneal transparency loss are linked to the absence of COL-V. In contrast, an abundance of COL-V has been associated with several health issues including atherosclerosis, inflammation, the production of granulation tissue, cancer, and fibrosis in organs such as the lungs, liver, adipose tissue, kidneys, and skin. COL-V is a stable ECM component found in the liver, and its levels rise in correlation with both early and severe liver fibrosis. COL-V neoepitopes have emerged as potential blood biomarkers for evaluating the course and remission of fibrosis in experimental liver fibrosis, without any adverse effects [12].

5.3 Non-collagenous Proteins

Non-collagenous proteins extracted from bone have been discovered to possess biological activity, despite their functions remaining incompletely comprehended. Each of these proteins may have distinct activities based on their localization patterns. Hence, it is crucial to gain a deeper comprehension of the characteristics and functionalities of these proteins, with the objective of developing novel approaches for the field of bone tissue engineering. There is speculation that non-collagenous proteins may play a significant role in regulation of hydroxyapatite mineral formation, and differentiation and attachment of cells. Some of these proteins may possess many functions, serving various roles inside the bone. Consequently, it may not be adequate to define a singular function for these proteins. Additionally, some of these proteins may collaborate, producing a synergistic impact on mechanical properties and cellular behavior of bone. Alternatively, they may counterbalance the effects caused by the disruption of other NCPs found in the bone matrix. Nevertheless, NCPs may have several disadvantages, including limited availability, high solubility in water, and lack of comprehensive understanding regarding their mode of operation [2].

5.3.1 Osteocalcin

Osteocalcin is a highly prevalent non-collagenous protein found in bone, which is solely synthesized by osteoblasts. Osteocalcin is primarily known for its role in inhibiting bone mineralization. However, current research indicates that it also has a broader impact on regulating body metabolism, cognition, and reproduction. Osteocalcin primarily functions to restrict bone growth while having little impact on bone mineralization or resorption. Furthermore, it stimulates testosterone synthesis in the testes and enhances neurotransmitter production in the brain. Although the pace of bone creation has increased, there is no comparable increase in the surface area of osteoblasts, suggesting that each osteoblast is depositing a more considerable amount of matrix. These findings indicate that osteocalcin, like other proteins in its family, likely interacts with a particular receptor to carry out its tasks. Identifying osteocalcin as a suppressor of bone development suggests that substances that inhibit osteocalcin's action or production, in addition to estrogen replacement treatment, could be used for therapeutic purposes [13]. BGLAP, which encodes an 11 kD pre-pro-protein with 98 amino acids, is a human osteocalcin gene and is localized on chromosome 1 at 1q25-q31 (Fig. 5.2) [14].

Osteocalcin, often known as bone γ-carboxyglutamic acid (Gla) protein or BGP, is a compact protein weighing 5.6 kDa, composed of 46 to 50 amino acids. Osteoblasts are the main producers of this substance, which they release into the extracellular space [16]. Hypertrophic chondrocytes and odontoblasts in teeth also generate smaller quantities [15]. Price et al. first extracted osteocalcin from human and bovine bone, establishing its substantial Gla content and protein existence inside bone [17, 18]. Subsequently, a second Gla protein, known as matrix Gla protein (MGP), was identified. Both proteins are classified under a distinct subset of the broader family of vitamin K-dependent proteins, with their main function being related to blood clotting [15]. Osteocalcin is often used as a blood biomarker to signify the process of bone production carried out by osteoblasts. Initially believed to only control mineralization in the bone matrix, modern pharmacological and genetic research has shown that osteocalcin also operates as a hormone. The recent discoveries establish a link between the energy requirements of bones and the general balance of the body, filling in various gaps in our understanding of the hormonal influences of insulin [19, 20], adiponectin [21], leptin [22, 23], and nutritional availability [24, 25] on skeletal metabolism [15].

Undercarboxylated osteocalcin in the circulation, which is controlled by insulin, functions in a positive feedback loop to stimulate the development and function of β-cells, as well as the synthesis and secretion of insulin. Moreover, osteocalcin enhances insulin sensitivity in both adipose tissue and skeletal muscle [15]. Several research groups have recently provided compelling evidence regarding the function of osteocalcin in the control of glucose metabolism. It has been supposed that osteocalcin, like adiponectin and leptin, acts as a hormone due to its exclusive production by osteoblasts and could participate in the circulation (Fig. 5.3) [13, 15, 19].

A study was conducted to ascertain the undercarboxylated osteocalcin receptor in the target tissues and its importance in a variety of physiological processes. Gprc6a is a protein that may serve as an osteocalcin receptor in the Leydig cells of the testes, adipose tissue, and skeletal muscle. It appears to be involved in the transmission of osteocalcin's effects through these tissues. Gprc6a is a receptor that is associated with an unidentified G protein. It has been discovered that osteocalcin is expressed by a variety of cell types that have responded to it in the circulation [26, 27]. Pancreatic islet β cells also express Gprc6a, and osteocalcin binds to this receptor to initiate the MAP kinase pathway [28]. The genetic inactivation of the receptor in the β-cell lineage, which prevents osteocalcin-stimulated β-cell proliferation, results in

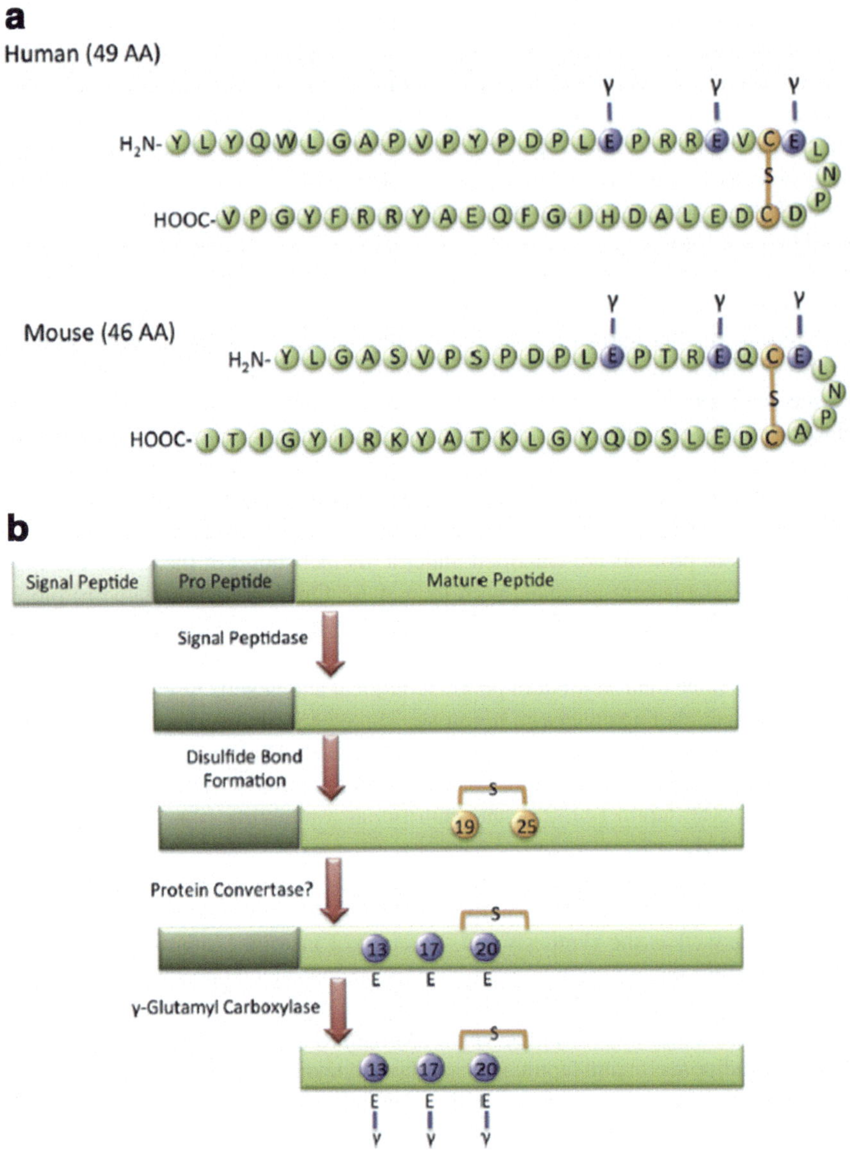

Fig. 5.2 Processing, translation, and structure of osteocalcin. **a** The osteocalcin structure is shown by a disulfide bond and three γ-carboxyglutamic acid residues. The human protein comprises 49 amino acids with a disulfide bond among cysteine residues at positions 23 and 29 and three γ-carboxyglutamic acid residues at positions 17, 21, and 24. The mouse protein has 3 amino acids less than human protein with γ-carboxyglutamic acid residues at positions 13, 17, and 20. **b** Osteocalcin translates as a pre-pro-peptide. After the signal sequence cleavage peptide is transported to the endoplasmic reticulum, where a disulfide bond develops. Osteocalcin undergoes γ-carboxylation on three glutamic acid residues after treatment in a potential protein convertase stage. Reprinted with permission from [15]

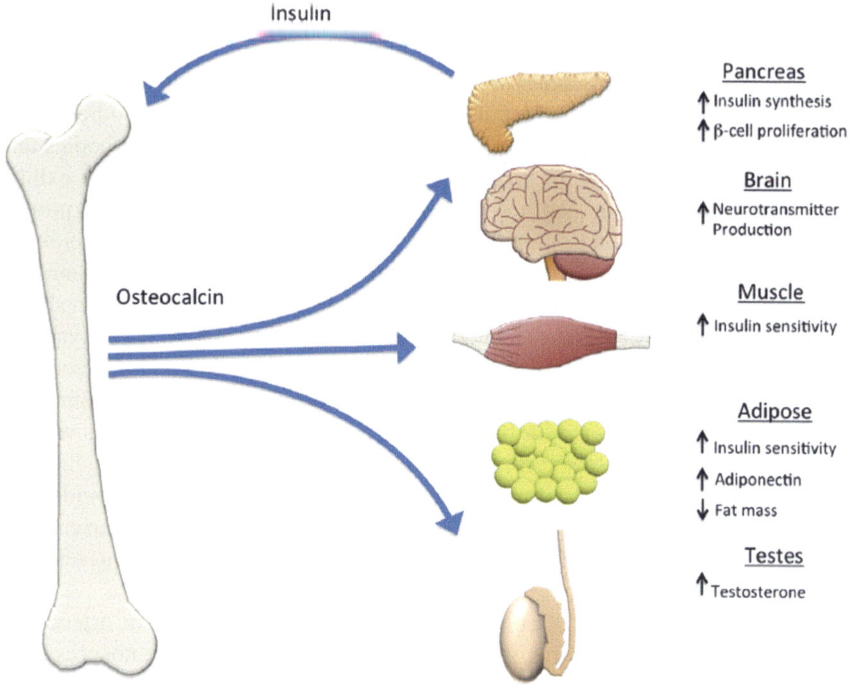

Fig. 5.3 Osteocalcin hormonal influences the metabolism of the entire body. In response to insulin, osteoblasts produce undercarboxylated osteocalcin or due to the low pH of the osteoclast resorption pit, it is released from the bone matrix and enters the circulation where it performs as a hormone. In the pancreas, osteocalcin increases the proliferation of β-cells as well as amplified insulin fabrication. Osteocalcin generates an extensive rise in sensitivity of insulin, which might be partly mediated by the amplified fabrication of adiponectin by adipose tissue. Further investigations have proposed that osteocalcin performs in the testes to enhance production of testosterone and in the brain to promote production of neurotransmitters. Reprinted with permission from [15]

glucose intolerance, low insulin, and elevated blood sugar levels [29]. Osteopenia is the consequence of the genetic deletion of the osteoblast-produced protein Gprc6a [15, 30]. The results of these investigations indicate that Gprc6a is involved in the promotion of osteocalcin's actions in a variety of tissues, including bone, muscle, and adipose.

5.3.2 Osteonectin

Osteonectin is a 32 kDa protein that is only found in bones and has the ability to concurrently bind to hydroxyapatite (HAP) and collagen. Osteonectin is believed to be unique to bone due to its specific biochemical properties compared to other non-collagenous components of bone. Insoluble collagen type I (COL-I) forms a strong attachment with osteonectin, while osteonectin has an affinity for both free calcium ions and manufactured apatite crystals. The development of mineral deposits is initiated by collagen–osteonectin complexes derived from unstable balanced salt solutions. The protein has a higher abundance in the matrix compared to bone cells and is only found inside the mineralized bone trabeculae. Although antibodies that specifically target osteonectin do not interact with other tissues, they do exhibit cross-reactivity with bone and, to a lesser extent, dentin. Osteonectin is a protein that is exclusive to certain tissues and acts as a connection between the mineral and collagen components of bone, as shown by studies. It has the potential to enhance the mineralization process of natural skeletal tissues [31]. Osteonectin has been shown to possess several key roles. Osteonectin has been shown to bind to hydroxyapatite crystals and collagen type I, most likely via several areas of COL-I. Osteonectin likely contributes to the facilitation of HAP's adherence to collagen-coated surfaces. Osteonectin has a preference for binding calcium ions over phosphate ions when it is attached to collagen. Recent findings reveal that osteonectin has both sialic acid and organic phosphate, indicating a notable attraction to calcium. The binding of hydroxyapatite to osteonectin likely occurs via similar interactions between calcium ions on the surface of the HAP crystal and osteonectin, as shown by previous research [31].

Immunocytochemically, osteonectin may be detected in immature osteocytes, osteoprogenitor cells, and active osteoblasts. Conversely, the protein is not present in resident and elderly osteocytes, indicating that it serves as an indicator of the functional differentiation of bone cells known as osteoblasts [32]. Osteonectin, a protein found only in bones [31], is produced solely by osteoblasts. On the other hand, this indicates that the visual characteristics of intracellular osteonectin by immunocytochemistry indicate the process of osteoblastic development in the cells it is linked with [32]. Extracellular matrix mineralization is a critical and indispensable phase of bone remodeling and regeneration. Extracellular mineralization is the process by which a portion of the water in the tissue is mineralized after apatite crystals are initially formed at specific locations along collagen fibers. During matrix mineralization, osteoblasts produce and discharge non-collagenous proteins and COL-I. The arrangement of collagen fibers is the fundamental structure that powers bone mineralization, despite the fact that non-collagenous proteins have contributed to the deposition of hydroxyapatite. Osteonectin, also referred to as BM-40 or secreted protein acidic and rich in cysteine (SPARC) [33, 34] has a high affinity for both HA and COL-I [31] and is significant in the complete mineralization of the bone matrix [35].

In order to elucidate the onset of mineralization, numerous investigations have been implemented. These include the direct formation of hydroxyapatite or other calcium phosphate minerals on the collagen fibers of the bone, the enzymatic removal of materials that obstruct mineral deposition, and the localized elevation of phosphate levels caused by alkaline phosphatase's breakdown of organic phosphates [31]. Recent research has demonstrated the advantageous effects of osteonectin on myocarditis and brain injury [36, 37]. It also serves as a substantial indicator for the diagnosis, treatment, and prognosis of malignant malignancies [38–40]. Osteonectin, a modular extracellular matrix protein, is responsible for regulating cell activity. It is not a structural component of the matrix; rather, it is involved in processes such as tissue remodeling, cell formation, turnover, and repair [41]. HOG1 or P38 MAPK are terms that are occasionally used to refer to MPK2, a critical signaling pathway. The protein's subgroup of the mitogen-activated protein kinases (MAPKs) superfamily consists of four members: p38a, p38b, p38c, and p38d [42]. In response to various environmental stimuli, the P38 MAPK signaling pathway facilitates a variety of biological responses from cells [43]. Various signals that promote bone formation, such as mechanical stimuli [44], Wnt proteins [45], parathyroid hormone [46], epinephrine [47], TGF-β [48], and BMP-2 [48, 49], regulate ECM mineralization and alkaline phosphatase (ALP) activity through the p38 MAPK pathway. This has been demonstrated in prior laboratory studies that utilized primary osteoblasts or cell lines that were treated with specific p38 inhibitors. In addition, numerous studies have demonstrated that the p38 pathway influences the expression and functionality of critical transcription factors that are involved in osteoblastogenesis, such as Osx, Runx2, and Dlx5 [48, 50–52]. Fra-2 is the primary regulator of bone matrix synthesis and osteoblast proliferation. The Fra-2/AP-1 complex modulates the formation of new bone by modulating collagen production and osteoblast development [53]. Fos/AP-1 proteins, such as FosB and Fra-1, have been associated in the regulation of bone mass through their impact on osteoblast activity and bone matrix production in previous research [54–57]. According to a study, the protein AP-1 is the primary target of P38 in the regulation of the mineralization process. The synthesis of non-collagenous materials is facilitated by P38 [58]. P38 is an essential regulator of osteoblastogenesis during skeletogenesis and is essential for the regulation of mineralization by osteonectin [59].

5.3.3 Bone Sialoproteins

Bone sialoprotein (BSP) comprises approximately 8–12% of the non-collagenous proteins in bone and cementum [60–62]. Conversely, dentin contains approximately 1% of BSP, which is significantly less than the amount found in bone [63]. The preponderance of the highly glycosylated and sulfated phosphoprotein known as BSP is found in mineralized connective tissues. Young osteocytes and osteoblasts are significant contributors to the production of BSP during the process of human bone growth [64]. The 23 kDa glycoprotein with a high quantity of sialic acid in the cortical bone of cows was first discovered by Herring et al. [65] and described as such. Although the first BSP to undergo comprehensive study, a complete description was hindered by difficulties in separating the protein's intact form [66]. Subsequently, Fisher and colleagues [61] determined that this molecule, known as BSP II [64], was promptly recognized as the result of the degradation of a bigger molecule with an estimated molecular weight of around 70,000 M_r (as predicted by cDNA sequence analysis [62], the expected M_r was around 33,000). The freshly formed BSP was obtained from a thin layer (1–2 mm) of the subperiosteal bone in fetal calves. This area is known for its fast bone development [61]. Further investigation revealed that osteopontin, another sialoprotein with similar characteristics to BSP, may have contaminated the sample [67].

BSP provides a less intense osteoinductive effect due to its membership in the SIBLING protein family. Through the vitronectin receptor ($\alpha_v\beta_3$), it initiates the mineralization process and enhances the differentiation and adhesion of osteoblasts [68]. Within a controlled laboratory setting, the addition of BSP enhanced the capacity of primary human osteoblasts to attach, expand, and survive on scaffolds made of calcium phosphate cement [69]. Additionally, research has demonstrated that it facilitates the migration and adhesion of endothelial cells, which leads to the formation of new blood vessels [70, 71]. Nevertheless, the numerous functions that BSP performs in bone remodeling and a variety of disorders are outlined by its characteristics related to cell adhesion and signaling. Arginine-glycine-aspartate (RGD) motifs and polyglutamic acid are frequently observed in protein sequences. These motifs have the potential to bind to hydroxyapatite and integrins on the surface of cells, respectively. Stimulatory cytokines and hormones that promote bone production, as well as inhibitory substances that impair bone development, modulate the activation of the bone sialoprotein gene in newly formed osteoblasts. Consequently, bone sialoprotein is indispensable for initiating the process of generating new minerals in the cementum and bone. Furthermore, BSP is essential for the formation of mineral crystals in a variety of diseases, such as breast cancer. The combination of RGD and hydroxyapatite-binding polyglutamic acid sequences could result in the formation of dual-functional entities that could assist in the targeting and attachment of normal and metastatic cells to bone surfaces, potentially with the assistance of BSP. The fact that BSP has the potential to enhance cellular adhesion and interact with the RGD motif suggests that its actions are multifaceted and necessitate additional research [72].

BSPs have an important function in the formation of new crystals and have a significant attraction to both calcium and HAP. Studies have shown that UMR cells are responsible for producing BSP, which has a dissociation constant (K_d) of around 2.6×10^{-9} for its binding to HAP. When exposed to a concentration of 60 mM potassium chloride (KCl), BSP has a binding constant range of 0.5 to 1.0 mM and has the ability to bind 83 calcium ions (Ca^{2+}) [73]. By comparing recombinant BSP with native BSP, one may see how post-translational modifications affect the inhibitory effect of BSP on HAP crystal formation. The specific mechanism behind the inhibitory action of BSP is yet unclear, in contrast to its nucleation activity. Consequently, BSP effectively inhibits the production of HAP crystals [72, 74]. The original BSP sequence was identified using rat cDNA sequencing [75]. Bovine spongiform encephalopathy (BSE) has been extracted from many animal species, including pig [76], rabbit [77], cow [61, 78], and human [60] bones. BSP typically has a M_r ranging from 60 to 80 K when analyzed using SDS-PAGE. This indicates the presence of a substantial carbohydrate component, which constitutes around 50% of its overall weight. This carbohydrate composition contains 6% galactosamine, 7% glucosamine, and 12% sialic acid [60, 61].

The amino acid composition of mammalian BSPs is approximately 327, which comprises a 16-residue signal sequence. The molecular weight of the mature protein is approximately 33 to 34 kDa. Residues of glutamic acid (Glu) and glycine (Gly) comprise approximately 33% of the total amino acid composition [72]. The N-terminal half of BSP contains two to three polyglutamic acid sequences that enhance HAP binding. These sequences demonstrate that Glu is present in significant quantities [75]. Additionally, the Arg-Gly-Asp (RGD) sequence at the carboxy-terminus of BSP is bound by integrins, which are cell membrane receptors for ECM proteins and promote cell adhesion [79]. The levels of BSP expression varied across various bone cell lines. UMR 106, specifically UMR-106-NIH, consistently demonstrates significant amounts of BSP expression, while ROS 17/2.8 only does so when stimulated [75, 80]. Bone sialoprotein is naturally produced by viable osteocytes, as shown by laboratory investigations using osteocytes derived from bovines and rodents [61, 81]. Mammalian bone sialoproteins have a protein sequence that is remarkably conserved, with 45% similarity and an additional 10–23% showing more conservative alterations. The regions of the protein that exhibit the highest degree of conservation include the carboxy- and amino-terminal parts, which are rich in aromatic amino acids, as well as the regions containing the polyglutamic acid and RGD sequences. Furthermore, there is a significant level of conservation in the phosphorylation, glycosylation, and sulfation domains. Nevertheless, the amino acid similarity diminishes to 39% when compared to avian BSP [72, 82].

Freshly synthesized BSP contains a low proportion of necessary amino acids and a high level of aspartic acid (Asp) and glutamic acid (Glu), resulting in an isoelectric point of around 3.9 [74]. Due to the presence of sialic acid, sulfate, and phosphate in its oligosaccharide side-chains, BSP exhibits anionic properties. This anionic nature poses challenges in reliably detecting and identifying BSP using commonly used protein stains such as Silver or Coomassie Blue. Consequently, cationic dyes such as "Stains-All" and Alcian Blue are used to perform BSP staining in acrylamide gels [61, 76]. BSP has mucin-like properties as a result of its elevated levels of glycosylation and sulfation [72, 83]. Utilizing ECM proteins is a technique to promote blood vessel formation and the development of bone inside hydroxyapatite scaffolds. The $\alpha_v\beta_3$ "vitronectin receptor" facilitates cell attachment via adhesion molecules like BSP, hence enhancing cell survival and promoting osteoblastic differentiation [75, 84]. Although physisorption may be used to coat calcium phosphate scaffolds with proteins, its effectiveness is limited by the abrupt release of the adsorbed protein [69, 85].

5.3.4 Proteoglycans

Proteoglycans (PGs) are intricate macromolecules composed of glycosaminoglycan (GAG) chains that are chemically bonded to a protein core. There are several established fundamental protein structures, and some groupings may be categorized based on the data included in their sequences. PGs have an important function in bone health since they influence the formation and growth of apatite crystals, which are responsible for regulating the mineralization process. The protein core may have a wide range of GAG chain attachments, ranging from one to almost one hundred, and these attachments are often linked to serine residues. Proteoglycans may be located on the cell surface, in the ECM, or within cells (often in secretory granules). Glycosaminoglycans are composed of hexosamine units, such as D-galactosamine (GalN) or D-glucosamine (GlcN), and hexuronic acid units, such as L-iduronic acid (IdoA) or D-glucuronic acid (GlcA). Keratan sulfate contains galactose units. The units undergo sulfation at many locations and are organized in irregular, linear sequences. Dermatan sulfate (DS), chondroitin sulfate (CS), keratan sulfate (KS), heparin, and heparan sulfate (HS) are some common examples of glycosaminoglycans. Hyaluronan is a commonly found glycosaminoglycan that exists in its free form, but is not classified as a proteoglycan. Heparin, HS, and DS are composed of IdoA and GlcA units, but CS exclusively contains GlcA as its hexuronic acid component. The presence of sulfate substitution allows for a high degree of variation in the sequence of GAG chains, both within individual chains and across distinct GAGs. This is especially true for the copolymeric species of GlcA/IdoA [86].

HS is synthesized by both osteoblasts and osteoclasts and has an important function in cellular adhesion. It regulates crucial signaling molecules necessary for bone remodeling and repair, such as TGF-β and osteoprotegerin/tumor necrosis factor receptor superfamily member 11B (OPG). In addition, it also interacts with the fibroblasts growth factor (FGF) and may serve as a protein coreceptor for it [3]. Recombinant DNA methods are essential for comprehending these proteins due to the complexity introduced by the presence of GAGs and oligosaccharide substituents in their sequence analysis [86]. Proteins like a lymphocyte "homing" receptor are referred to as "glycanated variants" or "part-time proteoglycans" due to the varied replacement of GAG chains. Besides the GAG-substituted section, the core protein may contain other functional domains that participate in protein–protein interactions or act as anchors for the proteoglycan. This is typically achieved through transmembrane peptide sequences or phosphoinositol linkages, connecting the proteoglycan to macromolecules present in the ECM or on the cell surface [86]. Proteoglycans have a diverse range of biological roles, influencing complex cellular processes including proliferation, motility, and adhesion, as well as providing basic mechanical support. Additionally, they have a function in the specialization and development of tissues. The primary method by which these effects are mediated is by the attachment of proteins to GAG chains. This process includes both highly specific interactions with well-defined regions of the GAG molecules and non-specific charge-based interactions [86]. Proteoglycans are classified and named based on their structural or functional qualities, unlike collagens. Some examples of these names include versican, syndecan, decorin, and aggrecan [86].

Aggrecan is a vital component of the primary spongiosa formed during endochondral ossification. Aggrecan mRNA has robust expression in osteoblasts in high-throughput gene expression studies and is detected throughout bone development, contradicting the earlier belief that it was mostly found in cartilage [87, 88]. Versican, a kind of proteoglycan, is found in soft connective tissues. Despite its distinctions, it has a close association with another medicine. Versican is often present in the loose, interstitial mesenchyme during bone development. It is believed to occupy physical space and gradually transform into bone. As osteogenesis progresses, this particular proteoglycan gradually vanishes. During the osteogenesis process, Versican is substituted by two distinct proteoglycans belonging to the same group. Biglycan and decorin are two proteoglycans that have different core proteins with diverse chemical properties [89]. The versican core protein includes sequences that have a resemblance to epidermal growth factor (EGF) [90]. The release of these sequences may potentially affect the cellular metabolism of the osteoblastic lineage [88]. Versican, a kind of proteoglycan, consists of chondroitin sulfate, a crucial component for cartilage development. It is situated inside the developing skeletal structure as a consequence. It has been found in osteoids in adult bones, suggesting that it may have a role in regulating or impeding mineralization [3].

Biglycan and decorin, two small proteoglycans, are found throughout the bone matrix. In addition to the chondroitin sulfate chains seen in bone, these proteoglycans may also incorporate dermatan sulfate into soft connective tissues. The function and presence of TGF-β are controlled by the interaction between decorin and biglycan [91]. Leucine-rich repeat (LRR) sequences, found in several proteins that have a role in morphogenesis such as chaoptin, Drosophila toll protein, serum's leucine-rich protein, and adenylyl cyclase, are considered essential proteins [89]. Upon solving the crystal structures of biglycan and decorin, it was determined that they are members of the small leucine-rich proteoglycans (SLRPs) family. Unlike ribonuclease, these SLRPs have a less pronounced curvature and have the potential to form dimeric structures [88]. SLRPs govern the size and shape of collagen fibrils, and may affect the collagen matrix. Additionally, SLRPs may enhance the interactions between collagen and minerals. Biglycan and decorin play crucial functions in maintaining osteoblast populations. Biglycan expression in osteoblast precursors may induce apoptosis, although in mature osteoblasts, it increases. It may serve as a sensor for shear stress since it is found in the pericellular areas of the matrix and in the osteocytes. On the other hand, decorin is produced at an early stage of cell differentiation by preosteoblasts and is suppressed in fully formed osteoblasts. The two proteoglycans may together interact with collagen and TGF-β to regulate the functions of growth factors and maintain the number of osteoblasts [3].

Decorin is located in the interterritorial matrix, which is far from the chondrocytes. On the other hand, biglycan is found in the intraterritorial matrix of cartilage. The spatial distribution of decorin throughout the process of bone endochondral formation closely resembles that of COL-I. The expression levels of this gene decrease as preosteoblasts get embedded in the ECM and undergo osteocyte differentiation. Biglycan has a distinct distribution pattern, with higher expression in osteoblasts and retention in osteocytes. It is located in certain morphologically structured areas around cells, particularly in canaliculi and lacunae. Osteocytes in the bone matrix function as mechanical sensors, whereas proteoglycans such as biglycan or cell surface-associated molecules like HSPGs act as force transducers [88]. It is appropriate to attempt to categorize the PGs into families, although there is still a need for data on the structure of several key proteins. This data is currently insufficient, and as a result, the basis for such categorization may not be completely consistent. Nevertheless, the chart provides a concise overview and can serve as a valuable resource for making comparisons and references [86].

PG families can be classified into the following categories: (a) High-molecular extracellular chondroitin sulfate proteoglycans primarily bind to hyaluronan. Multiple extracellular matrix proteoglycans obtained from different connective tissues belong to this classification. The protein sequences of fibroblast/versican and cartilage/aggrecan have been completely discovered. (b) Proteoglycans (PGs) containing tiny core proteins that have homology and are capable of carrying either one or two GAG chains, such as KS, DS, or CS. The fibromodulin, biglycan, and decorin proteins, which belong to this family, have been sequenced at both the protein and cDNA levels. Another prospective candidate is PG 100, a chondroitin sulfate proteoglycan produced by osteosarcoma cells. (c) ECM/basement membrane HSPGs, are similar to the large-molecular HSPG produced by EHS tumor. The data for affiliation between different members of this family are indefinite and are primarily derived from the immunological cross-reactivity of core proteins of similar size. If different types of proteins are encoded by more than one gene, more specific comparison of them will be present. (d) Cell-surface PGs that are located within the membrane and are related with the hybrid PG made up of CS and HS, known as syndecan. At least one other member of this group has been acknowledged, a fibroblast HSPG with a 48-kDa core protein. (e) Serglycins, a family of mostly intracellular PGs, including core proteins with elongated sequences of irregular glycine and serine parts, which are highly exchanged with heparin and/or CS chains [86].

Hyaluronan is composed of a long chain of glycosaminoglycan molecules that do not include sulfate groups. Hyaluronan and its receptor, CD44, work together to guide the development of the skeleton. Osteoblasts are mostly located in the periosteum and inner surfaces of bones, although they may also be found in the bone matrix surrounding other cell types, such as osteocytes [3]. The N-terminal portion of the protein core interacts to a decasaccharide sequence on the hyaluronan molecule. A spherical region (G1) facilitates the linkage between aggrecan, cartilage, and the large CSPG of hyaluronan. A short but elongated peptide separates the second spherical domain (G2) of the core protein from the hyaluronan-binding G1 domain. Although G2 has resemblances to G1, it lacks the capability to attach to hyaluronan [86]. Versican is a kind of proteoglycan that is located in fibroblasts and relies on the presence of the G2 globular domain. The core protein of this structure has a hyaluronan-binding domain at its N-terminal, which is comparable to the one found in aggrecan [86]. The presence of GAG chains on different proteoglycans enables them to engage with ECM proteins. Core proteins also facilitate certain interactions. Human lung fibroblasts produce an ECM-associated heparan sulfate proteoglycan (HSPG) that has a high affinity for a particular region of the fibronectin molecule. This basement membrane HSPG, which also binds to fibronectin, may be similar in size to the one being referred to. Biglycan and decorin possess the capacity to attach to TGF-β, perhaps as a result of the leucine-rich domains found in their core proteins [86].

The core protein of decorin, also known as PG-S2, DSPG II, or PG 40, interacts with fibronectin and collagen to affect the process of collagen fibrillogenesis. Although biglycan (PG-S1, DSPG I) has similarities with decorin, it does not have the ability to bind collagen. In contrast, fibromodulin is capable of binding collagen [86]. Instead of being limited to mast cells, heparin, HS, or sometimes DS (such as heparin cofactor II) may provide binding sites for proteins in vivo under normal physiological conditions. The binding capacity of these GAGs is superior to that of CS, even when the charge density is the same. This is likely due to the presence of IdoA units, which enhance the flexibility of the polymer's conformation [86]. Cell-surface HSPGs may cross-link proteoglycans and induce actin cytoskeleton remodeling by interacting with ECM components such fibronectin, thrombospondin, and other collagens. Consequently, the integral membrane protein cores have the ability to regulate the formation and breakdown of certain cell adhesions [86]. Angiogenesis is believed to need the presence of basic FGF, which is bound to and stored by ECM heparan sulfate proteoglycans (HSPGs). The alteration in the shape of N-CAM caused by interaction with the GAG chains of cell-surface HSPGs is crucial for the attachment of nerve cells. The endo glucuronidase produced by tumor cells facilitates the degradation of ECM heparan sulfate proteoglycans during the process of metastasis [86].

5.3.5 Bone Morphogenetic Proteins

The TGF-β superfamily encompasses the bone morphogenetic proteins (BMPs). In the late 1980s, the differentiation of BMPs unique proteins took place via the replication of human BMP-2 and BMP-4, together with the isolation and sequencing of bovine BMP-3 (osteogenin) [92–94]. Despite the original identification of BMPs' potential to generate bone in the 1960s [95], it took many decades to precisely define the proteins that constitute BMPs. There have been lately identified over 20 members of the BMP family. The primary means by which BMPs transmit signals is via type I and II serine/threonine kinase receptors, including the type IA activin receptor (ALK-2 or ActR-IA), type IA BMP receptor (ALK-3 or BMPR-IA), and type IB BMP receptor (ALK-6 or BMPR-IB) [96–98]. Moreover, it is shown that three type II receptors, namely type II and IIB activin receptors (ActR-IIB and ActR-II) and type II BMP receptor (BMPR-II) [99–101]. ActR-IA, II, and IIB are also signaling receptors for activins, while BMPR-IA, IB, and II are particular to BMPs. These receptors are expressed differentially in several tissues. Both type I and type II receptors are required for the transmission of the BMP signal. Upon the interaction of a ligand, they make a complicated heterotetrameric-activated receptor consisting of two pairs of a type I and II receptor complex [102].

Smad proteins transmit signals in a ligand-dependent manner by acting as substrates for type I BMP receptors [103–105]. Smads regulate gene transcription by associating with Smad4 to create a complex, which then moves into the nucleus upon phosphorylation [106]. The molecules Smad1, Smad5, and Smad8 play a crucial role in BMP signaling. Research using knockout and transgenic mice, together with studies involving individuals and animals with BMP-related disorders, emphasize the significance of BMPs in the development of cartilage, brain tissue, the heart, and postnatal bone. BMP-2 has shown therapeutic potential in clinical and preclinical studies for treating osteoporosis, non-union fracture repair, spinal fusion, root canal therapy, and bone structural abnormalities [106]. Furthermore, BMPs inhibit the process of myogenesis, hence impacting the production of somites. To be more precise, BMP-2 hinders the growth of the limb bud by interacting with FGF-4 and sonic hedgehog. At the same time, it promotes the formation of precursor cells for osteoblasts and chondrocytes [107, 108]. Confirmed evidence from null mutations in Smad genes, BMP receptors, and ligands has shown the role of BMP signaling proteins in both embryonic and postnatal development. Metanephric mesenchymal cells in mouse embryos that lack BMP-2 and BMP-4 fail to undergo differentiation, leading to the absence of glomeruli in the developing kidneys [106]. Consequently, the embryos fail to survive. The main location of BMP-6 expression is in hypertrophic cartilage, whereas animals that do not have BMP-7 only have slight skeletal defects [106, 109, 110].

Growth/differentiation factor-8 (GDF-8), often known as myostatin, is mainly produced by skeletal muscle throughout both the development and maturity stages. GDF-8 is produced at the early embryonic stage, namely in the myotome region of growing somites, and later in various adult muscles [106]. Embryonic development in mice with a null mutation in the BMPR-IA gene results in death. Given GDF-5's role in cartilage development and its strong binding to BMPR-IB, it is plausible that BMPR-IB may have a supplementary role in the production of cartilage. The appendicular defects seen in BMPR-IB mutant animals are indistinguishable with those shown in mice with a homozygous GDF - 5^{bp-j} null gene. During the process of bone and cartilage formation, BMP ligands have the capability to transmit signals across several type I BMP receptors [106]. Due to its ability to bind to both ActR-IA (Alk2) and BMPR-IB, it is believed that BMP-7 has complementary or overlapping roles in the in vivo formation of bone and cartilage [98, 106]. Embryos with mutated Smad5 genes had blood arteries that were longer and had a reduced number of vascular smooth muscle cells around them [111]. This indicates that Smad5 plays a role in regulating the connections between endothelium and mesenchyme during the process of angiogenesis [106].

Substantial quantities of lipids were inadvertently found in association with guanidine hydrochloride-soluble non-collagenous proteins during the extraction of BMPs from the bone matrix. The Sudan Black B-stained microspherules were formed by these protein-lipid complexes. Similar to the findings seen in previous experiments using unstained delipidized BMPs combined with non-collagenous proteins, the present study identified bone formation after the implantation of BMPs and non-collagenous protein-lipid microspherules that were labeled with Sudan Black B. Sudan Black B did not exhibit binding affinity toward lipid-depleted osteocalcin, albumin, osteonectin, or other bone matrix proteins, and it also did not possess the ability to stimulate bone formation. The presence or absence of BMPs does not affect the correlation between the refinement of non-collagenous proteins and a decrease in

Sudan Black B attachment. The bone matrix lipids that are soluble in acetone consist mostly of saturated short-chain fatty acids, cholesterol, and triglycerides. Monounsaturated fatty acids and phospholipids are either insignificant or not present. Implants containing acetone-soluble lipids in addition to recombinant BMP-2 resulted in a greater amount of bone formation compared to implants having just recombinant BMP-2 [112].

The separation of BMP from bone involves three phases of lipid extraction: (1) isolation of non-collagenous proteins in chaotropic solvents under dissociative conditions; (2) extraction before demineralization of the bone matrix; and (3) extraction after demineralization of the bone matrix [113, 114]. The chloroform–methanol mixture is used in this procedure. The chloroform–methanol mixture was subjected to acidification and heating at 60°C to extract lipids that are closely associated with skin collagen in a quantitative manner [115]. However, under such situations, BMPs undergo denaturation as a result of the combination of low pH and organic–inorganic solvents, which causes the irreversible disruption of hydrophilic and hydrophobic interactions [113, 116]. After delipidizing BMP-NCP and combining it with lipids extracted using different solvents, both acetone-soluble lipids and lipids soluble in chloroform–methanol showed equivalent effectiveness in regenerating new bone volume in muscle. Unlike chloroform–methanol, acetone has minimal amounts of phospholipids, which are often seen near the calcification front [117, 118]. The removal of phospholipid-containing lipids using chloroform–methanol did not have any impact on the level of bone formation that was stimulated. When chloroform–methanol is made acidic and heated to 50 °C [115], all BMPs undergo denaturation and the staining of Sudan Black B is completely erased, leading to the full removal of lipids [113].

Research has shown that lipids obtained from different sources such as heterotopic bone, healing fractures, the mineralizing front in developing teeth and bones, and demineralized bone, differ from lipids associated with the bone matrix, including BMP [117–119]. Within a span of seven days after the injury, Lane et al. [119] detected a notable rise in lipids within the fracture callus. They attributed this rise to the generation of triglycerides in the bone marrow. The fatty acid, triglyceride, diglyceride, and phospholipid quantities were similar to those seen in natural, intact bone both before and after a week. The focus was on lipids that are connected to minerals and the capacity of acidic phospholipids to bond with calcium. These phospholipids act as a central point for the formation of crystal apatite [117]. Prior studies on Sudan Black B binding indicate that the process of removing lipids from bones actually obscures rather than reveals the role of lipids in the integration of bone grafts [120–122]. In order to remove lipids, it is necessary to separate the BMP-containing non-collagenous proteins from bone collagen using chaotropic solvents. Under these conditions, the presence of a BMP-NCP lipid molecule may be detected using Sudan Black B dye binding. The complex may exhibit dye binding due to the presence of lipids, BMP, or NCP. However, this binding does not inhibit BMP activity in muscle implants; on the contrary, it may potentially enhance it [112].

5.3.5.1 Negative Regulation of Bone Morphogenetic Protein Signaling

Smad6 strongly binds to type I BMP receptors and prevents the phosphorylation of Smad1 and Smad5. This limits the formation of heteromers with Smad4, hence exerting a negative regulatory function in BMP signaling. Tob, a constituent of the anti-proliferative protein family which comprises BTG1, BTG2, BTG3, Tob, and Tob2, engages with Smad1 and Smad5 to inhibit BMP-induced, Smad-dependent transcription in osteoblasts. Smurf1, discovered by the yeast two-hybrid assay, regulates the degradation rate of Smad1 and Smad5 [106, 123, 124]. Noggin is an excreted polypeptide that attaches to BMP-2, BMP-4, and BMP-7 in order to restrict their function. The cocrystal structures of BMP-7 and noggin reveal that noggin hinders the activity of the type I and type II BMP receptors on BMP-7, hence impeding BMP signaling [106].

Osteocytes and osteoblasts both secrete sclerostin, which acts as an antagonist to BMP. The remaining individuals of this antagonistic family are Noggin, Dan, Gremlin, and Chordin. Sclerostin suppresses osteoblastogenesis in multipotent fibroblastic C3H10T1/2 cells by forming a robust interaction with BMP-5, BMP-6, and BMP-7, hence impeding the formation of bone-producing cells [125]. Osteoporosis is a condition where bones become weak and brittle. It is caused by an overproduction of sclerostin by osteoblasts, which is triggered by the osteocalcin promoter in transgenic mice. These findings emphasize the need of regulating BMP activity after birth to facilitate optimal bone development, and suggest that enhancing natural BMP signaling may aid in promoting bone formation [106]. Furthermore, the process of ubiquitin-mediated proteasomal degradation, which relies on three enzymes and is crucial for numerous biological processes such as gene transcription, signal transmission, and cell cycle progression, controls the activity of BMP signaling proteins [126, 127]. The three enzymes are referred to as ubiquitin-activating enzyme (E1), ubiquitin-conjugating enzyme (E2), and ubiquitin ligase (E3). E3 ensures the selectivity of protein ubiquitination and the subsequent degradation of the 26S proteasomes [128, 129]. A research on gene therapy suggests that the use of bone marrow MSCs implanted with Adenovirus BMP-2, together with a bioresorbable polymer, may be beneficial in treating bone defects [130]. New experimental findings indicate that recombinant human BMP-2 (rhBMP-2) may outperform autogenous bone transplants in some situations, providing a possible alternative to bone grafts in spinal fusion surgeries [106].

5.3.6 Growth Factors

Transforming growth factor-beta (TGF-β), acidic and basic fibroblast growth factors (FGFs), bone morphogenetic proteins (BMPs), platelet-derived growth factors (PDGFs), and insulin-like growth factors I and II (IGFs) are among the several growth factors present in the bone matrix [131]. Heparan sulfate (HS) and extracellular matrix proteins form connections with TGF-β, FGF, IGFs, and hepatocyte growth factor (HGF), as well as other growth factors. These growth factors regulate the synthesis of ECM, cellular differentiation, and tissue remodeling. Therefore, by processing necessary information, the ECM impacts the development of complex multicellular organisms by modifying the signaling of growth factors [132].

The interaction between growth factors and ECM has a dynamic impact on many cellular processes. Growth factors are securely enclosed, protected from deterioration, and transported to their specific receptors in a controlled way by attaching to the ECM. Furthermore, this interaction not only stabilizes the growth factors but also generates gradients that have an important function in embryonic development, tissue repair, and disease progression. Furthermore, altering the ECM may release bound growth factors and alter their availability for biological processes. Matrix metalloproteinases (MMPs) have an important function in the process of remodeling, which subsequently impacts the distribution and function of growth hormones. Hence, the ECM and its associated growth factors constitute a complex regulatory system that is essential for maintaining tissue balance and facilitating the process of tissue repair. Acquiring comprehension of this intricate system may aid in the development of tissue engineering, cancer therapy, and regenerative medicine treatment approaches. While polypeptide growth factors are molecules that are specified for the intercellular interaction, peptide and non-peptide hormones transmit signals between various organ systems. There is expanding evidence that growth factor binding to the ECM is an essential mechanism controlling growth factor activity, though they were initially expressed as soluble molecules (Table 5.1). While growth factors facilitate transmission of the data needed for formation of complicated cellular structures, the ECM prepares a structural basis for multicellularity [132].

In contrast to heparin, HS has a lower degree of sulfation and is synthesized on a core protein. Heparan sulfate proteoglycans (HSPGs) are present in the ECM and are also connected to the cell membrane [133]. HS has certain structural features that enable it to strongly bind to HGF and FGF-2 [134, 135]. Protein sequences that bind heparin in growth factors like PDGF-A and VEGF usually have a high number of basic residues. Various bonds have the capacity to alter these binding sites. FGF-2 [135] and HGF [136] may have different specificities for HS substructures compared to other members of the FGF family [137, 138].

Specific growth factors form an indirect bond with HS by using other proteins. As an example, follistatin, a protein that binds to heparin, interacts with activin, whereas TGF-β binds to heparan sulfate (HS) via a 60 kDa protein [139, 140]. Hepatocyte growth factor/scatter factor (HGF/SF) has domains that aid in binding and shares a similar structure with plasminogen [141]. Upon tissue damage, the inactive form of HGF, which is stored in the ECM of organs including the kidney, spleen, and liver, becomes active to stimulate the regeneration of epithelial cells and the restructuring of the matrix [132]. Hematopoietic stem cells (HSCs) undergo differentiation into several types of blood cells in response to the spatial arrangement of microenvironments inside the body. The importance of the ECM heparan sulfate proteoglycans (HSPGs) in transporting these substances to hematopoietic cells is emphasized by

Table 5.1 ECM-associated growth factors[a] [132]

Binding of growth factors to matrix proteins		Growth factors associated with Heparan sulfate or Heparin	
Decorin	Active TGF-β	FGFs 1–9	GM-CSF
Fibronectin			IL-3
Thrombospondin			IL-4
COL-IV		TGF-β	
Microfibrillar structures	Latent TGF-β	VEGF	NT-6
		PDGF-A	HB-GAM
COL-IV	BMP-2, -7	PDGF-B	MK
Matrix form of FGF receptor	FGFs	HB-EGF	IP-10
160 kDa protein	LIF		PF-4
		HGF	MCP-1
SPARC	PDGF-BB and AB		RANTES
Fibronectin	TNF-α		IL-8
IGFBP-5 (and -3?)	IGF-I	Activin via follistatin	
		TGF-β via 60 kDa protein	
Binding of growth factors to unknown extracellular constituents			
	Sonic hedgehog		
	Wnt proteins		

[a]Heparan sulfate (HS), certain components of the ECM, and other elements of the matrix are often associated with growth factors. The following exchanges are exemplified: The ECM proteins shown on the right side establish interactions with the growth factors situated on the left. Specific growth factors, including HGF, TGF-β, and CSF-1, lose their functionality when binding to the matrix. The factors that bind to HS may be grouped into five groups: TGF-β-related factors, FGF family members (FGF 1–9), chemokines (such as IP-10), neural factors (such as NT-6), and hematopoietic growth factors (such as GM-CSF). HGF and HBEGF are uncategorized. CSF-1 (M-CSF) may be synthesized from a precursor that includes glycosaminoglycan. Activin and TGF-β interact with heparan sulfate (HS) via accessory proteins, rather than directly binding to HS. PDGF, VEGF, and PIGF binding is affected by alternative splicing.

the interaction between heparan sulfate (HS) and hematopoietic cytokines including GM-CSF and IL-3. The differentiation routes of HSCs may be affected by varied culture conditions [142].

HS enhances FGF signaling [143] by increasing the binding of FGF to its receptors [144]. Extracellular and cell-surface HSPGs control the FGF family. In addition, extracellular matrix HSPGs may alter the biological response to FGF by guiding complexes to cell–matrix or cell–cell adhesions, or by inhibiting endocytosis and the breakdown of ligand-receptor complexes. The pH level has an impact on the interaction between PDGF-AB and -BB and SPARC, a protein that is synthesized during the process of tissue remodeling and repair [132]. TGF-β latency-associated protein (LAP) and active TGF-β interact to create latent, inactive complexes which are subsequently released by TGF-βs. Referred to as tiny latent TGF-β1, this complex remains inactive unless exogenous proteases trigger its activation, leading to rapid and localized signaling that is crucial for angiogenesis and tissue healing. The regulation of growth factors ensures that signaling is activated in a timely and spatially controlled manner. This involves separating the mature growth factor from latency proteins during activation, and releasing it from its binding sites in the matrix [132]. The ECM's role in storing growth factors enables rapid signal transmission without the need for new protein synthesis. Localized signaling is crucial for immune response and tissue repair, and it is enhanced by the proteolytic activation and subsequent release of stored growth factors. Growth factors maintain organ function and form by establishing a chronological record of cellular activity via their interaction with the extracellular matrix. Growth factors such as HGF, TGF-β, and PDGF have an essential function in the accumulation and modification of the matrix in many clinical conditions [132].

5.3.7 Fibronectin

Fibronectin is a versatile protein found in the tissues and blood of vertebrates, as well as in fibroblast and astrocyte cell cultures that attach to surfaces [145]. Fibronectin, also known as CSP or LETS protein [146–150], is an essential constituent of several fibroblastic cell cultures. The concentration of fibronectin in human plasma is around 300 μg/ml [151]. Fibronectin is found in the extracellular connective tissue matrix and basement membranes of mouse and chicken embryos. Its role in cell adhesion has been extensively studied in cell culture settings [145, 152, 153]. Human fibronectin is present in about 20 different splicing isoforms. Structurally, fibronectin is a glycoprotein that has a large molecular weight of around 440 kDa. It is composed of two almost similar subunits, each weighing around 250 kDa, which are covalently connected by disulfide bonds in their C-terminal region (Fig. 5.4). Each of these subunits is composed of three distinct types of repeats: type I, II, and III. Type I repetitions consist of 12 modules, containing 40 amino acidic residues, which are arranged as stacked β-sheets, connected by a disulfide bond. Additionally, they possess a hydrophobic core composed of conserved aromatic residues. Type II repetitions, consisting of 2 modules, composed of perpendicular antiparallel β-sheets that are connected by two intra-chain disulfide bonds. These repetitions are 60 residues in length. Finally, type III repetitions consist of 90 residues and typically contain 15–17 modules. The repetitions lack disulfide bonds and the antiparallel β-sheets are connected by flexible loops and maintained by hydrogen bonding. Type III repetitions exhibit the greatest sensitivity to potential unfolding of fibronectin [154].

Collagen bundles and fibronectin fibers have a tight relationship, particularly when they form thin fibers in non-muscular arteries or are positioned near smooth muscle cells in muscular arteries [145]. Fibronectin is present on the outer surfaces of individual adipocytes in adipose tissues. It is also often seen in the loose connective tissue of the adventitia, as well as in the loose connective tissues of the epididymis and testis [145]. Other proteins included in reticulin fibers consist of many glycoproteins

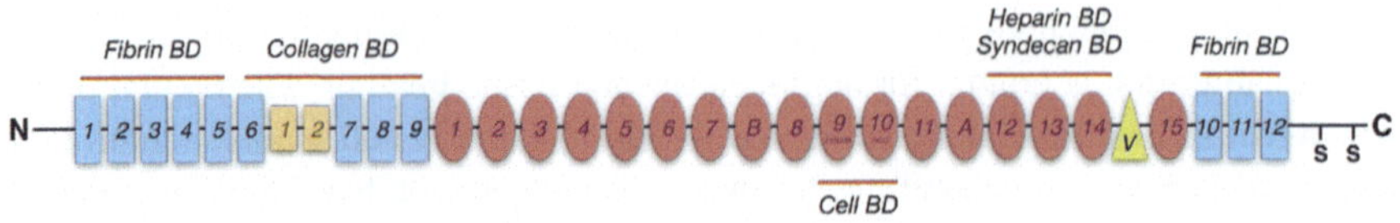

Fig. 5.4 Structure of fibronectin. Illustration depicting the organization of fibronectin individual subunits. The presence of binding domains (BD) and repeats are specified. Reprinted with permission from [154]

that are not collagenous [155, 156], as well as COL-III [157]. Soluble fibronectin binds to the glycosaminoglycan heparin and has a notable attraction to collagenous proteins under laboratory conditions. Plasma transglutaminase, an active blood coagulation factor XIII, has the ability to bind both cell-associated [149] and soluble [150] fibronectin to fibrin or fibrinogen. This process potentially forms connections between collagenous proteins and cell surfaces in connective tissue matrices in living organisms [145].

Cells that are rapidly multiplying around injury sites or inside clots are linked to the connective tissue matrix via interactions involving cell surfaces, collagen, fibrin, and either serum or cellular fibronectin. Fibronectin has a role in guiding the growth of epithelial cells that are undergoing regeneration on basement membranes [145, 158]. Collagen molecules adapt their molecular composition to engage with other elements in the biological structure of connective tissue for certain functions [159]. Fibronectin was first identified as a cell-surface protein that is absent in transformed cells. Over time, it has been known as a prototype adhesive protein, with a similar protein detected in plasma. Plasma fibronectin may be derived from the liver [160]. Fibronectin, a versatile protein found in both plasma and the extracellular matrix, plays a crucial role in facilitating cell attachment. Fibronectin was first characterized as an ECM protein [161], but it also has the ability to interact with other proteins, including fibrinogen, collagen [162], proteoglycans, and glycosaminoglycans [163, 164]. The ability of fibronectin to interact with cells and promote attachment is emphasized by the separate discovery of its cell attachment-stimulating activity [160, 165, 166].

The synthesis of fibronectin involves two steps: the cleavage of a signal peptide and a short propeptide, resulting in the production of the mature polypeptide [167, 168]. The fibronectin polypeptide consists of about 2500 amino acids and is characterized by three distinct types of repeats: type I, type II, and type III [160]. The molecule consists of two polypeptides that combine to form a dimer by disulfide connections at the COOH terminus. The dimer has a molecular weight of around 550,000 [169].

The glycoprotein fibronectin contains carbohydrates, which make up around 4 to 10% of its composition, varying depending on its origin. Tumor cells and embryonic sources often have higher carbohydrate content compared to adult plasma and fibronectins obtained from cultured normal adult skin fibroblasts [160]. The glycosylation of fibronectin is crucial [160] because carbohydrates provide protection to the fibronectin polypeptide from proteolysis [170]. Possible changes of fibronectin include phosphorylation at a serine residue at the COOH-terminus [169], acylation

[171], and the presence of tyrosine-linked sulfate [172]. Fibronectin has a strong attraction to collagen, namely denatured collagen (gelatin). Very few proteins possess the same capacity as fibronectin to bind to fibronectin, resulting in an almost exclusive binding phenomenon [173, 174]. Denaturation of acetylcholinesterase and Clq results in the breakdown of their triple helix structures. It has been shown that fibronectin can bind to all known forms of collagen. The collagen-binding site on the fibronectin polypeptide has been identified using polypeptide segments derived from fibronectin cDNA in bacteria and fragmentation tests. This site has been localized to two type II repeats [175].

Immunostaining reveals that fibronectin and collagen fibrils line precisely at a consistent interval of 67 nm, indicating an inherent connection between fibronectin and collagen [176]. The cell attachment-stimulating region of fibronectin is situated inside the central segment of the fibronectin polypeptide. This segment contains the RGD sequence, which is crucial for cell adhesion to surfaces [177, 178]. The RGD sequence may be identified by ECM and platelet adhesion proteins, such as osteopontin [179], von Willebrand factor [180], fibrinogen [180, 181], vitronectin [182, 183], thrombospondin [184], and perhaps laminin [185]. By examining fibronectin fragments and synthetic peptides that mimic certain amino acid sequences, it has been shown that the RGD region plays a role in cell attachment. Application of peptides containing the RGD sequence enhances cell adherence to surfaces. However, when peptides or fibronectin-coated surfaces are dissolved in a solution, they effectively inhibit cell attachment [160].

The collagen-binding domain and the DNA-binding domain are located next to each other at the C-terminus [186, 187]. The binding affinity of fibronectin enhances the interaction between fibronectin and DNA [160, 188]. Fibronectin may bind to both large aggregating chondroitin sulfate proteoglycans prevalent in cartilage and fibroblasts, as well as small chondroitin/dermatan sulfate PGs present in bone and fibroblasts [189, 190]. The binding of proteoglycans to fibronectin affects its interactions with collagen and cells. Cell-surface PGs may enhance the interactions between fibronectin and cells by binding to the heparin-binding sites on fibronectin [191, 192]. In contrast, soluble polymers have the ability to hinder cell attachment by disrupting these processes. Fibroblast proteoglycans typically consist of a solitary glycosaminoglycan chain and may directly interact with fibronectin via their core protein [193]. In addition, the extracellular matrix contains heparan/chondroitin sulfate proteoglycans [194], and the cell surface is covered with heparan sulfate proteoglycans [195, 196], which are connected to fibronectin [160].

5.3.7.1 Fibronectin Receptors

Multiple receptors exist for fibronectin. Fibronectin receptors, which bind to the RGD cell binding site, have been identified in tissues and cultured cells using antibodies that inhibit cell binding to fibronectin. Alternatively, these receptors have been characterized through affinity chromatography on specific regions of fibronectin that are involved in cell attachment. These approaches have been used to obtain fibronectin receptors from various cell types. The changing requirements for fibronectin interaction in different cells may need this variance [160].

5.3.7.2 Fibronectin Deposition into ECM

Regarding the molecular mechanics of matrix synthesis, three steps are essential: The processes involved include the self-binding of fibronectin, the crosslinking of fibronectin, and the binding of fibronectin to cell-surface receptors [160]. Fibronectin mostly exists in the form of an insoluble protein. This is the specific form in which fibronectin interacts with cells. Fibronectin exhibits a weak attraction to its receptor, with a dissociation constant (K_d) in the micrometer range. Consequently, it is probable that only a tiny fraction of fibronectin molecules in solution will bind to its receptors, and this fraction is expected to be lower than the concentration of soluble fibronectin in plasma [151]. Conversely, the tissue fibronectin that cannot be dissolved might work along with cells to affect their ability to adhere. The process of fibronectin insolubilization is crucial. The inhibition of matrix deposition in cancerous cells highlights the importance of this therapy [160].

5.3.7.3 Fibronectin Biological Effects

Fibronectin and its receptors regulate several cellular characteristics. The polarity, anchoring, differentiation stage, and mobility exhibit a high degree of complexity. In addition, the cellular phenotypic and growth properties may be affected by the interactions between cells and matrix molecules, such as fibronectin [160]. Both fibronectin RGD receptor subunits include cytoplasmic tails, and their specific sequences have been identified. The cytoplasmic tail of the β subunit is recognized for its interaction with cytoskeletal proteins [197] and may undergo phosphorylation at a tyrosine residue [198]. In addition, these sequences do not provide significant insights into the mechanism by which a growth or differentiation signal is sent to the cell. Furthermore, the fibronectin molecule [160] might potentially include supplementary sequences that have the ability to interact with cells. Studies have shown that TGF-β stimulates the production of fibronectin, leading to an increase in cell proliferation on soft agar [199]. Examining the possible role of fibronectin in the anchorage dependence of cell growth is crucial due to its impact on cell proliferation. Cell growth seems to need the occupation of fibronectin receptors. It is possible to imagine that a comparable signal may lead to unrestricted cell growth without attachment, as seen in cancerous cells [160].

5.3.7.4 Fibronectin and Biomaterials

The bioactivity of biomaterials is intricately linked to the biological response at the interface between the biomaterial and the cells. Nevertheless, once materials are inserted, they rapidly become covered by proteins found in biological fluids, rendering direct surface detection by cells nearly difficult. Hence, regulating the attachment of specific proteins at the interface of biomaterials could potentially serve as a technique in tissue engineering to create optimally functioning scaffolds that can effectively modulate cellular activities. Regrettably, the adsorption of proteins on biomaterials is typically an unpredictable process, primarily influenced by the physical and chemical characteristics of the material, as well as the interactions between the material and proteins. As a result of a weaker structure, protein adsorption may occur without the necessary stimulus for cells [200, 201]. ECM proteins play a crucial role in this context since they are well recognized for their ability to modify cellular behavior. Fibronectin is crucial for supporting the regeneration and repair of both healthy and injured tissues [154].

The design of the adsorbed protein film at the interface is critical in determining important elements of cell response, such as adhesion, migration, spreading, differentiation, and proliferation. This configuration has an essential function in revealing the nature of the reciprocal destiny between the material and the tissue. Protein adsorption, specifically fibronectin, has an essential function in the interaction between biomaterials and tissues. Body fluids contain a diverse range of biological components, and their attachment to biomaterials is a complex process. Blood contains around 150 proteins, along with carbohydrates, lipids, and other compounds that compete to be absorbed at the material contact. When considering diffusion, smaller molecules present in the bulk solution and at high concentration deposit more quickly than larger molecules with lower concentration Table 5.2 [154].

Another important factor is the biomaterial's adhesive strength. Indeed, molecules with higher surface affinities may cause molecules with lower affinities that were previously adsorbed to be displaced. According to the Vroman effect, exchanges start when all of a substrate's binding sites are filled and go on until molecules and proteins with a higher affinity and contact with the material occupy the surface [203]. Clearly,

Table 5.2 Hierarchical exchange of plasma proteins on the surface [154, 202]

Protein	Molecular weight (kDa)	Blood plasma concentration (mg/ml)	
Factor XII	80	0.015–0.047	Lower affinity to adsorb
Fibronectin	220	0.2	
Fibrinogen	340	3	
Immunoglobulin-G	150	15	
Albumin	66	40	Higher affinity to adsorb

desirable proteins, like fibronectin, may be specifically immobilized to trigger certain responses, therefore blocking the interchange of proteins that compete with one another on the surface. Table 5.3 lists the methods for improving the biomaterials–fibronectin interaction and highlights the main factors to be taken into account [154].

The process of fibronectin immobilization on scaffold surfaces may occur by either physical adsorption or surface functionalization, followed by covalent immobilization [204]. Despite its increased difficulty, covalently immobilized fibronectin effectively addresses and overcomes many challenges associated with physical adsorption. These concerns include undesired changes in conformation and spontaneous detachment of fibronectin, both of which eventually lead to loss of fibronectin's functionality [205, 206]. An evident constraint linked to immobilizing whole fibronectin is the

Table 5.3 Techniques to enrich the interface of biomaterials with fibronectin [154]

Source of fibronectin	Biomaterial interface enrichment approach with fibronectin	Disadvantages
Autologous	Aptamer immobilization	High production costs
	Monoclonal antibody immobilization	High production costs;
		High molecular size;
		Probable host immune response
Recombinant cell binding domains	Functionalization of the surface and subsequent covalent immobilization	Absence of particular binding sites interaction
Recombinant fragments	Functionalization of the surface and subsequent covalent immobilization	High production costs;
		Lack of complete availability of protein
	Direct immobilization through physical adsorption	High production costs;
		Lack of complete availability of protein;
		Probable spontaneous desorption
Heterologous	Functionalization of the surface and subsequent covalent immobilization	Probable loss of protein mobility;
		Probable adsorption in an undesired conformation;
		Probable host immune response
	Direct immobilization through physical adsorption	Probable adsorption in an undesired conformation;
		Probable spontaneous desorption;
		Probable host immune response

difficulty in effectively controlling the protein's conformational adsorption, which is greatly influenced by the substrate [154]. Given that the proper interaction with cell integrins necessitates the interaction of many fibronectins' individual domains (such as RGD, IKVAV, and PHSRN), it is better to anchor the fibronectin section rather than immobilizing single binding domains [207–210]. The high molecular weight of fibronectin restricts the quantity of biomaterial that may be enhanced with it because of its negative impact on the stability and bioavailability of the material. Hence, it is crucial to develop materials that exhibit specific affinity toward fibronectin. The concept of introducing tiny molecules onto the surface of scaffolds to catch and hold fibronectin from the extracellular space is an innovative notion [154, 211].

The idea of enhancing biomaterials with targeted binding capabilities was first proposed to enhance the retention of BMP-2 by attaching monoclonal antibodies (mABs) to scaffolds designed for bone regeneration. Oliveira et al. [212] later used a similar approach. Nevertheless, apart from compatibility concerns, the considerable dimensions, expensive production expenses, and incapacity to detect minuscule molecules restrict the use of mABs. Consequently, aptamer might serve as a viable substitute to mitigate this issue. The notion was originally introduced by Hoffmann and his colleagues in 2011. To protect endothelial cells from blood on vascular grafts, aptamers were used [213]. Chen et al. [214] used aptamers to decorate synthetic biomaterials, aiming to enhance cell colonization and investigate the impact of aptamers on cell behavior in a dose-dependent manner. The aptamer's capacity to generate biomaterials with selective binding was shown in both studies [154].

The aptamer-modified hyaluronic acid hydrogel significantly enhanced the quantity of adsorbed fibronectin, hence increasing the accessibility of adhesive proteins for cells. However, unlike its basic counterpart, chitosan enriched with aptamers demonstrated a similar amount of fibronectin adsorption. Nevertheless, the preservation of its function and structure was only seen when aptamers were present [215, 216].

5.3.8 *Osteopontin*

Osteopontin (OPN) is a non-collagenous protein that accounts for 1–2% of the NCPs in bone. It is essential for the biomineralization of bone tissue [217]. It is imperative to maintain the mechanical characteristics and structural integrity of bone. Depalle et al. [218] conducted research that illustrates the importance of OPN in the biomineralization processes and the preservation of the structural integrity of collagen fibrils in bone. Due to its high affinity for calcium, capacity for phosphorylation, and highly charged state, OPN serves a multitude of functions in the regulation of bone mineral density [218–220]. OPN plays a substantial function in various pathological and physiological processes, including apoptosis, wound healing, biomineralization, bone remodeling, and tumor metastasis [221, 222]. It is essential for the mineralized ECM that is observed in teeth and bones [221]. It functions as an "adhesive" at the interface between minerals and collagen, establishing sacrificial connections that fortify the structure of bones and prevent the separation of mineralized collagen fibrils. As a result, the absence of OPN may reduce bone toughness and facilitate the propagation of fractures [218].

Studies conducted on mice lacking OPN ($Opn^{-/-}$) have demonstrated a substantial decrease in bone hardness in comparison to healthy bone, underscoring the essential function of OPN in the prevention of fracture propagation [223, 224]. A quantitative investigation utilizing 2D small-angle X-ray scattering (SAXS) spectra has demonstrated that an increase in crystal misorientation is a consequence of OPN deficiency [225, 226]. Additionally, OPN is essential for the formation and reorganization of the collagen framework within the bone. The arrangement of collagen fibrils at the tissue level is significantly altered, and bones lacking OPN contain less collagen in comparison to normal bones. Although the mineral's size remains constant, its absence evidently impacts the collagen matrix, resulting in a reduction in tissue structure and the presence of collagen fragments that are not fibrillar in the $Opn^{-/-}$ bone [218].

The administration of exogenous OPN results in an increase in CD44 levels in OPN-deficient osteoclasts, which implies that autocrine OPN is crucial for osteoclast function, including bone resorption and migration [227, 228]. The injection of exogenous OPN has been demonstrated to partially restore osteoclast resorptive activity, indicating its critical role in this process [227]. Nevertheless, external OPN is unable to access the osteopontin secreted by osteoclasts in the resorption lacuna [222, 227]. The OPN molecule contains numerous conserved regions, such as a putative calcium-binding site, a location where thrombin may cleave the molecule, two sites where heparin can engage, a site where serine/threonine phosphorylation can occur, and a domain that can bind to (RG). Cell-interacting domains include the arginine-glycine-aspartic acid (RGD) cell-binding sequence and the serine-valine-valine-tyrosine-glutamate-leucine-arginine (SVVYGLR) motif. These specific locations enable osteopontin to perform its numerous functions in bone and other tissues [229, 230]. The cleavage sites contain thrombin and cleavage sites of matrix metalloproteinase (MMP) [229]. SVVYGLR site is exposed and directed to the development of two sections in response to cleavage by thrombin: C-terminal fragment and N-terminal fragment (Fig. 5.5) [222]. The structural similarities and diverse post-translational modifications that the protein family undergoes have an impact on their activities in mineralization processes [218].

In addition, it has been hypothesized that the binding of OPN to integrin $\alpha v\beta 3$ influences the levels of intracellular calcium (Ca^{2+}) by regulating the entry of calcium from the extracellular environment and promoting the release of Ca^{2+} from intracellular compartments through the Ca^{2+}-ATPase pump [231, 232]. The production of cytosolic calcium ions (Ca^{2+}) further regulates the activity of osteoclasts [233, 234]. In conclusion, OPN influences mineralization, collagen matrix synthesis, and osteoclast function, among other components of bone physiology, by interacting with a variety of receptors and signaling pathways. The absence or deficiency of OPN has a substantial impact on bone structure and function, rendering it essential for the

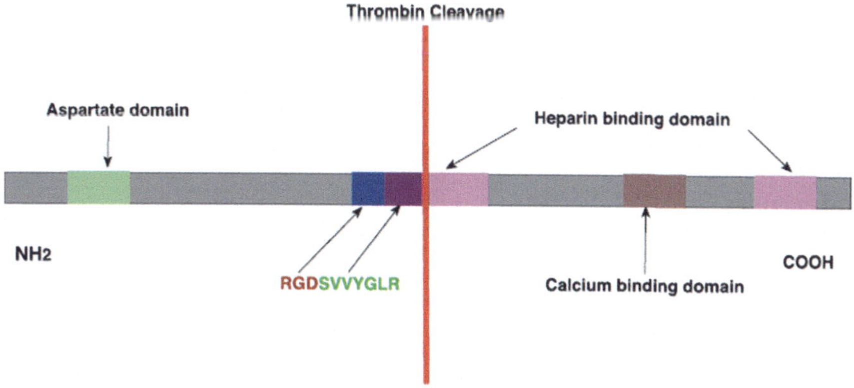

Fig. 5.5 A schematic illustration of thrombin cleavage site and OPN structure. SVVYGLR and RGD binding domains are shown. Reprinted with permission from [222]

health and development of the skeleton. The movement of osteoclasts, their survival, and the formation of podosomes are all dependent on OPN. Periodontal remodeling is facilitated by the involvement of numerous OPN-dependent signaling pathways, which in turn facilitate the movement of orthodontic teeth [222].

5.4 Conclusion

In this comprehensive chapter, we have examined the intricate realm of nano-bioceramics and its prospective applications in regenerative medicine and bone tissue engineering. COL-I, which is the structural component of bone tissue, imparts bone with its tensile strength and endurance. Conversely, COL-V controls the formation and structure of COL-I fibrils, which impacts the mechanical properties of bone. Aside from collagen, the bone microenvironment consists of a complex network of cellular connections regulated by non-collagenous proteins such as growth factors, proteoglycans, bone sialoproteins, osteocalcin, osteonectin, bone morphogenetic proteins (BMPs), and osteopontin. Osteonectin has a role in controlling cell adhesion and the formation of the extracellular matrix, whereas osteocalcin is responsible for mineralization and maintaining calcium balance. Proteoglycans facilitate the hydration and signaling of the ECM, bone sialoproteins assist in the process of mineralization, and bone morphogenetic proteins stimulate the formation of new bone tissue (osteogenesis). Osteopontin mediates the link between inflammation and bone remodeling, fibronectin facilitates cellular adhesion to collagen, and growth factors regulate cellular function and tissue regeneration. Despite the significant potential of nano-bioceramics, there are still unresolved challenges such as achieving clinical translation, tailoring features for specific applications, and ensuring biocompatibility. Through a meticulous examination of individual ceramic nanoparticles, we aim to uncover the specific biological characteristics that may be modified in nanobioceramics, leading to groundbreaking advancements in bone tissue engineering and regenerative medicine.

References

1. Kianfar, E. (2021). Protein nanoparticles in drug delivery: Animal protein, plant proteins and protein cages, albumin nanoparticles. *Journal of Nanobiotechnology, 19*(1), 159.
2. Carvalho, M. S., et al. (2021). Bone matrix non-collagenous proteins in tissue engineering: Creating new bone by mimicking the extracellular matrix. *Polymers, 13*(7), 1095.
3. Burr, D. B. (2019). Bone morphology and organization. *Basic and applied bone biology* (pp. 3–26). Elsevier.
4. Avila Rodríguez, M. I., Rodriguez Barroso, L. G., & Sánchez, M. L. (2018). Collagen: A review on its sources and potential cosmetic applications. *Journal of Cosmetic Dermatology, 17*(1), 20–26.
5. Rezvani Ghomi, E., et al. (2021). Collagen-based biomaterials for biomedical applications. *Journal of Biomedical Materials Research Part B: Applied Biomaterials, 109*(12), 1986–1999.
6. Andrianarivo, A. G., et al. (1992). Growth on type I collagen promotes expression of the osteoblastic phenotype in human osteosarcoma MG-63 cells. *Journal of Cellular Physiology, 153*(2), 256–265.
7. Lynch, M. P., et al. (1995). The influence of type I collagen on the development and maintenance of the osteoblast phenotype in primary and passaged rat calvarial osteoblasts: Modification of expression of genes supporting cell growth, adhesion, and extracellular matrix mineralization. *Experimental Cell Research, 216*(1), 35–45.
8. Green, J., et al. (1995). Cell-matrix interaction in bone: Type I collagen modulates signal transduction in osteoblast-like cells. *American Journal of Physiology-Cell Physiology, 268*(5), C1090–C1103.
9. KN, A., Bhatia, P., & Singh, S. (2021). Review on collagen type 1 & type 2.
10. Mizuno, M., Fujisawa, R., & Kuboki, Y. (2000). Type I collagen-induced osteoblastic differentiation of bone-marrow cells mediated by collagen-$\alpha 2\beta 1$ integrin interaction. *Journal of Cellular Physiology, 184*(2), 207–213.
11. Niyibizi, C., & Eyre, D. R. (1994). Structural characteristics of cross-linking sites in type V collagen of bone: Chain specificities and heterotypic links to Type I collagen. *European Journal of Biochemistry, 224*(3), 943–950.
12. Mak, K. M., Png, C. Y. M., & Lee, D. J. (2016). Type V collagen in health, disease, and fibrosis. *The Anatomical Record, 299*(5), 613–629.
13. Ducy, P., et al. (1996). Increased bone formation in osteocalcin-deficient mice. *Nature Materials, 382*(6590), 448–452.
14. Puchacz, E., et al. (1989). Chromosomal localization of the human osteocalcin gene. *Endocrinology, 124*(5), 2648–2650.
15. Zoch, M. L., Clemens, T. L., & Riddle, R. C. (2016). New insights into the biology of osteocalcin. *Bone, 82*, 42–49.
16. Hauschka, P. V., et al. (1989). Osteocalcin and matrix Gla protein: Vitamin K-dependent proteins in bone. *Physiological Reviews, 69*(3), 990–1047.
17. Price, P. A., et al. (1976). Characterization of a gamma-carboxyglutamic acid-containing protein from bone. *Proceedings of the National Academy of Sciences, 73*(5), 1447–1451.
18. Price, P. A., Poser, J. W., & Raman, N. (1976). Primary structure of the gamma-carboxyglutamic acid-containing protein from bovine bone. *Proceedings of the National Academy of Sciences, 73*(10), 3374–3375.
19. Lee, N. K., et al. (2007). Endocrine regulation of energy metabolism by the skeleton. *Cell, 130*(3), 456–469.

20. Clemens, T. L., & Karsenty, G. (2011). The osteoblast: An insulin target cell controlling glucose homeostasis. *Journal of Bone Mineral Research, 26*(4), 677–680.
21. Kajimura, D., et al. (2013). Adiponectin regulates bone mass via opposite central and peripheral mechanisms through FoxO1. *Cell Metabolism, 17*(6), 901–915.
22. Takeda, S., et al. (2002). Leptin regulates bone formation via the sympathetic nervous system. *Cell, 111*(3), 305–317.
23. Ducy, P., et al. (2000). Leptin inhibits bone formation through a hypothalamic relay: A central control of bone mass. *Cell, 100*(2), 197–207.
24. Nussbaum, M., et al. (1985). Short stature in anorexia nervosa patients. *Journal of Adolescent Health Care, 6*(6), 453–455.
25. Bolton, J. G., et al. (2005). A prospective study of changes in bone turnover and bone density associated with regaining weight in women with anorexia nervosa. *Osteoporosis International, 16*(12), 1955–1962.
26. Pi, M., et al. (2005). Identification of a novel extracellular cation-sensing G-protein-coupled receptor. *Journal of Biological Chemistry, 280*(48), 40201–40209.
27. Kuang, D., et al. (2005). Cloning and characterization of a Family C orphan G-protein coupled receptor. *Journal of Neurochemistry, 93*(2), 383–391.
28. Pi, M., Wu, Y., & Quarles, L. D. (2011). GPRC6A mediates responses to osteocalcin in β-cells in vitro and pancreas in vivo. *Journal of Bone Mineral Research, 26*(7), 1680–1683.
29. Wei, J., et al. (2014). Osteocalcin promotes β-cell proliferation during development and adulthood through Gprc6a. *Diabetes, 63*(3), 1021–1031.
30. Pi, M., et al. (2008). GPRC6A null mice exhibit osteopenia, feminization and metabolic syndrome. *PloS One, 3*(12).
31. Termine, J. D., et al. (1981). Osteonectin, a bone-specific protein linking mineral to collagen. *Cell, 26*(1), 99–105.
32. Jundt, G., et al. (1987). Osteonectin—A differentiation marker of bone cells. *Cell Tissue Research, 248*(2), 409–415.
33. Bradshaw, A. D. (2012). Diverse biological functions of the SPARC family of proteins. *The International Journal of Biochemistry Cell Biology, 44*(3), 480–488.
34. Murphy-Ullrich, J. E., & Sage, E. H. (2014). Revisiting the matricellular concept. *Matrix Biology, 37*, 1–14.
35. Nakase, T., et al. (1994). Alterations in the expression of osteonectin, osteopontin and osteocalcin mRNAs during the development of skeletal tissues in vivo. *Bone Mineral, 26*(2), 109–122.
36. Strunz, M., et al. (2019). Modulation of SPARC/Hevin proteins in Alzheimer's disease brain injury. *Journal of Alzheimer's Disease, 68*(2), 695–710.
37. Deckx, S., et al. (2019). Extracellular SPARC increases cardiomyocyte contraction during health and disease. *PloS One, 14*(4).
38. Fukushima, Y., et al. (2007). Induction of glioma cell migration by vitronectin in human serum and cerebrospinal fluid. *Journal of Neurosurgery, 107*(3), 578–585.
39. Luo, H.-L., et al. (2019). Methylation of SPARCL1 is associated with oncologic outcome of advanced upper urinary tract urothelial carcinoma. *International Journal of Molecular Sciences, 20*(7), 1653.
40. Ribeiro, N., et al. (2014). Role of SPARC in bone remodeling and cancer-related bone metastasis. *Journal of Cellular Biochemistry, 115*(1), 17–26.
41. Bradshaw, A. D., & Sage, E. H. (2001). SPARC, a matricellular protein that functions in cellular differentiation and tissue response to injury. *The Journal of Clinical Investigation, 107*(9), 1049–1054.
42. Zarubin, T., & Jiahuai, H. (2005). Activation and signaling of the p38 MAP kinase pathway. *Cell Research, 15*(1), 11–18.
43. Cuadrado, A., & Nebreda, A. R. (2010). Mechanisms and functions of p38 MAPK signalling. *Biochemical Journal, 429*(3), 403–417.
44. You, J., et al. (2001). Osteopontin gene regulation by oscillatory fluid flow via intracellular calcium mobilization and activation of mitogen-activated protein kinase in MC3T3–E1 osteoblasts. *Journal of Biological Chemistry, 276*(16), 13365–13371.

45. Caverzasio, J., & Manen, D. (2007). Essential role of Wnt3a-mediated activation of mitogen-activated protein kinase p38 for the stimulation of alkaline phosphatase activity and matrix mineralization in C3H10T1/2 mesenchymal cells. *Endocrinology, 148*(11), 5323–5330.
46. Rey, A., et al. (2007). Evidences for a role of p38 MAP kinase in the stimulation of alkaline phosphatase and matrix mineralization induced by parathyroid hormone in osteoblastic cells. *Bone, 41*(1), 59–67.
47. Suzuki, A., et al. (1999). Regulation of alkaline phosphatase activity by p38 MAP kinase in response to activation of Gi protein-coupled receptors by epinephrine in osteoblast-like cells. *Endocrinology, 140*(7), 3177–3182.
48. Lee, K.-S., Hong, S.-H., & Bae, S.-C. (2002). Both the Smad and p38 MAPK pathways play a crucial role in Runx2 expression following induction by transforming growth factor-β and bone morphogenetic protein. *Oncogene, 21*(47), 7156–7163.
49. Guicheux, J., et al. (2003). Activation of p38 mitogen-activated protein kinase and c-Jun-NH2-terminal kinase by BMP-2 and their implication in the stimulation of osteoblastic cell differentiation. *Journal of Bone Mineral Research, 18*(11), 2060–2068.
50. Ortuño, M. J., et al. (2010). P38 regulates expression of osteoblast-specific genes by phosphorylation of osterix. *Journal of Biological Chemistry, 285*(42), 31985–31994.
51. Greenblatt, M. B., et al. (2010). The p38 MAPK pathway is essential for skeletogenesis and bone homeostasis in mice. *The Journal of Clinical Investigation, 120*(7), 2457–2473.
52. Ulsamer, A., et al. (2008). BMP-2 induces Osterix expression through up-regulation of Dlx5 and its phosphorylation by p38. *Journal of Biological Chemistry, 283*(7), 3816–3826.
53. Bozec, A., et al. (2010). Fra-2/AP-1 controls bone formation by regulating osteoblast differentiation and collagen production. *Journal of Cell Biology, 190*(6), 1093–1106.
54. Kveiborg, M., et al. (2002). The increased bone mass in ΔFosB transgenic mice is independent of circulating leptin levels. *Endocrinology, 143*(11), 4304–4309.
55. Jochum, W., et al. (2000). Increased bone formation and osteosclerosis in mice overexpressing the transcription factor Fra-1. *Nature Medicine, 6*(9), 980–984.
56. Sabatakos, G., et al. (2000). Overexpression of ΔFosB transcription factor(s) increases bone formation and inhibits adipogenesis. *Nature Medicine, 6*(9), 985–990.
57. Eferl, R., et al. (2004). The Fos-related antigen Fra-1 is an activator of bone matrix formation. *The EMBO Journal, 23*(14), 2789–2799.
58. Hipskind, R. A., & Bilbe, G. (1998). MAP kinase signaling cascades and gene expression in osteoblasts. *Frontiers in Bioscience, 3*, d804–d816.
59. Zhu, Y. S., et al. (2020). Osteonectin regulates the extracellular matrix mineralization of osteoblasts through P38 signaling pathway. *Journal of Cellular Physiology, 235*(3), 2220–2231.
60. Fisher, L. W., et al. (1987). Purification and partial characterization of small proteoglycans I and II, bone sialoproteins I and II, and osteonectin from the mineral compartment of developing human bone. *Journal of Biological Chemistry, 262*(20), 9702–9708.
61. Fisher, L. W., et al. (1983). Matrix sialoprotein of developing bone. *Journal of Biological Chemistry, 258*(20), 12723–12727.
62. Fisher, L., et al. (1990). Human bone sialoprotein. Deduced protein sequence and chromosomal localization. *Journal of Biological Chemistry, 265*(4), 2347–2351.
63. Fujisawa, R., et al. (1993). Differences in composition of cell-attachment sialoproteins between dentin and bone. *Journal of Dental Research, 72*(8), 1222–1226.
64. Bianco, P., et al. (1991). Expression of bone sialoprotein (BSP) in developing human tissues. *Calcified Tissue International, 49*(6), 421–426.
65. Herring, G. (1972). The organic matrix of bone. In *The biochemistry and physiology of bone*. Academic Press.
66. Termine, J. D., et al. (1981). Mineral and collagen-binding proteins of fetal calf bone. *Journal of Biological Chemistry, 256*(20), 10403–10408.
67. Sodek, J. (1992). Sialoprotein in bone remodeling. In *The biological mechanisms of tooth movement craniofacial adaptation* (p. 127–136).

68. Tye, C. E., et al. (2003). Delineation of the hydroxyapatite-nucleating domains of bone sialoprotein. *Journal of Biological Chemistry, 278*(10), 7949–7955.
69. Klein, A., et al. (2018). Effect of bone sialoprotein coated three-dimensional printed calcium phosphate scaffolds on primary human osteoblasts. *Journal of Biomedical Materials Research Part B: Applied Biomaterials, 106*(7), 2565–2575.
70. Bellahcene, A., et al. (2000). Bone sialoprotein mediates human endothelial cell attachment and migration and promotes angiogenesis. *Circulation Research, 86*(8), 885–891.
71. Klein, A., et al. (2019). Effect of bone sialoprotein coating on progression of bone formation in a femoral defect model in rats. *European Journal of Trauma Emergency Surgery,* 1–10.
72. Ganss, B., Kim, R. H., & Sodek, J. (1999). Bone sialoprotein. *Critical Reviews in Oral Biology Medicine, 10*(1), 79–98.
73. Chen, Y., Bal, B., & Gorski, J. (1992). Calcium and collagen binding properties of osteopontin, bone sialoprotein, and bone acidic glycoprotein-75 from bone. *Journal of Biological Chemistry, 267*(34), 24871–24878.
74. Stubbs, J. T., III., et al. (1997). Characterization of native and recombinant bone sialoprotein: Delineation of the mineral-binding and cell adhesion domains and structural analysis of the RGD domain. *Journal of Bone Mineral Research, 12*(8), 1210–1222.
75. Oldberg, A., Franzen, A., & Heinegaard, D. (1988). The primary structure of a cell-binding bone sialoprotein. *Journal of Biological Chemistry, 263*(36), 19430–19432.
76. Zhang, Q., et al. (1990). Characterization of fetal porcine bone sialoproteins, secreted phosphoprotein I (SPPI, osteopontin), bone sialoprotein, and a 23-kDa glycoprotein. Demonstration that the 23-kDa glycoprotein is derived from the carboxyl terminus of SPPI. *Journal of Biological Chemistry, 265*(13), 7583–7589.
77. Kinne, R., & Fisher, L. (1987). Keratan sulfate proteoglycan in rabbit compact bone is bone sialoprotein II. *Journal of Biological Chemistry, 262*(21), 10206–10211.
78. Franzen, A., & Heinegård, D. (1985). Isolation and characterization of two sialoproteins present only in bone calcified matrix. *Biochemical Journal, 232*(3), 715–724.
79. Ruoslahti, E., & Pierschbacher, M. D. (1987). New perspectives in cell adhesion: RGD and integrins. *Science, 238*(4826), 491–497.
80. Midura, R. J., et al. (1990). A rat osteogenic cell line (UMR 106–01) synthesizes a highly sulfated form of bone sialoprotein. *Journal of Biological Chemistry, 265*(9), 5285–5291.
81. Ecarot-Charrier, B., Bouchard, F., & Delloye, C. (1989). Bone sialoprotein II synthesized by cultured osteoblasts contains tyrosine sulfate. *Journal of Biological Chemistry, 264*(33), 20049–20053.
82. Yang, R., et al. (1995). Characterization of an avian bone sialoprotein (BSP) cDNA: Comparisons to mammalian BSP and identification of conserved structural domains. *Journal of Bone Mineral Research, 10*(4), 632–640.
83. Midura, R. J., & Hascall, V. C. (1996). Bone sialoprotein—A mucin in disguise? *Glycobiology, 6*(7), 677–681.
84. Hynes, R. O. (2002). Integrins: bidirectional, allosteric signaling machines. *Cell, 110*(6), 673–687.
85. Wernike, E., et al. (2010). Long-term cell-mediated protein release from calcium phosphate ceramics. *Journal of Biomedical Materials Research Part A: An Official Journal of The Society for Biomaterials, The Japanese Society for Biomaterials, The Australian Society for Biomaterials, the Korean Society for Biomaterials, 92*(2), 463–474.
86. Kjellén, L., & Lindahl, U. (1991). Proteoglycans: Structures and interactions. *Annual Review of Biochemistry, 60*(1), 443–475.
87. Wong, M., et al. (1992). Aggrecan core protein is expressed in membranous bone of the chick embryo. Molecular and biomechanical studies of normal and nanomelia embryos. *Journal of Biological Chemistry, 267*(8), 5592–5598.
88. Gorski, J. P., & Hankenson, K. D. (2020). Secreted noncollagenous proteins of bone. *Principles of bone biology* (pp. 359–378). Elsevier.
89. Fisher, L. W., Termine, J. D., & Young, M. F. (1989). Deduced protein sequence of bone small proteoglycan I (biglycan) shows homology with proteoglycan II (decorin) and several

nonconnective tissue proteins in a variety of species. *Journal of Biological Chemistry, 264*(8), 4571–4576.
90. Wight, T. N. (2002). Versican: A versatile extracellular matrix proteoglycan in cell biology. *Current Opinion in Cell Biology, 14*(5), 617–623.
91. Markmann, A., et al. (2000). Influence of decorin expression on transforming growth factor-β-mediated collagen gel retraction and biglycan induction. *Matrix Biology, 19*(7), 631–636.
92. Luyten, F. P., et al. (1989). Purification and partial amino acid sequence of osteogenin, a protein initiating bone differentiation. *Journal of Biological Chemistry, 264*(23), 13377–13380.
93. Wozney, J. M., et al. (1988). Novel regulators of bone formation: Molecular clones and activities. *Science, 242*(4885), 1528–1534.
94. Wozney, J. M. (1992). The bone morphogenetic protein family and osteogenesis. *Molecular Reproduction Development, 32*(2), 160–167.
95. Urist, M. R. (1965). Bone: Formation by autoinduction. *Science, 150*(3698), 893–899.
96. Ten Dijke, P., et al. (1994). Identification of type I receptors for osteogenic protein-1 and bone morphogenetic protein-4. *Journal of Biological Chemistry, 269*(25), 16985–16988.
97. Koenig, B. B., et al. (1994). Characterization and cloning of a receptor for BMP-2 and BMP-4 from NIH 3T3 cells. *Molecular Cellular Biology, 14*(9), 5961–5974.
98. Macías-Silva, M., et al. (1998). Specific activation of Smad1 signaling pathways by the BMP7 type I receptor, ALK2. *Journal of Biological Chemistry, 273*(40), 25628–25636.
99. Rosenzweig, B., et al. (1995). Cloning and characterization of a human type II receptor for bone morphogenetic proteins. *Proceedings of the National Academy of Sciences, 92*(17), 7632–7636.
100. Yamashita, H., et al. (1995). Osteogenic protein-1 binds to activin type II receptors and induces certain activin-like effects. *The Journal of Cell Biology, 130*(1), 217–226.
101. Kawabata, M., Chytil, A., & Moses, H. L. (1995). Cloning of a novel type II serine/threonine kinase receptor through interaction with the type I transforming growth factor-β receptor. *Journal of Biological Chemistry, 270*(10), 5625–5630.
102. Moustakas, A., & Heldin, C.-H. (2002). From mono-to oligo-Smads: The heart of the matter in TGF-β signal transduction. *Genes Development, 16*(15), 1867–1871.
103. Chen, Y., Bhushan, A., & Vale, W. (1997). Smad8 mediates the signaling of the receptor serine kinase. *Proceedings of the National Academy of Sciences, 94*(24), 12938–12943.
104. Hoodless, P. A., et al. (1996). MADR1, a MAD-related protein that functions in BMP2 signaling pathways. *Cell, 85*(4), 489–500.
105. Nishimura, R., et al. (1998). Smad5 and DPC4 are key molecules in mediating BMP-2-induced osteoblastic differentiation of the pluripotent mesenchymal precursor cell line C2C12. *Journal of Biological Chemistry, 273*(4), 1872–1879.
106. Chen, D., Zhao, M., & Mundy, G. R. (2004). Bone morphogenetic proteins. *Growth Factors, 22*(4), 233–241.
107. Niswander, L., & Martin, G. R. (1993). FGF-4 regulates expression of Evx-1 in the developing mouse limb. *Development, 119*(1), 287–294.
108. Wall, N. A., & Hogan, B. L. (1994). TGF-β related genes in development. *Current Opinion in Genetics Development, 4*(4), 517–522.
109. Luo, G., et al. (1995). BMP-7 is an inducer of nephrogenesis, and is also required for eye development and skeletal patterning. *Genes Development, 9*(22), 2808–2820.
110. Dudley, A. T., Lyons, K. M., & Robertson, E. J. (1995). A requirement for bone morphogenetic protein-7 during development of the mammalian kidney and eye. *Genes Development, 9*(22), 2795–2807.
111. Yang, X., et al. (1999). Angiogenesis defects and mesenchymal apoptosis in mice lacking SMAD5. *Development, 126*(8), 1571–1580.
112. Urist, M., et al. (1997). Neutral lipids facilitate transfer of bone morphogenetic proteins and other noncollagenous proteins. *Medical Hypotheses, 49*(6), 465–475.
113. Urist, M.R., et al., Inductive substrates for bone formation. Clinical Orthopaedics Related Research®, 1968. 59: p. 59–96.

114. Urist, M. R., et al. (1987). [28] Preparation and bioassay of bone morphogenetic protein and polypeptide fragments. *Methods in enzymology* (pp. 294–312). Elsevier.
115. Melcher, A. (1969). Histologically demonstrable bound lipid apparently associated with relatively stable, mature collagen fibres. *Gerontology, 15*(2–3), 217–232.
116. Urist, M. R., et al. (1969). 23 Osteogenetic competence. *Clinical Orthopaedics Related Research®, 64*, 194–220.
117. Boskey, A. L. (1992). Mineral-matrix interactions in bone and cartilage. *Clinical Orthopaedics Related Research, 281*, 244–274.
118. Boyan-Salyers, B., & Boskey, A. (1980). Relationship between proteolipids and calcium-phospholipid-phosphate complexes in Bacterionema matruchotii calcification. *Calcified Tissue International, 30*(1), 167–174.
119. Lane, J., et al. (1979). A temporal study of collagen, proteoglycan, lipid and mineral constituents in a model of endochondral osseous repair. *Metabolic Bone Disease Related Research, 1*(4), 319–324.
120. Thorén, K., Aspenberg, P., & Thorngren, K.-G. (1995). Lipid extracted bank bone. Bone conductive and mechanical properties. *Clinical Orthopaedics Related Research, 1995*(311), 232–246.
121. Prolo, D. J., et al. (1982). Superior osteogenesis in transplanted allogeneic canine skull following chemical sterilization. *Clinical Orthopaedics Related Research, 168*, 230–242.
122. Urist, M. R., Mikulski, A., & Boyd, S. D. (1975). A chemosterilized antigen-extracted autodigested alloimplant for bone banks. *Archives of Surgery, 110*(4), 416–428.
123. Yoshida, Y., et al. (2000). Negative regulation of BMP/Smad signaling by Tob in osteoblasts. *Cell, 103*(7), 1085–1097.
124. Zhu, H., et al. (1999). A SMAD ubiquitin ligase targets the BMP pathway and affects embryonic pattern formation. *Nature, 400*(6745), 687.
125. Winkler, D. G., et al. (2003). Osteocyte control of bone formation via sclerostin, a novel BMP antagonist. *The EMBO Journal, 22*(23), 6267–6276.
126. Hershko, A., & Ciechanover, A. (1998). The ubiquitin system. *Annual Review of Biochemistry, 67*(1), 425–479.
127. Weissman, A. M. (2001). Themes and variations on ubiquitylation. *Nature Reviews Molecular Cell Biology, 2*(3), 169–178.
128. Ciechanover, A., Orian, A., & Schwartz, A. L. (2000). The ubiquitin-mediated proteolytic pathway: Mode of action and clinical implications. *Journal of Cellular Biochemistry, 77*(S34), 40–51.
129. Hershko, A. (1983). Ubiquitin: Roles in protein modification and breakdown. *Cell, 34*(1), 11–12.
130. Chang, S. C., et al. (2003). Ex vivo gene therapy in autologous bone marrow stromal stem cells for tissue-engineered maxillofacial bone regeneration. *Gene Therapy, 10*(24), 2013–2019.
131. Mohan, S., & Baylink, D. J. (1991). Bone growth factors. *Clinical Orthopaedics Related Research, 263*, 30–48.
132. Taipale, J., & Keski-Oja, J. (1997). Growth factors in the extracellular matrix. *The FASEB Journal, 11*(1), 51–59.
133. Hardingham, T. E., & Fosang, A. J. (1992). Proteoglycans: Many forms and many functions. *The FASEB Journal, 6*(3), 861–870.
134. Lyon, M., et al. (1994). Interaction of hepatocyte growth factor with heparan sulfate. Elucidation of the major heparan sulfate structural determinants. *Journal of Biological Chemistry, 269*(15), 11216–11223.
135. Faham, S., et al. (1996). Heparin structure and interactions with basic fibroblast growth factor. *Science, 271*(5252), 1116–1120.
136. Mizuno, K., et al. (1994). Hairpin loop and second kringle domain are essential sites for heparin binding and biological activity of hepatocyte growth factor. *Journal of Biological Chemistry, 269*(2), 1131–1136.
137. Park, J. E., Keller, G.-A., & Ferrara, N. (1993). The vascular endothelial growth factor (VEGF) isoforms: Differential deposition into the subepithelial extracellular matrix and bioactivity of extracellular matrix-bound VEGF. *Molecular Biology of the Cell, 4*(12), 1317–1326.

138. Raines, E. W., & Ross, R. (1992). Compartmentalization of PDGF on extracellular binding sites dependent on exon-6-encoded sequences. *The Journal of Cell Biology, 116*(2), 533–543.
139. Bützow, R., et al. (1993). A 60-kD protein mediates the binding of transforming growth factor-beta to cell surface and extracellular matrix proteoglycans. *The Journal of Cell Biology, 122*(3), 721–727.
140. Nakamura, T., et al. (1991). Follistatin, an activin-binding protein, associates with heparan sulfate chains of proteoglycans on follicular granulosa cells. *Journal of Biological Chemistry, 266*(29), 19432–19437.
141. Rosen, E. M., Nigam, S. K., & Goldberg, I. D. (1994). Scatter factor and the c-met receptor: A paradigm for mesenchymal/epithelial interaction. *The Journal of Cell Biology, 127*(6 Pt 2), 1783–1787.
142. Roberts, R., et al. (1988). Heparan sulphate bound growth factors: A mechanism for stromal cell mediated haemopoiesis. *Nature, 332*(6162), 376–378.
143. Schlessinger, J., Lax, I., & Lemmon, M. (1995). Regulation of growth factor activation by proteoglycans: What is the role of the low affinity receptors? *Cell, 83*(3), 357–360.
144. Aviezer, D., et al. (1994). Perlecan, basal lamina proteoglycan, promotes basic fibroblast growth factor-receptor binding, mitogenesis, and angiogenesis. *Cell, 79*(6), 1005–1013.
145. Stenman, S., & Vaheri, A. (1978). Distribution of a major connective tissue protein, fibronectin, in normal human tissues. *The Journal of Experimental Medicine, 147*(4), 1054–1064.
146. Hogg, N. M. (1974). A comparison of membrane proteins of normal and transformed cells by lactoperoxidase labeling. *Proceedings of the National Academy of Sciences, 71*(2), 489–492.
147. Hynes, R. O. (1973). Alteration of cell-surface proteins by viral transformation and by proteolysis. *Proceedings of the National Academy of Sciences, 70*(11), 3170–3174.
148. Gahmberg, C. G., & Hakomori, S.-I. (1973). Altered growth behavior of malignant cells associated with changes in externally labeled glycoprotein and glycolipid. *Proceedings of the National Academy of Sciences, 70*(12), 3329–3333.
149. Hunt, R. C., Gold, E., & Brown, J. C. (1975). Cell cycle dependent exposure of a high molecular weight protein on the surface of mouse L cells. *Biochimica et Biophysica Acta-Biomembranes, 413*(3), 453–458.
150. Keski-Oja, J., Mosher, D. F., & Vaheri, A. (1976). Cross-linking of a major fibroblast surface-associated glycoprotein (fibronectin) catalyzed by blood coagulation factor XIII. *Cell Biology International, 9*(1), 29–35.
151. Mosesson, M., & Umfleet, R. (1970). The cold-insoluble globulin of human plasma I. Purification, primary characterization, and relationship to fibrinogen and other cold-insoluble fraction components. *Journal of Biological Chemistry, 245*(21), 5728–5736.
152. Wartiovaara, J., Stenman, S., & Vaheri, A. (1976). Changes in expression of fibroblast surface antigen (SFA) during cytodifferentiation and heterokaryon formation. *Differentiation, 5*(2–3), 85–89.
153. Linder, E., et al. (1975). Distribution of fibroblast surface antigen in the developing chick embryo. *The Journal of Experimental Medicine, 142*(1), 41–49.
154. Parisi, L., et al. (2020). A glance on the role of fibronectin in controlling cell response at biomaterial interface. *Japanese Dental Science Review, 56*(1), 50–55.
155. Pras, M., et al. (1974). Antigenic properties of a non-collagenous reticulin component of normal connective tissue. *Immunology, 27*(3), 469.
156. Pras, M., & Glynn, L. (1973). Isolation of a non-collagenous reticulin component and its primary characterization. *British Journal of Experimental Pathology, 54*(4), 449.
157. Nowack, H., et al. (1976). Preparation and use in immunohistology of antibodies specific for type I and type III collagen and procollagen. *Journal of Immunological Methods, 12*(1–2), 117–124.
158. Vracko, R. (1974). Basal lamina scaffold-anatomy and significance for maintenance of orderly tissue structure: A review. *The American Journal of Pathology, 77*(2), 313.
159. Gallop, P., Blumenfeld, O., & Seifter, S. (1972). Structure and metabolism of connective tissue proteins. *Annual Review of Biochemistry, 41*(1), 617–672.

160. Ruoslahti, E. (1988). Fibronectin and its receptors. *Annual Review of Biochemistry, 57*(1), 375–413.
161. Wartiovaara, J., et al. (1974). Distribution of fibroblast surface antigen: Association with fibrillar structures of normal cells and loss upon viral transformation. *The Journal of Experimental Medicine, 140*(6), 1522–1533.
162. Engvall, E., & Ruoslahti, E. (1977). Binding of soluble form of fibroblast surface protein, fibronectin, to collagen. *International Journal of Cancer, 20*(1), 1–5.
163. Perkins, M. E., Ji, T. H., & Hynest, R. O. (1979). Cross-linking of fibronectin to sulfated proteoglycans at the cell surface. *Cell, 16*(4), 941–952.
164. Stathakis, N. E., & Mosesson, M. W. (1977). Interactions among heparin, cold-insoluble globulin, and fibrinogen in formation of the heparin-precipitable fraction of plasma. *The Journal of Clinical Investigation, 60*(4), 855–865.
165. Kleinman, H. K., McGoodwin, E. B., & Klebe, R. J. (1976). Localization of the cell attachment region in types I and II collagens. *Biochemical Biophysical Research Communications, 72*(2), 426–432.
166. Pearlstein, E. (1976). Plasma membrane glycoprotein which mediates adhesion of fibroblasts to collagen. *Nature, 262*(5568), 497–500.
167. Dean, D. C., Bowlus, C. L., & Bourgeois, S. (1987). Cloning and analysis of the promotor region of the human fibronectin gene. *Proceedings of the National Academy of Sciences, 84*(7), 1876–1880.
168. Gutman, A., Yamada, K. M., & Kornblihtt, A. (1986). Human fibronectin is synthesized as a pre-propolypeptide. *FEBS Letters, 207*(1), 145–148.
169. Skorstengaard, K., et al. (1986). Complete primary structure of bovine plasma fibronectin. *European Journal of Biochemistry, 161*(2), 441–453.
170. Bernard, B., Yamada, K., & Olden, K. (1982). Carbohydrates selectively protect a specific domain of fibronectin against proteases. *Journal of Biological Chemistry, 257*(14), 8549–8554.
171. Lange-Mutschler, J. (1986). Acylated fibronectin: A new type of posttranslational modification of cellular fibronectin. *FEBS Letters, 201*(2), 210–214.
172. Paul, J. I., & Hynes, R. (1984). Multiple fibronectin subunits and their post-translational modifications. *Journal of Biological Chemistry, 259*(21), 13477–13487.
173. Johansson, S., & Smedsrød, B. (1986). Identification of a plasma gelatinase in preparations of fibronectin. *Journal of Biological Chemistry, 261*(10), 4363–4366.
174. Vartio, T., Hovi, T., & Vaheri, A. (1982). Human macrophages synthesize and secrete a major 95,000-dalton gelatin-binding protein distinct from fibronectin. *Journal of Biological Chemistry, 257*(15), 8862–8866.
175. Owens, R. J., & Baralle, F. E. (1986). Mapping the collagen-binding site of human fibronectin by expression in Escherichia coli. *The EMBO Journal, 5*(11), 2825–2830.
176. Furcht, L. T., Mosher, D. F., & Wendelschafer-Crabb, G. (1978). Immunocytochemical localization of fibronectin (LETS protein) on the surface of L6 myoblasts: Light and electron microscopic studies. *Cell, 13*(2), 263–271.
177. Pierschbacher, M. D., & Ruoslahti, E. (1984). Cell attachment activity of fibronectin can be duplicated by small synthetic fragments of the molecule. *Nature, 309*(5963), 30–33.
178. Ruoslahti, E., & Pierschbacher, M. D. (1986). Arg-Gly-Asp: A versatile cell recognition signal. *Cell, 44*(4), 517–518.
179. Oldberg, A., & Ruoslahti, E. (1986). Evolution of the fibronectin gene. Exon structure of cell attachment domain. *Journal of Biological Chemistry, 261*(5), 2113–2116.
180. Plow, E. F., et al. (1985). The effect of Arg-Gly-Asp-containing peptides on fibrinogen and von Willebrand factor binding to platelets. *Proceedings of the National Academy of Sciences, 82*(23), 8057–8061.
181. Gartner, T. K., & Bennett, J. S. (1985). The tetrapeptide analogue of the cell attachment site of fibronectin inhibits platelet aggregation and fibrinogen binding to activated platelets. *Journal of Biological Chemistry, 260*(22), 11891–11894.

182. Hayman, E., Pierschbacher, M., & Ruoslahti, E. (1985). Detachment of cells from culture substrate by soluble fibronectin peptides. *The Journal of Cell Biology, 100*(6), 1948–1954.
183. Suzuki, S., et al. (1985). Complete amino acid sequence of human vitronectin deduced from cDNA. Similarity of cell attachment sites in vitronectin and fibronectin. *The EMBO Journal, 4*(10), 2519–2524.
184. Lawler, J., & Hynes, R. O. (1986). The structure of human thrombospondin, an adhesive glycoprotein with multiple calcium-binding sites and homologies with several different proteins. *The Journal of Cell Biology, 103*(5), 1635–1648.
185. Horwitz, A., et al. (1985). The cell substrate attachment (CSAT) antigen has properties of a receptor for laminin and fibronectin. *The Journal of Cell Biology, 101*(6), 2134–2144.
186. Skorstengaard, K., et al. (1986). Purification and complete primary structures of the heparin-, cell-, and DNA-binding domains of bovine plasma fibronectin. *European Journal of Biochemistry, 154*(1), 15–29.
187. Pande, H., et al. (1985). Primary structure of a glycosylated DNA-binding domain in human plasma fibronectin. *Journal of Biological Chemistry, 260*(4), 2301–2306.
188. Zardi, L., et al. (1979). Fibronectin: A chromatin-associated protein? *Cell, 18*(3), 649–657.
189. Rich, A., et al. (1981). Cartilage proteoglycans inhibit fibronectin-mediated adhesion. *Nature, 293*(5829), 224–226.
190. Knox, P., & Wells, P. (1979). Cell adhesion and proteoglycans. I. The effect of exogenous proteoglycans on the attachment of chick embryo fibroblasts to tissue culture plastic and collagen. *Journal of Cell Science, 40*(1), 77–88.
191. Laterra, J., Silbert, J. E., & Culp, L. A. (1983). Cell surface heparan sulfate mediates some adhesive responses to glycosaminoglycan-binding matrices, including fibronectin. *The Journal of Cell Biology, 96*(1), 112–123.
192. Woods, A., et al. (1986). Adhesion and cytoskeletal organisation of fibroblasts in response to fibronectin fragments. *The EMBO Journal, 5*(4), 665–670.
193. Schmidt, G., et al. (1987). Interaction of small dermatan sulfate proteoglycan from fibroblasts with fibronectin. *The Journal of Cell Biology, 104*(6), 1683–1691.
194. Isemura, M., et al. (1987). Isolation and characterization of fibronectin-binding proteoglycan carrying both heparan sulfate and dermatan sulfate chains from human placenta. *Journal of Biological Chemistry, 262*(18), 8926–8933.
195. Koda, J. E., Rapraeger, A., & Bernfield, M. (1985). Heparan sulfate proteoglycans from mouse mammary epithelial cells. Cell surface proteoglycan as a receptor for interstitial collagens. *Journal of Biological Chemistry, 260*(13): p. 8157–8162.
196. Koda, J. E., & Bernfield, M. (1984). Heparan sulfate proteoglycans from mouse mammary epithelial cells. Basal extracellular proteoglycan binds specifically to native type I collagen fibrils. *Journal of Biological Chemistry, 259*(19), 11763–11770.
197. Horwitz, A., et al. (1986). Interaction of plasma membrane fibronectin receptor with talin—A transmembrane linkage. *Nature, 320*(6062), 531–533.
198. Hirst, R., et al. (1986). Phosphorylation of the fibronectin receptor complex in cells transformed by oncogenes that encode tyrosine kinases. *Proceedings of the National Academy of Sciences, 83*(17), 6470–6474.
199. Ignotz, R. A., & Massague, J. (1986). Transforming growth factor-beta stimulates the expression of fibronectin and collagen and their incorporation into the extracellular matrix. *Journal of Biological Chemistry, 261*(9), 4337–4345.
200. Morris, A. H., & Kyriakides, T. R. (2014). Matricellular proteins and biomaterials. *Matrix Biology, 37*, 183–191.
201. Clause, K. C., & Barker, T. H. (2013). Extracellular matrix signaling in morphogenesis and repair. *Current Opinion in Biotechnology, 24*(5), 830–833.
202. Dee, K. C., Puleo, D. A., & Bizios, R. (2002). Protein-surface interactions, in *An introduction to tissue-biomaterial interactions* (p. 37–52).
203. Hirsh, S. L., et al. (2013). The Vroman effect: Competitive protein exchange with dynamic multilayer protein aggregates. *Colloids Surfaces B: Biointerfaces, 103*, 395–404.

204. Parisi, L., et al. (2018). Tailoring the interface of biomaterials to design effective scaffolds. *Journal of functional biomaterials, 9*(3), 50.
205. Vallieres, K., et al. (2007). AFM imaging of immobilized fibronectin: Does the surface conjugation scheme affect the protein orientation/conformation? *Langmuir, 23*(19), 9745–9751.
206. García, A. J., Vega, M. D., & Boettiger, D. (1999). Modulation of cell proliferation and differentiation through substrate-dependent changes in fibronectin conformation. *Molecular Biology of the Cell, 10*(3), 785–798.
207. Nuttelman, C. R., et al. (2001). Attachment of fibronectin to poly (vinyl alcohol) hydrogels promotes NIH3T3 cell adhesion, proliferation, and migration. *Journal of Biomedical Materials Research: An Official Journal of The Society for Biomaterials, The Japanese Society for Biomaterials, The Australian Society for Biomaterials, the Korean Society for Biomaterials, 57*(2), 217–223.
208. Gao, M., et al. (2006). Molecular mechanisms of cellular mechanics. *Physical Chemistry Chemical Physics, 8*(32), 3692–3706.
209. Jongpaiboonkit, L., King, W. J., & Murphy, W. L. (2009). Screening for 3D environments that support human mesenchymal stem cell viability using hydrogel arrays. *Tissue Engineering Part A, 15*(2), 343–353.
210. Arnold, M., et al. (2004). Activation of integrin function by nanopatterned adhesive interfaces. *ChemPhysChem, 5*(3), 383–388.
211. Parisi, L., et al. (2017). Aptamers improve the bioactivity of biomaterials. *Aptamers, 1*, 3–12.
212. Oliveira, C., et al. (2014). Biofunctional nanofibrous substrate comprising immobilized antibodies and selective binding of autologous growth factors. *Biomacromolecules, 15*(6), 2196–2205.
213. Hoffmann, J., et al. (2008). Immobilized DNA aptamers used as potent attractors for porcine endothelial precursor cells. *Journal of Biomedical Materials Research Part A: An Official Journal of The Society for Biomaterials, The Japanese Society for Biomaterials, The Australian Society for Biomaterials, the Korean Society for Biomaterials, 84*(3), 614–621.
214. Chen, N., et al. (2012). Cell adhesion on an artificial extracellular matrix using aptamer-functionalized PEG hydrogels. *Biomaterials, 33*(5), 1353–1362.
215. Saccani, M., et al. (2019). Surface modification of chitosan films with a fibronectin fragment–DNA aptamer complex to enhance osteoblastic cell activity: A mass spectrometry approach probing evidence on protein behavior. *Rapid Communications in Mass Spectrometry, 33*(4), 336–342.
216. Parisi, L., et al. (2017). Anti-fibronectin aptamers improve the colonization of chitosan films modified with D-(+) Raffinose by murine osteoblastic cells. *Journal of Materials Science: Materials in Medicine, 28*(9), 136.
217. Sroga, G. E., et al. (2011) Biochemical characterization of major bone-matrix proteins using nanoscale-size bone samples and proteomics methodology. *Molecular Cellular Proteomics, 10*(9).
218. Depalle, B., et al. (2020). Osteopontin regulates type I collagen fibril formation in bone tissue. *Acta Biomaterialia.*
219. Addison, W. N., et al. (2010). Phosphorylation-dependent inhibition of mineralization by osteopontin ASARM peptides is regulated by PHEX cleavage. *Journal of Bone Mineral Research, 25*(4), 695–705.
220. Boskey, A., et al. (2002). Osteopontin deficiency increases mineral content and mineral crystallinity in mouse bone. *Calcified Tissue International, 71*(2).
221. Sodek, J., Ganss, B., & McKee, M. (2000). Osteopontin. *Critical Reviews in Oral Biology Medicine, 11*(3), 279–303.
222. Singh, A., et al. (2018). Role of osteopontin in bone remodeling and orthodontic tooth movement: A review. *Progress in Orthodontics, 19*(1), 1–8.
223. Poundarik, A. A., et al. (2012). Dilatational band formation in bone. *Proceedings of the National Academy of Sciences, 109*(47), 19178–19183.
224. Thurner, P. J., et al. (2010). Osteopontin deficiency increases bone fragility but preserves bone mass. *Bone, 46*(6), 1564–1573.

225. Poundarik, A. A., et al. (2018). Biomolecular regulation, composition and nanoarchitecture of bone mineral. *Scientific Reports, 8*(1), 1–8.
226. Poundarik, A. (2013). The role of non-collagenous proteins. In *Particular, osteocalcin and osteopontin, in the determination of bone matrix quality*. Rensselaer Polytechnic Institute.
227. Chellaiah, M. A., et al. (2003). Osteopontin deficiency produces osteoclast dysfunction due to reduced CD44 surface expression. *Molecular Biology of the Cell, 14*(1), 173–189.
228. Chellaiah, M., & Hruska, K. (2003). The integrin alpha v beta 3 and CD44 regulate the actions of osteopontin on osteoclast motility. *Calcified Tissue International, 72*(3), 197–205.
229. Kahles, F., Findeisen, H. M., & Bruemmer, D. (2014). Osteopontin: A novel regulator at the cross roads of inflammation, obesity and diabetes. *Molecular metabolism, 3*(4), 384–393.
230. Wolak, T. (2014). Osteopontin–a multi-modal marker and mediator in atherosclerotic vascular disease. *Atherosclerosis, 236*(2), 327–337.
231. Miyauchi, A., et al. (1991). Recognition of osteopontin and related peptides by an alpha v beta 3 integrin stimulates immediate cell signals in osteoclasts. *Journal of Biological Chemistry, 266*(30), 20369–20374.
232. Zimolo, Z., et al. (1994). Soluble alpha v beta 3-integrin ligands raise $[Ca^{2+}]i$ in rat osteoclasts and mouse-derived osteoclast-like cells. *American Journal of Physiology-Cell Physiology, 266*(2), C376–C381.
233. Kikuta, J., & Ishii, M. (2013). Osteoclast migration, differentiation and function: Novel therapeutic targets for rheumatic diseases. *Rheumatology, 52*(2), 226–234.
234. Tanabe, N., et al. (2011). Osteopontin signals through calcium and nuclear factor of activated T cells (NFAT) in osteoclasts A Novel RGD-dependent pathway promoting cell survival. *Journal of Biological Chemistry, 286*(46), 39871–39881.

Chapter 6
Bone ECM Proteins—Part II

Abstract The functions of bones depend on the proteins involved in it and in ossification. In addition to collagenous and non-collagenous proteins, there are other proteins that play a significant role in bone function. The proteins can be categorized into two groups: plasmatic proteins and cytoskeleton proteins. Plasmatic proteins include transferrin, immunoglobulin, albumin, α2HS, and phospholipid transfer protein. Tensin, vinculin, talin, and paxillin are proteins that make up the cytoskeleton. Transferrin plays a vital role in the process of attaching to and carrying iron throughout living organisms. Antibodies, also known as immunoglobulins, recognize antigens present on the surfaces of foreign entities such as viruses and bacteria. They then trigger mechanisms like phagocytosis and cell lysis to destroy these infectious particles and cells. Albumin, the predominant protein in plasma, is an essential constituent of the fluid located outside of blood arteries, referred to as extravascular tissue fluid. Albumin and plasma α_2HS glycoprotein exhibit elevated concentrations within the bone matrix, while immunoglobulin G is mostly associated with the blood capillaries in the bone. Phospholipid transfer proteins (PLTPs) are proteins present in the plasma that can transfer phospholipids between lipoproteins and liposomes. Tensin has a crucial role in linking the actin cytoskeleton, extracellular matrix (ECM), and signal transduction. Vinculin regulates adhesion via recruiting actin remodeling proteins, facilitating actin polymerization, and directly binding to actin, despite its limited enzymatic activity. Paxillin is a versatile protein that acts as an adaptor in focal adhesions, playing a crucial role in cell migration and movement. Talin is an essential protein that connects the cytoplasmic parts of integrin β subunits to actin filaments. Understanding the activity of these cells can have a significant impact on the development of materials used in bone tissue engineering. This chapter offers extensive details on the composition and functions of these proteins.

Abbreviations

ABD Actin-binding domain
BCR Breakpoint cluster region

R. Gholami and S. M. Naghib, *Nanobioceramics for Bone Tissue Engineering and Regenerative Biomedicine*, Biological and Medical Physics, Biomedical Engineering,
https://doi.org/10.1007/978-981-96-0041-0_6

BMPs	Bone morphogenetic proteins
ECM	Extracellular matrix
EM	Electron microscopy
FAB	Focal adhesion-binding
F-actin	Filamentous actin
FAK	Focal adhesion kinase
Fc	Fragment crystallizable
FOP	Fibrodysplasia ossificans progressiva
HDL	High-density lipoprotein
HSA	Human serum albumin
Ig	Immunoglobulins
ITP	Idiopathic thrombocytopenic purpura
kDa	Kilodaltons
KSF	Keratinocyte serum-free
PLTPs	Phospholipid transfer proteins
PTB	Phosphotyrosine-binding
SH2	Src homology 2
SPECT	Single photon emission computed tomography
Tfr2	Transferrin receptor 2
VASP	Vasodilator-stimulated phosphoprotein
VLDL	Very low-density lipoproteins

6.1 Plasmatic Proteins

6.1.1 Transferrin

Transferrin, a collection of glycoproteins that are nearly identical, is essential for the binding and transportation of iron throughout organisms [1, 2]. In 1947, Laurell and Ingelman made the initial discovery and description of this phenomenon. Since that time, its essential function in facilitating the transport of iron across a variety of tissues in the body has been widely acknowledged [3–10]. The methodologies for isolating transferrin from a variety of sources have made significant progress [1, 4]. Some of the identified variants of transferrin in animals include ovotransferrin [11], central nervous system transferrin [4, 12–14], testicular transferrin [15], and other forms found in serum and mucosal tissues [16]. Despite the fact that plasma transferrin is the primary source of iron for the majority of tissues, there are other types of transferrin, including mucosal transferrin, which is produced through biliary secretion and is involved in the absorption of iron in the human intestine. The levels of iron in specific compartments that are not accessible by plasma are locally regulated by these transferrin types. There is a negative connection between total iron levels and bone mineral density [17, 18]. Iron overload diseases, such as familial hemochromatosis, result in the development of osteoporosis at an early age. Iron is essential for the processes of erythropoiesis, bacterial immunity, and cellular respiration [19]. Excessive iron may have detrimental effects on cells. Iron overload-induced multi-organ damage may be prevented by maintaining iron concentrations within a certain range. Transferrins have a crucial function in this process since they effectively transfer and attach to iron, which protects against microbial invasion that relies on iron [4].

The molecular volume of transferrin has been determined to be 144 ± 10^3 Å [20] using small-angle neutron scattering investigations. X-ray analysis [21] has shown that its structure exhibits ellipsoidal characteristics in both its solution and crystalline states. While other tissues, including the central nervous system, testicular Sertoli cells, fibroblasts, macrophages, lymphocytes, and others, also produce transferrin, the liver is the main organ responsible for synthesizing transferrin in the bloodstream. The hydrodynamic properties of human transferrin are altered by the presence of iron, which causes changes in its structure [22, 23]. In addition, the binding of metal ions helps to maintain the structure of the protein by preventing deuterium exchange or proteolytic degradation, unlike the unbound form [4, 24, 25]. Human plasma transferrin has a molecular weight of 79,570 and consists of 678 amino acids with two *N*-linked oligosaccharide chains. Carbohydrates constitute around 2–12% of its composition [4, 26, 27]. Transferrin has a notable attraction to bicarbonate or carbonate ions, which facilitate the internal binding of iron and prevent its oxidation in polynuclear complexes. This trait is recognized as one of the distinctive features of transferrin [28, 29].

Transferrin has the ability to bind several metals, including platinum, indium, plutonium, europium, and others, when there is a low amount of iron present. Within a living organism, it further facilitates the transportation of plutonium (IV), americium, curium, as well as magnesium (III), chromium (III), and zinc (II). The transferrin gene is located on human chromosome 3. The receptor for this gene was first identified on reticulocytes in the 1960s and subsequently characterized chemically in the 1980s [4, 30–35]. Transferrin receptor 2 (Tfr2), mostly found in the liver, plays a crucial role in regulating iron balance. By engaging with bone morphogenetic proteins (BMPs) and Wnt inhibitors like sclerostin, it influences the process of bone formation and regulates the signaling pathway of BMP-p38 MAPK [36]. Tfr2 exists in two isoforms: Tfr2α, responsible for regulating iron levels in the liver, and Tfr2β, which is involved in exporting iron from the spleen [37]. Through the management of ferroportin-mediated iron export, it interacts with holo-transferrin to extend its half-life. This interaction is connected to the regulation of hepcidin, which is crucial for maintaining iron balance. Disorders characterized by excessive accumulation of iron, such as hereditary hemochromatosis, may occur due to the malfunctioning of various biological mechanisms [36, 38].

BMP signaling is essential for maintaining bone balance and also plays a critical role in regulating iron metabolism via its interaction with hepcidin and other components of the BMP pathway. Imbalances in BMP signaling are associated with conditions such fibrodysplasia ossificans progressiva (FOP), which is characterized by abnormal heterotopic ossification. Tfr2 exerts its influence on these pathways via its interaction with BMPs, hence suggesting the potential for therapeutic use in bone-related disorders [36]. Transferrin and its receptors have several roles in cellular and systemic iron regulation, BMP signaling, and bone metabolism, and they are crucial for maintaining iron balance. Further investigation into the molecular interrelationships and treatment possibilities of these conditions is anticipated to enhance our understanding and capacity to manage bone disorders and iron-related diseases.

6.1.2 *Immunoglobulin*

Antibodies or immunoglobulins (Ig) have a dual role in immunity: they both identify antigens on the foreign body surfaces including viruses and bacteria, and activate removal mechanisms including phagocytosis and lysis of cells, to eliminate these infecting particles and cells [39]. The immunoglobulins expose a basic four-chain assembly consisting of two equal light chains of 25 kDa, and two equal heavy chains of 50–70 kDa [39]. The detection of the action of proteolytic enzymes on IgG1 by Porter in 1959 [40] presented the first eminent comprehension into the structure of antibodies. The typical Y shaped arrangement of an immunoglobulin relates to a fragment crystallizable (Fc) section and two fragment antigen binding arms [39]. Immunoglobulin light chain amyloidosis is described by a clonal group of bone marrow plasma cells, which generates a λ or κ type monoclonal light chain as a fragment or a complete molecule [41]. The light chain protein erroneously folds to make a β-pleated sheet instead of adapting to the α-helical structure of most proteins [42]. When observed under polarized light, the β-pleated sheet structure causes positive staining with Congo red; this staining is needed for the identification of immunoglobulin light chain amyloidosis [43, 44]. An increase in the level of free light chain serum leads to the progression of immunoglobulin light chain amyloidosis by several years [45]. This insoluble protein interferes with organ function when deposited in tissues.

Immunoglobulin-secreting cells appear in all lymphoid tissues, such as the bone marrow. There are significant differences between the various organs regarding the heavy chain isotype distribution of the secreted immunoglobulins and their number of immunoglobulin-secreting cells. Additionally, both distribution patterns rely on age. In adults, most of all immunoglobulin-secreting cells of the individual are localized in the bone marrow. Primary in life, however, the majority of immunoglobulin-secreting cells are localized in lymph nodes and spleen [46]. It is notable that contrary to lymph nodes, spleen, bronchus-associated and gut-associated lymphoid tissues, the bone marrow has never been commonly acknowledged as an antibody formation site. Several reports in which the bone marrow was explored as an antibody formation site returned negative outcomes [46]. This might be because of the fact that lymph nodes and spleen are quite easy organs to manipulate, contrary to the bone marrow. In addition, the bone marrow of mice, the most common investigational animal in immunological studies, was discovered to present little, if any, antibody-producing cells after immunization. Nevertheless, there are now convincing proofs that the bone marrow is actually a key antibody production site in humans as well as in investigational animals [46].

Immunoglobulins are generated and released into the culture medium through short-term cultures of monkey [47], rabbit [48, 49], guinea-pig [50, 51], and human [52, 53] bone marrow cells. The production was observed to be the greatest in the bone marrow, wherever the immunoglobulin fabrication by various lymphoid tissues from adult organisms was quantitated, particularly when calculated per whole organ [46, 48, 50, 52]. Immunoglobulin production also occurs in murine bone marrow. The number of immunoglobulin-secreting cells [54] and the number of cytoplasmic immunoglobulin-positive plasmablasts and plasma cells (CIg cells) [55–57] in this organ is considerable in compare to the numbers in the other lymphoid organs. The bone marrow contribution to the entire CIg cells is reliant on age [46].

The localization of most IgA- and IgG-secreting cells in the bone marrow of an adult mouse and their defect in germ-free mice [54] propose that the IgA and IgG generated are primarily directed against local antigens, which mostly diffuse via the digestive and respiratory tract and mucous membranes [46]. In humans, since the IgA generated outside the secretory sites is almost exclusively of the monomeric form, the contribution of the bone marrow to the secretory IgA is possibly insignificant [46, 58]. Vossen has shown the shift of the majority of CIg cells to the bone marrow for humans as a function of age [59]. There is an outstanding relationship between the levels of the different immunoglobulin classes and subclasses in the serum and the heavy chain isotype distribution profile of the CIg cells in the human bone marrow. This has been displayed for the subsequent combinations: IgA1-IgA2 [60], IgG1-IgG2-IgG3-IgG4 [61], IgM-IgG-IgA [62, 63], monomeric IgA-polymeric IgA [58] and IgD versus all other isotypes together [64]. In addition, the κ/λ ratio of the CIg cells in the bone marrow correlated with the percentage distribution of the levels of Ig(κ) and Ig(λ) in the serum [62, 63]. However, the pattern of CIg cell distribution in other lymphoid organs did not match with serum levels of different immunoglobulin classes and subclasses. This information supports the idea that the human bone marrow is the main source of all classes and subclasses of immunoglobulins in the serum [46].

Deutsch [65] displayed that guinea-pigs immunization with typhoid vaccine directed to the presence of anti-typhoid antibodies in the bone marrow. Almost simultaneously, Pfeiffer [66] stated that bone marrow extracts of rabbits immunized with *Vibrio cholerae* include particular antibodies. These pioneers did not properly verify that these antibodies were actually generated in the bone marrow, but their records were redolent for this vision. Reiter [67] and Lüdke [68] displayed that immunized animal bone marrow cells release particular antibodies through in vitro cultures. But the ultimate evidence that the bone marrow actually generates antibodies upon immunization was provided by Thorbecke and Keuning [69]. Presently, antibody formation in the bone marrow has been determined in man [70, 71], rabbit [48, 72], guinea-pig [50, 51], mole rat [73], rat [74], mouse [75, 76], chicken [77], and frog [78]. In mice, antibodies of the IgE [79, 80], IgA, IgG, and IgM, [75, 76] classes are fabricated by the bone marrow [46]. For evident reasons, humans are usually not available for investigating formation of antibodies in bone marrow. Some data, however, on the fabrication of autoantibody in different lymphoid tissues of patients suffering from idiopathic thrombocytopenic purpura (ITP) and rheumatoid arthritis are accessible. The bone marrow was discovered to be the primary source of autoantibodies in chronic ITP. In this disease, the production of IgG anti-platelet antibodies in the bone marrow was estimated to be tenfold more than that in the other lymphoid tissues [46, 70]. Also, in rheumatoid arthritis, plaque-forming cells that produce rheumatoid factors are much more abundant in the bone marrow than in other lymphoid tissues [71].

6.1.3 Albumin

Albumin, the most common plasma protein, is a vital component of fluid found outside of blood vessels, known as extravascular tissue fluid [81]. A study was conducted on rabbits using a dual isotope approach with ^{125}I-labeled albumin to investigate the transport of albumin from the blood vessels to the surrounding tissues (extravascular space) and its concentration in various tissues such as muscle, skin, gut, kidney, and bone fluid [82]. While the calcified matrix component of albumin in bone remains rather steady, the proportion of extravascular tissue fluid may be easily replenished. However, plasma and the majority of albumin outside of blood vessels may be exchanged with each other in soft tissues. Plasma has the ability to substitute albumin alone in the fluid found in bone tissue [81]. The plasma proteins are present in the interstitial fluid that surrounds the endosteal and periosteal surfaces, as well as the walls of the Haversian canal, and the lacunae and canaliculi [83–85]. These proteins serve as buffers to maintain the pH of the microenvironment around the bones, control the osmotic pressure inside the bone tissue, and transport hormones, vitamins, and other metabolites. The extensive mineral surface area present on the surfaces of bones and the walls of canaliculi facilitates the interchange and movement of tissue fluid, hence contributing to the maintenance of mineral homeostasis [81, 86]. OWEN et al. used radial immunodiffusion to assess the quantity of albumin in the plasma of young rabbits, as well as in the extravascular tissue fluid of several organs including the stomach, kidneys, bone, muscle, and skin [81]. The extravascular albumin volume per unit mass of bone is similar to that of soft tissues. The calcified matrix in juvenile bone consists of around 57% albumin, with tissue fluid accounting for 27% and blood vessels making up the remaining 16% [81].

Human serum albumin (HSA) is an unglycosylated transport protein present in plasma. It is distinguished by its many binding sites, which are located in particular subdomains. As a result, it possesses the capability to attach to a wide range of ligands, including some medicinal drugs. In 1975, the original sequencing of HSA revealed the presence of a signal peptide (residues 1–18), a propeptide (residues 19–22), and the mature protein (residues 23–609). In 1989, Carter et al. identified the first structure of HSA with a resolution of 6.0 Å. In 1999, the Sugio group improved the 3D configuration of HSA to a precision of 2.5 Å. Helices make up around 67% of the secondary structure of HSA. The molecular weight of HSA is

66.5 kilodaltons (kDa). Before being released into the bloodstream, it undergoes the cleavage of a 6-amino acid N-terminal oligopeptide in the trans-Golgi apparatus. It is often found in plasma at a concentration ranging from 35 to 50 g/L. HSA plays a crucial role in the distribution, metabolism, transportation, and binding of ligands. Reduced quantities of albumin might hinder the movement of substances, metabolism, and distribution of fluids in tissues. The esterase-like activity of HSA is essential for the transformation of prodrugs into active medicinal molecules, making it a significant resource in pharmaceutical production. The entire structure of HSA is composed of three homologous α-helical domains: I (residues 1–195), II (residues 196–383), and III (residues 384–585). Each domain is composed of ten antiparallel helices, which are then separated into two subdomains. Subdomain A (h1-h6) has six helices, while subdomain B (h7-h10) contains four helices. HSA contributes about 75% of the average oncotic pressure and acts as a chemical scavenger by attaching to various tiny molecules, therefore protecting the body from dangerous substances like free radicals. Due to the presence of charged amino acid residues, it functions well as a plasma buffer. The solubility of hydrophobic medicines in plasma is increased as a consequence of their interaction, leading to reduced toxicity and prolonged half-life. The crystal structures of HSA complexed with a range of chemicals, both endogenous (such as bilirubin, thyroxine, and heme) and exogenous (such as warfarin, phenylbutazone, and aspirin), have been widely recorded. The presence of the Cys34 residue in HSA sets it apart from other extracellular proteins. This residue is crucial in capturing free radicals because of its distinct free thiol group. Antioxidant molecules have the ability to delay cellular damage and improve human health by preventing free radical reactions. HSA is presently obtained from human plasma. However, scientists have successfully produced recombinant HSA utilizing other expression systems, including *Pichia pastoris* [87], *Kluyveromyces lactis* [88], *Saccharomyces cerevisiae* [89], *Escherichia coli* [90], transgenic animals [91], and transgenic plants [92–96]. HSA plays a crucial role in eliminating naturally occurring toxins, such as bilirubin, bile acids, and uremic toxins that may lead to malfunction in the brain, kidneys, and cardiovascular system. The ability of HSA to bind and neutralize toxins is especially advantageous in sepsis-related disorders such hepatitis, liver failure, and renal failure. The buildup of these toxins leads to cell death, tissue damage, and harm to the cells lining blood vessels.

Although intravenous injection of HSA has its benefits, it may also lead to negative outcomes such as circulatory failure and hypersensitivity responses. Possible adverse reactions include erythema, edema, pruritus, rash, skin rash, vertigo, hypotension, urticaria, asthma, local allergic responses, vomiting, and increased salivation (Table 6.1) [92].

Utilizing human serum albumin (HSA) as a medicine delivery method has several advantages. The ability to reduce toxicity by preventing an excessive buildup of the active drug form is a notable benefit, particularly for drugs with a narrow therapeutic range. Moreover, HSA has shown advantageous enzymatic activity in some therapeutic situations [92].

Recent research has shown that Bone-Albumin, which refers to bone allografts coated with serum albumin, enhances the formation of new bone in rat models with

Table 6.1 Intravenous injection of human serum albumin and its side effects [92]

Organ	Symptoms
Renal	Clearance and filtered load reduced, osmotic clearance reduced GFR, sodium removal reduced
Dermatologic	Pruritus, skin rash, urticarial, erythema, and edema
Gastrointestinal	Increased salivation, vomiting, and nausea
Cardiovascular	Hypotension
Respiratory	Bronchospasm

critical size and non-union defects [97–99]. For example, research has shown that Bone-Albumin may enhance bone formation and reduce complications at the donor site after anterior cruciate ligament repair using bone-patellar tendon-bone autografts. Moreover, studies conducted on different surfaces coated with serum albumin have shown that this coating enhances the attachment and growth of stem cells. This suggests that serum albumin-coated biomaterials have the potential to be valuable in cell transplantation [99–102]. Furthermore, albumin is crucial for bone regeneration. The Bone-Albumin approach enhances the production of albumin in the local area, hence promoting the activation of stem cells that naturally occur in the body. Therefore, progenitor cells are highly localized at the site of injury, resulting in almost complete bone repair, even in circumstances of significant size shortages [103]. Through the activation of internal precursor cells, Bone-Albumin has shown effectiveness in many experimental models. Based on in vitro binding assays, it has been shown that the presence of albumin coating enhances the ability of stem cells to stick to demineralized bone surfaces in aged cell populations. Additional investigation is required to comprehensively comprehend the mechanism by which Bone-Albumin's regenerative characteristics influence tissue aging and stem cell functionality [103].

Utilizing allogeneic cells in stem cell therapy has ethical, immunologic, and regulatory challenges due to its reliance on a young donor cell pool for optimal regeneration [104]. Therefore, it seems to be more feasible to focus on endogenous stem cells. Horvathy et al. [103] conducted many experiments and showed that Bone-Albumin effectively enhances bone regeneration by increasing the levels of albumin in the affected area. This increase in albumin attracts more naturally occurring progenitor cells, which aid in the process of tissue remodeling. An initial human experiment used sizable albumin-coated structural allografts to enhance recuperation following total joint revision arthroplasty, therefore providing evidence in favor of their efficacy. One-year post-surgery, a single photon emission computed tomography (SPECT) investigation showed increased osteoblast activity. Based on this information, albumin might be a beneficial supplement for the elderly population to counteract the reduced ability to repair caused by a drop in stem cell function [103]. Through the activation of the body's own stem cells, Bone-Albumin has demonstrated efficacy in regenerating missing bone tissue. This approach may be advantageous for patients who are old or immunocompromised, since stem cell function often declines with age. It promotes the process of healing, stimulates faster and more complete reconstruction, and perhaps reduces the risk of re-injury [103].

6.1.4 α_2HS Glycoprotein

Adult hepatocytes are the primary producers of α_2HS glycoprotein, and its production is regulated by systemic inflammation. This glycoprotein is present in dentine and bone in numbers comparable to plasma albumin, but not exceeding its plasma levels in other tissues. It plays a crucial role in the calcified cortical bone matrix. Albumin and plasma α_2HS glycoprotein are found in high levels inside the bone matrix, whereas immunoglobulin G is mostly linked with the blood capillaries in the bone [105]. The human α_2HS glycoprotein, found in the mineralizing regions of normal human bone matrix, has resemblance to the α-glycoprotein found in rabbit bone [106]. The idea that the G2B glycoprotein found in rabbit and cow bones is analogous to the human α_2HS glycoprotein is corroborated by research undertaken by Ashton and colleagues [107]. This glycoprotein is mostly located in the extravascular regions of bone; however, it may also be derived from plasma [106].

Just like epidermal growth factor and bovine pituitary extract in keratinocyte serum-free (KSF) medium, α_2HS glycoprotein has been shown to enhance the pace at which wounds heal and the speed at which epidermal keratinocytes move. The study on intestinal epithelial cells and breast cancer demonstrates that the ability of α_2HS glycoprotein to stick to other cells is facilitated by annexins A2 and A6, respectively. The glycoprotein in question is the human equivalent of fetuin A, which is a cellular adhesion protein used in tissue culture and is found in fetal calf serum [108–110]. Ultracentrifugal investigations indicate that the G2B glycoprotein has an estimated molecular weight of around 49,000. Additionally, it exhibits α globulin electrophoretic mobility. According to the Herring method for separating EDTA-soluble proteins from the bone matrix, the main activity was seen in the glycoprotein fraction, which is a tiny part of the blood and has low acidity [107].

The research conducted by Quelch et al. [111] investigated the concentrations of albumin, sialic acid, and α_2HS glycoprotein in both healthy and sick human bones. Patients with osteogenesis imperfecta had significantly higher levels of these proteins compared to those with normal rates. The concentration of α_2HS glycoprotein in newborn bones was thrice more than in pediatric bones and sevenfold greater than in adult bones. The levels of these proteins were significantly increased in Paget's disease. The lowest levels were seen in osteopetrotic bones, while there was a significant increase in α_2HS glycoprotein in renal osteodystrophy [111]. Immunofluorescent antibody labeling reveals that α_2HS glycoprotein is mostly found outside of blood vessels in the natural bone matrix, particularly in areas where mineralization is occurring. This protein may also be identified in human plasma at a concentration of around 60 mg L^{-1} [107]. It is present in both the bone matrices of fetuses and adults and is likely taken up by the bone via interactions with other components of the mineralizing matrix. α_2HS glycoprotein aids in the phagocytosis of Staphylococcus aureus and Escherichia coli by human neutrophils in a laboratory setting, due to its opsonic properties [107].

The quantity of α_2HS glycoprotein seen in adult human bone is about one-tenth of the amount observed in fetal bone [112]. According to Dickson and Bagga [113], the levels of α_2HS glycoprotein decrease with aging. Furthermore, Quelch et al. discovered that neonatal bone has seven times the amount of α_2HS glycoprotein compared to adult bone [111]. α_2HS glycoprotein is more prevalent in bone compared to albumin [111]. While the specific factors that affect the levels of albumin and α_2HS glycoprotein in bone are not well understood, both substances are incorporated into the bone structure via the process of mineralization. α_2HS glycoprotein has shown a significant attraction to the bone mineral phase, as reported in studies. Research has shown that 96% of α_2HS glycoprotein is associated with the mineral phase, in contrast to 58% of albumin [113]. Based on the fact that sialic acid levels in bone are less affected by age compared to α_2HS glycoprotein levels, it is likely that α_2HS glycoprotein does not significantly contribute to the total concentration of sialic acid in bone extracts [111]. Patients diagnosed with osteogenesis imperfecta may exhibit elevated rates of bone turnover in types I and III of the disorder, perhaps accounting for the increased presence of sialic acid, albumin, and α_2HS glycoprotein in their bones. Histological studies indicate that osteogenesis imperfecta bone turnover rates remain constant with age, maintaining elevated protein levels from infancy to early adulthood. In addition, Dickson and colleagues [114] observed increased levels of α_2HS glycoprotein in the skeletal systems of people with osteogenesis imperfecta, as well as abnormally high levels of this glycoprotein in some urine samples [115].

6.1.5 Phospholipid Transfer Protein

Phospholipid transfer proteins (PLTPs) are proteins found in the plasma that have the ability to transfer phospholipids between lipoproteins and liposomes under laboratory conditions [116–119]. Additionally, they exhibit efficacy as conversion factors for high-density lipoprotein (HDL) and may have a function in transporting vitamin E [120–122]. HDLs are often regarded as anti-atherogenic because they play a part in reverse cholesterol transport, which includes the removal of cholesterol from the body. HDLs have been shown to possess additional anti-atherosclerotic capabilities, which emphasize their ability to protect against cardiovascular illnesses [123, 124]. Due to the correlation between PLTPs and HDL function, it has been proposed that PLTPs could impact the progression of atherosclerosis. In living organisms, the transport of phospholipids via PLTP is made easier by the breakdown of lipoproteins that are high in triglycerides, such as chylomicrons and very low-density lipoproteins (VLDL), by the enzyme lipoprotein lipase. Triglyceride-rich lipoproteins transfer their phospholipids to HDLs in this process [125, 126]. PLTPs have the ability to bind to several types of phospholipids, including phosphatidylcholine, phosphatidylethanolamine, phosphatidylinositol, and sphingomyelin, facilitating their movement. PLTPs in plasma are often linked to HDLs, underscoring their significance in lipid metabolism and cardiovascular well-being [125–128].

6.2 Cytoskeleton Proteins

6.2.1 Tensin

The αβ integrin heterodimer forms a transmembrane core that encloses focal adhesions, which are specialized portions of the plasma membrane. These focal adhesions are the site where the 220 kDa protein tensin is found. The cytoplasmic tail of the integrin extracellular domain serves as an anchor for actin filaments, binding to various components of the extracellular matrix [129]. Tensin is mostly expressed in human tissues such as the ovary, testis, prostate, colon, liver, small intestine, lung, kidney, skeletal muscle, and heart. Expression in the brain, thymus, and circulating leukocytes was either non-existent or very minimal [130]. Current research suggests that Tensin has an important function in connecting the actin cytoskeleton, extracellular matrix (ECM), and signal transduction. Upon ligation of integrin, tensin rapidly relocates to the assembling focal adhesion sites [131]. In 1991, Davis and colleagues [132] discovered the primary cDNA sequence of the tensin protein. The ability to characterize the structure and function of the encoded protein was facilitated [133].

Tensin is a phosphoprotein found in the cytoplasm, especially in focal adhesions that are regulated by integrins. The protein has a phosphotyrosine-binding (PTB) domain that interacts with the cytoplasmic termini of beta integrins and attaches to actin filaments. Tensin has the capacity to link integrin receptors to actin filaments via these interactions. In addition, tensin has a Src Homology 2 (SH2) domain that attaches to tyrosine-phosphorylated proteins involved in signal transduction and undergoes tyrosine phosphorylation [133, 134]. The tensin molecular structure is characterized by gel filtration, electron microscopy, and light scattering. The outcomes specified that tensin can make a dimer [135]. Tensin is a multi-domain protein with ability to bind to various signal transduction and structural molecules (Fig. 6.1) [133]. While the calculated molecular weight was 185 kDa, recombinant and native tensin protein transferred about 220 kDa on SDS gels. The difference is because of abnormally slow migration characteristics of the central section [130].

Tensin acts as a structural support inside focal adhesions, facilitating the formation and breakdown of signaling complexes by providing attachment points for other proteins having SH2 domains. Tensin's scaffolding role facilitates the interaction between proteins containing SH2 domains and signaling molecules that have been phosphorylated on tyrosine residues. Tensin's involvement in downstream signaling cascades makes it a promising biomarker and target for therapeutic interventions in

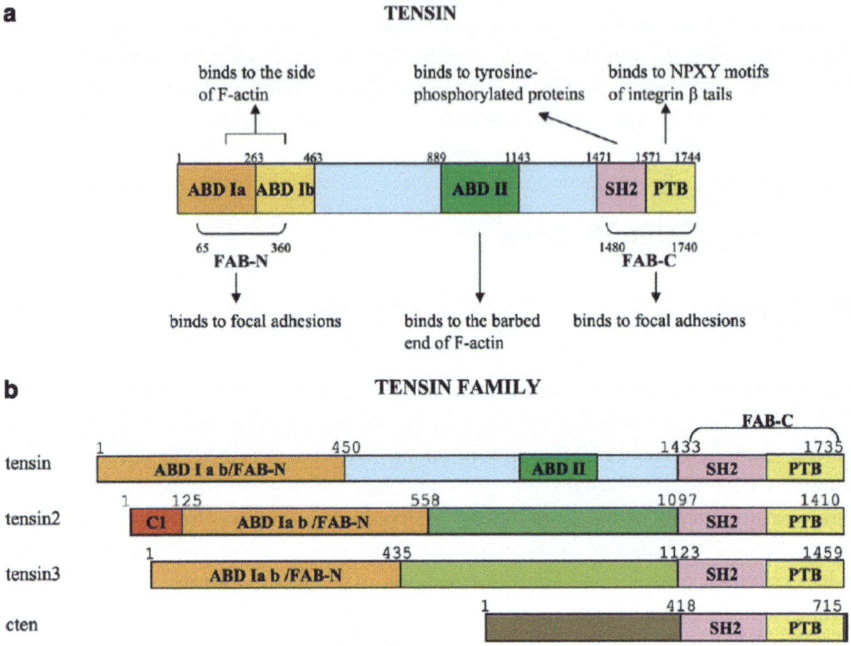

Fig. 6.1 Tensins' structural and functional domains. **a** Tensin can be separated into three sections, each of which is occupied in mediating protein–protein interactions. The C-terminus includes a focal adhesion-binding (FAB) site, PTB domain, and SH2 domain. The PTB domain interacts with β1, 3, 5, and 7 integrin tails. The SH2 domain connects to p130Cas, FAK, and tyrosine phosphorylated PI3 kinase. The middle section of tensin includes the ABD II, which delays the rate of actin polymerization. The N-terminus includes actin-binding domain (ABD) Ia and Ib, which interact with actin filaments. This section also includes the PTEN homologous sequence and another FAB activity. The number of amino acids is based on chicken tensin sequence. **b** Human tensin, tensin2, tensin3, and cten are highly conserved at C- and N- terminals. The middle sections of members of the tensin family display no sequence homology. A C1 domain (protein kinase C conserved Sect. 6.1) is discovered only in tensin2. Reprinted with permission from [133]

illnesses requiring wound healing, renal disease, and cancer. Research done on genetically modified mice has shown the essential roles of tensin in muscle regeneration, cell migration, and kidney function regulation. For instance, the ability of fibroblasts from mouse embryos that do not have tensin is much reduced compared to that of fibroblasts with normal levels of tensin [133, 134]. The sequence homology between tensin and many receptors suggests its role in structural maintenance and signal transduction. Tensin displays sequence homologies with the breakpoint cluster region (BCR), several tyrosine-phosphorylated receptors, and other structural molecules like actin, synapsin, insertin, and catenin-α. The concept that tensin mediates the connection between signal transduction processes and the cytoskeleton is substantiated by these commonalities, which emphasize its dual role in promoting signal transduction and preserving cellular structure [136].

6.2.2 Vinculin

Vinculin is a protein with a molecular weight of 116 kDa. It is mainly found in areas where cells come into contact with each other or with their surrounding substrate. Keith Burridge and Benjamin Geiger made the discovery over forty years ago, each working separately [137, 138]. Vinculin was characterized via cDNA cloning, revealing its composition of 1,066 amino acids. Vinculin consists of eight antiparallel α-helical filaments that are arranged into five different domains, as determined by crystallization studies [139]. The crystal structure of vinculin provides a clear understanding of how several ligand-binding sites are regulated. The formation of a 3-lobe head, consisting of domains 1-3 (D1-3), results in a molecular mass of 95 kDa [140] and a diameter of 80 Å [141]. Domain D1 encompasses the specific locations where talin [142] and α-actinin bind [143]. The head and tail are separated by a linker sequence consisting of 61 amino acids (residues 837–878) that are rich in proline [139]. The tail region, spanning amino acids 879-1066, has a helical bundle consisting of five helices joined by short loops. This helical bundle has a molecular weight of 30 kDa [139, 144]. The vinculin head binds to proteins such as α-actinin, α-catenin, β-catenin, IpaA, and talin. The linker molecule forms complexes with Vasodilator-stimulated phosphoprotein (VASP), ponsin, vinexin, and Arp2/3. The tail section of vinculin contains binding sites for the head domain of vinculin, paxillin, acidic phospholipids, and actin [145–150].

The Arp2/3 complex generates dendritic actin networks in lamellipodia. The observed problems in cells lacking vinculin may be due to the fast interaction between the Arp2/3 complex and the proline-rich region of vinculin during cell spreading [151]. These cells exhibit traits such as diminished stability of focal adhesions, decreased cell proliferation, adhesion, and migration, as well as resistance to apoptosis [152]. Vinexin-β, a crucial component for the cell spreading triggered by EGF, is connected to the proline-rich region of vinculin. However, it is absent in the focal adhesions of cells that do not have vinculin [150, 153]. The ligand-binding sites of vinculin, including the talin binding site, are hidden by an internal connection between the head and tail domains, which has been well recognized. These data suggest that vinculin exists in a state of dynamic equilibrium between its active and inactive forms. Endothelial and epithelial cells form connections with adjacent cells and build connections with the basement membrane via interactions with the extracellular matrix and direct contact with adjoining cells, respectively. Cell migration, tissue remodeling, embryonic development, and other processes that maintain internal stability rely on both kinds of adhesions [152, 154, 155].

Sticky receptors serve as the main method by which adhesion receptors attach to the actin cytoskeleton. Vinculin, despite its low enzymatic activity, controls adhesion by recruiting actin remodeling proteins, promoting actin polymerization, and directly attaching to actin. Vinculin depletion significantly impairs cell–cell and cell–matrix adhesion, highlighting its crucial role in human physiology [146]. Integrins are often present in focal adhesions, also known as cell–matrix adhesions, and may attach to almost fifty different proteins via their cytoplasmic ends [156]. The proteins are organized into nano-domains that have three-dimensional structures, as shown by recent study using super-resolution microscopy. The nano-domains consist of a signaling layer located next to the integrin membrane, an intermediate layer that transfers force and binds actin, and an upper layer that controls the activity of actin [157]. Vinculin first localizes in the signaling layer, but quickly moves to the actin-binding layer when its inhibitory head–tail interactions decrease. Vinculin controls the movement and stability of focal adhesions, which are important for cell migration. It does this by promoting the recruitment of other proteins in this layer [146].

In order to participate in cell–matrix and cell–cell adhesion, vinculin has to form a link with F-actin. Changes in this connection primarily result in a reduced ability to create and transmit pressures, which in turn affects cell shape, adhesion, stiffness, motility, and the vinculin-F-actin link. Vinculin overexpression decreases the migration of cells, while downregulation of vinculin increases the motility of cells. Motile cells have heightened signaling via FAK and paxillin. In addition, cells lacking vinculin exhibit enhanced motility and decreased adhesion compared to normal cells, as shown by the presence of smaller and fewer focal adhesions. This factor may clarify the role of vinculin as a tumor suppressor in model systems [146, 158–161]. Electron microscopy (EM) and biochemical analyses, along with, determining the main sequence of vinculin, display that vinculin consists of a spherical head connected to a tail domain by a proline-rich section, and interaction sites for several binding associates have been charted onto all three sections of the protein (Table 6.2) [162].

Recent study has revealed two previously unknown and unexpected roles of vinculin. Researchers [163] have shown that the activity of the mitogen-activated protein kinases ERK1 and ERK2 is much higher in cells lacking vinculin when the caspase-9 initiator is blocked. The proposal suggests that there is a rivalry between the tail domain (Vt) of paxillin and vinculin for binding to paxillin. When vinculin is not present, the ability to resist apoptosis is increased by the activation of paxillin or FAK via ERK [163]. Mouse F9 embryonal cancer cells lacking vinculin [164] exhibit resistance to caspase-3 activation caused by serum deprivation, treatment with camptothecin (a cytotoxic anticancer medication that inhibits DNA topoisomerase I), or dissociation from the ECM. Although the exact role of tyrosine 822 phosphorylation is not fully understood, the presence of full-length vinculin may make cells more susceptible to death. However, a mutation in vinculin that replaces Tyr822Phe does not have the same effect. Apoptosis susceptibility in cells lacking vinculin is regained by suppressing ERK [163].

Paxillin phosphorylation on Tyr118 and FAK on Tyr397 is unexpectedly elevated in cells lacking vinculin. Introducing a vinculin Vt polypeptide (residues 811–1066) decreases phosphorylation levels and restores the sensitivity of these cells to apoptotic stimuli. FAK and vinculin share certain binding sites in paxillin [163, 165]. The crystallographic analysis of full-length vinculin has shown its auto-inhibited conformation, which consists of five distinct domains. Domains D1-D3 function as a clamp, effectively anchoring the tail (Vt). The proline-rich region forms a covalent link between the head of vinculin and Vt. The isolated D1 and Vt domains have a low affinity of around 10^{-6} M. However, the head and tail have an unusually strong

Table 6.2 Vinculin binding partners[a] [152]

Vinculin domain	Ligand	Ligand site	Methods	Function and regulation
D1	IpaA	Not determined	ELISA, IP, OV	F-actin depolymerization; bacterial entry, vinculin activation
D1	Talin	Rod, up to 10 talin VBSs	SPOT, PD, Y2H, X-ray structure (D1/tVBS)	Link to membrane or F-actin, FA strength; vinculin activation, helix-bundle conversion
D1	α-Actinin	CH domain; Spectrin R4	cryoEM, X-ray structure (D1/aVBS)	FA strength; vinculin activation, helix-bundle conversion
D1-D3	β-catenin	Not determined	IP	No activation of F-actin binding, localization to cell–cell junctions
D1-D3	α-catenin	aa 326–509 of αE-catenin; aa 697–906 of α-catenin	ITC, SPR, PD, OV	PTEN stability (F9/γ229 cells); Localization to cell–cell junctions (adherence junctions), link to F-actin (?)
D1 and D4	(intra.)	Vt (D5)	FRAP, ITC, FRET, D5-Ala-mutants, X-ray structure (vinculin)	Release from FAs, auto-inhibited cytoplasmic conformation
PRR	Ponsin	SH3–1/-2	OV, IP	
PRR	Vinexin	SH3–1/-2	SPR, IP, Y2H	Binds activated vinculin; promotes spreading, enhanced FA size and stress fibers
PRR, P876/P878	Arp 2/3	Not determined	PD, IP	Lamellipodial extension, spreading
PRR, FPPPP	VASP	EVH1, EVH2	PD, IP	Link to F-actin
Vt (D5)	Acidic phospholipids	Not determined	Bilayer insertion, OV, PD	FA turnover; competition of F-actin binding
Vt (D5)	F-actin	Two sites, separate actin monomers	cryoEM, D5-Ala-mutants, Co-sedimentation	Link to actin cytoskeleton; vinculin dimerization, actin filament bundling

(continued)

Table 6.2 (continued)

Vinculin domain	Ligand	Ligand site	Methods	Function and regulation
Vt (D5)	Paxillin	LD1, LD2, LD4	PD, OV	Interaction of FAK/paxillin affects ERK activity, motility, and apoptosis (F9/ γ229 cells); constitutive
Vt (D5), C-terminal arm	PKC-α	Regulatory domain	Kinase assay; IP, OV	PKCα recruitment; D5 phosphorylation S1033/ S1045

[a]*Y2H* yeast two-hybrid; *SPR* surface plasmon resonance; *SPOT* SPOT-peptide assay; *SH3* Src homology domain 3; *PRR* proline-rich region; *PD* pull down; *OV* blot/gel overlay; *LD* paxillin-LD motif; *ITC* isothermal titration calorimetry; *IP* immunoprecipitation; *FRET* Förster resonance energy transfer; *FPPPP* PheProProProPro motif; *FA* focal adhesion; *ELISA* enzyme-linked immunosorbent assay; *cryoEM* cryo electron microscopy; *CH* calponin homology domain; *aa* amino acid

binding with a K_d of less than 10^{-9} M [139, 166]. This is mostly because of the additional contacts between D4 and Vt. The proline-rich region located between Vt and D4 binds the Arp2/3 complex [148], ponsin [149], vinexin [150], and VASP [145, 167].

Cells react to mechanical pressure on integrins that are attached to ligands from within the cell (e.g., RhoA stimulation of actomyosin contractility) or externally (e.g., shear stress, pulling force, stretching) [168–171]. This response is accompanied by an increase in focal adhesions and enlarged recruitment of vinculin (along with paxillin and talin) to focal adhesions [172]. The recruitment of vinculin [152] may be dependent on actomyosin contractility rather than actin polymerization. This is because Gd^{3+} (gadolinium) treatment, did not influence Rac signaling, which controls cell spreading and membrane-protrusive activities via Arp2/3-mediated actin polymerization [168], was not affected by Gd^{3+} treatment [173]. However, inhibiting the Ca^{2+} signal with Gd^{3+} is linked to a decrease in vinculin and phosphotyrosine at focal adhesions, along with a reduction in cell migration. Mechanical stretch in fibroblasts promotes Ca^{2+} entrance into the cell's leading edge, hence regulating local traction forces [173].

6.2.3 Paxillin

Paxillin is a multi-domain and multifunctional focal adhesion adapter protein that performs a significant scaffolding function in focal adhesions through the use of signaling and structural molecules engaged in migration and movement of cells, when phosphorylated on particular Ser and Tyr residues. It is located in the intracellular surface of cell adhesion sites to the ECM [174]. Paxillin is an adapter protein including four LIM domains enriched with cysteine-histidine at the C-terminus and five aspartate- and leucine-rich LD (Leu-Asp) motifs found at the N-terminus (LD1 to LD5), which are protein-binding segments that have the consensus sequence LDxL-LxxL [174]. The paxillin N-terminus regulates most of its signaling activity. Furthermore, it comprises a proline-rich sequence that binds SH3-containing proteins, located within the N-terminus, in addition to several tyrosine and serine residues all over the protein that bind to SH2 domains (Fig. 6.2) [175].

The observation of an increased amount of phosphorylation of paxillin on Tyr118 and FAK on Tyr397 in vinculin-null cells is surprising. The phosphorylation levels are decreased and the cells become more sensitive to apoptotic stimuli when the

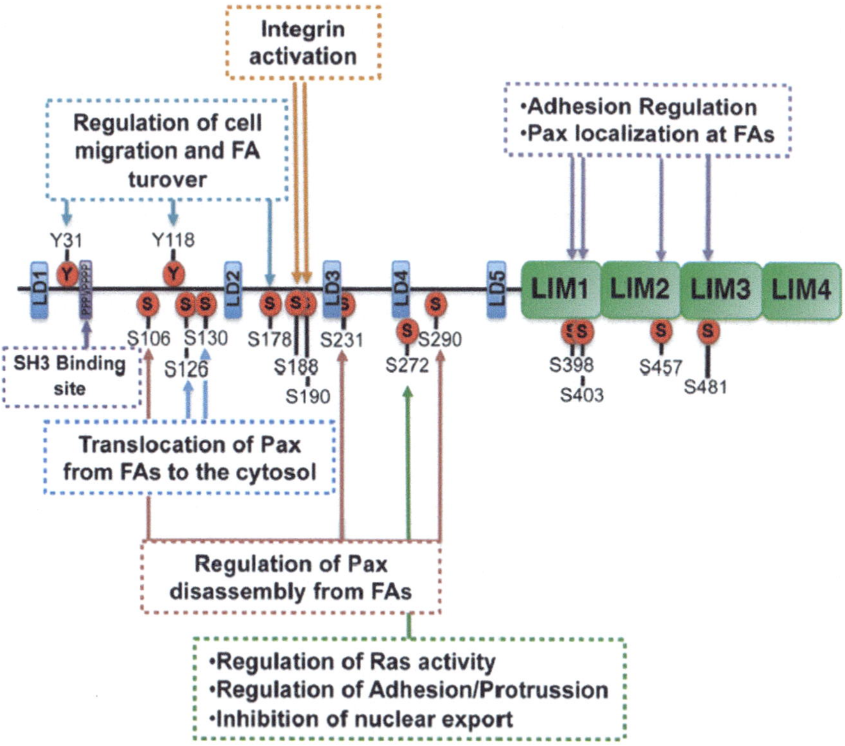

Fig. 6.2 Functional and structural organization of paxillin. Human paxillin protein comprises 591 amino acids. The C-terminus has four LIM domains enriched with cysteine-histidine, engaged in the paxillin anchorage to the plasma membrane and its targeting to focal adhesions. The N-terminus has five leucine-rich LD domains (LD1–LD5), which contain binding sequences for the employment of structural and signaling molecules including Crk, vinculin, and FAK, and a proline-rich section that anchors SH3-containing proteins. Serine phosphorylation causes some efficient adjustments engaged in the control of the cell movement and adhesion process. Paxillin is phosphorylated in response to cell adhesion and growth factor or cytokine stimulation: Tyrosine phosphorylation makes interaction sites for signaling molecules that in turn control targeting of paxillin to FAs. Reprinted with permission from [175]

vinculin Vt polypeptide (residues 811–1066) is introduced. FAK and vinculin have partial overlap in their binding sites inside paxillin.

6.2.4 Talin

Talin is a crucial protein that links the cytoplasmic sections of integrin β subunits to actin filaments. It belongs to a protein group that consists of skelemin, melusin, integrin-linked kinase, tensin, filamin, and α-actinin [176, 177]. Talin, with a molecular weight of 270 kilodaltons, is very prone to breakdown by the enzyme calpain 2. Talin consists of a spherical N-terminal head segment and a flexible rod region. The head region is responsible for the dynamics of focal adhesions and interacts with focal adhesion kinase (FAK) and PIPK1γ90, which is a variation of phosphatidylinositol (4)-phosphate 5-kinase type Iγ. Nevertheless, the precise connection to FAK remains uncertain. The FERM domain, consisting of F1, F2, and F3 subdomains, is also found in the skull. These subdomains contain binding sites for β-integrins, filamentous actin (F-actin), and layilin cytoplasmic domains (which serve as receptors for hyaluronan).

Talin's interaction with the cytoplasmic portions of β-integrin modifies the structure of the extracellular area. This contact increases the ability of β-integrin to bind to ECM proteins and promotes the creation of focal adhesions, which are connections between the ECM and cultured cells. However, studies on flies with altered integrin alleles indicate that talin and integrins do not solely depend on each other for their positioning at cell-ECM junctions. This implies that there are other processes at play in the recruitment of talin to these locations [178–180]. The talin rod domain has a minimum of two sites that bind to actin, numerous sites that bind to vinculin, and an extra site that binds to integrin. The exact arrangement of the subunits in this dimer is still unclear, even though the C-terminal helices of two talin monomers come together to create an antiparallel dimer. Proteomic and yeast two-hybrid screening approaches have shown the existence of several proteins that interact with talin and many sites of phosphorylation within the talin rod domain [180–182]. This domain consists of 62 amphipathic α-helices arranged in a sequence of helical bundles. Around 10 of these helices have the ability to attach to hydrophobic crevices in the vinculin head. Although talin is often found in a helical bundle, its major vinculin-binding residues are usually hydrophobic [180, 183].

Talin's interaction with F-actin is strongly linked to its ability to form dimers. The C-terminal section of the talin rod consists of a helix at the end and a bundle of five helices. These components are crucial for the creation of dimers. This combination generates the THATCH domain, which is essential for binding to actin [184]. The THATCH dimer binds to three actin monomers in a linear configuration on an actin filament, indicating that dimerization plays a role in the binding mechanism. Talin has a minimum of two extra actin-binding sites that enhance its ability to bundle actin [180]. Vertebrates possess two talin genes that have a 74% similarity. Talin 1 is mostly found in platelets, while Talin 2 is the principal type present in the cerebellum and striated muscle. The significance of the osteoclast cytoskeletal structure linked with integrin in bone disintegration is emphasized by the increased bone density and surplus of osteoclasts seen when talin is conditionally removed in mice. This is caused by a decrease in the ability of bones to break down and a lack of proper attachment of osteoclasts [180, 185]. The processes responsible for activating the various ligand-binding sites in talin are not fully understood. Although talin plays a crucial role in several integrin-mediated actions in both living organisms and laboratory settings, the mechanisms that control its activity are currently under investigation. Recent study indicates that vinculin, as opposed to talin, is crucial for the linkage between the actin cytoskeleton and focal adhesions [180, 186].

6.3 Conclusion

This chapter has explored the complex realm of nano-bioceramics and their prospective uses in bone tissue engineering and regenerative medicine. The investigations have taken us through the intriguing interaction of biological traits, material attributes, and medical consequences, emphasizing the revolutionary capacity of nanoparticles in the treatment of bone regeneration. Nanoparticles, possessing distinct physicochemical characteristics such as size, shape, composition, and charge, have significantly transformed methods in bone tissue engineering. They have great potential for medication administration, gene therapy, and cell labeling by controlling bone remodeling processes. Plasmatic and cytoskeleton proteins are essential in the bone extracellular matrix. Important plasma proteins involved in bone health include transferrin, which regulates mineralization and the formation of bone tissue; immunoglobulin, which affects the balance of bone in the body; albumin, which is involved in maintaining calcium levels and cell communication; α_2 HS glycoprotein, which plays a role in bone metabolism and repair; and phospholipid transfer protein (PLTP), which has an impact on overall bone health. Important proteins of the cytoskeleton include tensin, vinculin, paxillin, and talin. These proteins have a role in the construction of the extracellular matrix and connections between cells. The analysis of newly developed calcium-silicate nano-bioceramics demonstrates their favorable physicochemical and biological characteristics for the purpose of bone regeneration, as well as their potential use as scaffold fillers and grafts in bone cavities and periodontal diseases. The addition of nanoparticles such as hydroxyapatite to nanocellulose scaffolds with high porosity improves the ability of proteins to bind to the scaffold, promotes the spreading of cells, and increases the expression of genes related to bone formation. This demonstrates a promising potential for use in clinical applications. In the future, the main difficulty will be finding a balance between promoting bone regeneration and preventing negative consequences. Personalized methods will be crucial in achieving this. The future advancements in nano-bioceramics will be driven by tailoring them to meet the specific requirements of individual patients, achieved via surface changes, drug release kinetics, and cellular interactions. As we progress, the integration of nano-bioceramics persists, aligning science, medicine, and optimism, and directing us toward revolutionary treatments that cater to the requirements of those seeking rejuvenation and revitalization.

References

1. Aisen, P., & Listowsky, I. (1980). Iron transport and storage proteins. *Annual review of biochemistry, 49*(1), 357–393.
2. Morgan, E. (1981). Transferrin, biochemistry, physiology and clinical significance. *Molecular aspects of medicine, 4*(1), 1–123.
3. Laurell, C.-B., & Ingelman, B. (1985). The iron-binding protein of swine serum. *Nutrition Reviews, 43*(7), 209–211.
4. Huebers, H. A., & Finch, C. A. (1987). The physiology of transferrin and transferrin receptors. *Physiological Reviews, 67*(2), 520–582.
5. Aisen, P., & Brown, E. (1975). Structure and function of transferrin. *Program in Hematology, 9*, 28–56.
6. Huebers, H. A., & Finch, C. A. (1984). Transferrin: Physiologic behavior and clinical implications. *Blood, The Journal of the American Society of Hematology, 64*, 763–767.
7. Lane, R. (1976). Transferrin. *Structure and function of plasma proteins* (pp. 53–78). Springer.
8. Putnam, F. W. (1975). Transferrin. *The plasma proteins* (pp. 265–316). Elsevier.
9. Spik, G., et al. (1974). Amino acid sequences of two glycopeptides isolated from tryptic and chymotryptic hydrolysates of human lactotransferrin. *FEBS Letters, 38*(2), 213–216.
10. Young, S., & Bomford, A. (1984). Transferrin and cellular iron exchange. *Clinical Science, 67*(3), 273–278.
11. Thibodeau, S. N., Lee, D. C., & Palmiter, R. D. (1978). Identical precursors for serum transferrin and egg white conalbumin. *Journal of Biological Chemistry, 253*(11), 3771–3774.
12. Gallo, P., et al. (1985). The cerebrospinal fluid transferrin/Tau proteins: a study by two-dimensional polyacrylamide gel electrophoresis (2D) and agarose isoelectrofocusing (IEF) followed by double-antibody peroxidase labeling and avidin-biotin amplification. *Journal of the neurological sciences, 70*(1), 81–92.
13. Dickson, P. W., et al. (1985). High prealbumin and transferrin mRNA levels in the choroid plexus of rat brain. *Biochemical and Biophysical Research Communications, 127*(3), 890–895.
14. Bloch, B., et al. (1985). Transferrin gene expression visualized in oligodendrocytes of the rat brain by using in situ hybridization and immunohistochemistry. *Proceedings of the National Academy of Sciences, 82*(19), 6706–6710.
15. Skinner, M. K., Cosand, W. L., & Griswold, M. D. (1984). Purification and characterization of testicular transferrin secreted by rat Sertoli cells. *Biochemical Journal, 218*(2), 313–320.
16. Huebers, H., et al. (1976). Isolation and characterization of iron-binding proteins from rat intestinal mucosa. *European Journal of Biochemistry, 66*(3), 447–455.
17. Imel, E. A., et al. (2016). Serum fibroblast growth factor 23, serum iron and bone mineral density in premenopausal women. *Bone, 86*, 98–105.
18. Guggenbuhl, P., et al. (2005). Bone mineral density in men with genetic hemochromatosis and HFE gene mutation. *Osteoporosis International, 16*(12), 1809–1814.
19. Muckenthaler, M. U., et al. (2017). A red carpet for iron metabolism. *Cell, 168*(3), 344–361.
20. Martel, P., Kim, S., & Powell, B. (1980). Physical characteristics of human transferrin from small angle neutron scattering. *Biophysical Journal, 31*(3), 371–380.
21. Rosseneu-Motreff, M., et al. (1971). Size and shape determination of apotransferrin and transferrin monomers. *Biopolymers: Original Research on Biomolecules, 10*(6), 1039–1048.
22. Charlwood, P. (1971). Differential sedimentation-velocity and gel-filtration measurements on human apotransferrin and iron–transferrin. *Biochemical Journal, 125*(4), 1019–1026.
23. Jarritt, P. H. (1976). Effect of iron on sedimentation-velocity and gel filtration behaviour of transferrins from several vertebrates. *Biochimica et Biophysica Acta, 453*(2), 332–343.

24. Palmour, R. M., & Sutton, H. E. (1971). Vertebrate transferrins. Molecular weights, chemical compositions, and iron-binding studies. *Biochemistry, 10*(22), 4026–4032.
25. Yeh, Y., Iwai, S., & Feeney, R. E. (1979). Conformations of denatured and renatured ovotransferrin. *Biochemistry, 18*(5), 882 889.
26. Uzan, G., et al. (1984). Molecular cloning and sequence analysis of cDNA for human transferrin. *Biochemical and Biophysical Research Communications, 119*(1), 273–281.
27. MacGillivray, R., et al. (1983). The primary structure of human serum transferrin. The structures of seven cyanogen bromide fragments and the assembly of the complete structure. *Journal of Biological Chemistry, 258*(6), 3543–3553.
28. Aisen, P., Leibman, A., & Pinkowitz, R. (1974). The anion-binding functions of transferrin. *Protein-metal interactions* (pp. 125–140). Springer.
29. Schlabach, M. R., & Bates, G. W. (1975). The synergistic binding of anions and Fe3+ by transferrin. Implications for the interlocking sites hypothesis. *Journal of Biological Chemistry, 250*(6), 2182–2188.
30. Jandl, J. H., & Katz, J. H. (1963). The plasma-to-cell cycle of transferrin. *The Journal of Clinical Investigation, 42*(3), 314–326.
31. Booth, A. G., & Wilson, M. J. (1981). Human placental coated vesicles contain receptor-bound transferrin. *Biochemical Journal, 196*(1), 355–362.
32. Davies, M., Parry, J., & Sutcliffe, R. (1981). Examination of different preparations of human placental plasma membrane for the binding of insulin, transferrin and immunoglobulins. *Reproduction, 63*(2), 315–324.
33. Ecarot-Charrier, B., et al. (1980). Reticulocyte membrane transferrin receptors. *Canadian Journal of Biochemistry, 58*(5), 418–426.
34. Enns, C., & Sussman, H. (1981). Physical characterization of the transferrin receptor in human placentae. *Journal of Biological Chemistry, 256*(19), 9820–9823.
35. Seligman, P. A., Schleicher, R. B., & Allen, R. H. (1979). Isolation and characterization of the transferrin receptor from human placenta. *Journal of Biological Chemistry, 254*(20), 9943–9946.
36. Rauner, M., et al. (2019). Transferrin receptor 2 controls bone mass and pathological bone formation via BMP and Wnt signalling. *Nature Metabolism, 1*(1), 111–124.
37. Forejtnikova, H., et al. (2010). Transferrin receptor 2 is a component of the erythropoietin receptor complex and is required for efficient erythropoiesis. *Blood, The Journal of the American Society of Hematology, 116*(24), 5357–5367.
38. Powell, L. W., Seckington, R. C., & Deugnier, Y. (2016). Haemochromatosis. *The Lancet, 388*(10045), 706–716.
39. Lefranc, M.-P., & Lefranc, G. (2001). *The immunoglobulin factsbook*. Academic Press.
40. Porter, R. (1959). The hydrolysis of rabbit γ-globulin and antibodies with crystalline papain. *Biochemical Journal, 73*(1), 119–127.
41. Zhou, P., et al. (2008). CD32B is highly expressed on clonal plasma cells from patients with systemic light-chain amyloidosis and provides a target for monoclonal antibody–based therapy. *Blood, The Journal of the American Society of Hematology, 111*(7), 3403–3406.
42. Bhat, A., et al. (2010). Currents concepts on the immunopathology of amyloidosis. *Clinical Reviews in Allergy & Immunology, 38*(2–3), 97–106.
43. Picken, M. M. (2010). Amyloidosis—Where are we now and where are we heading? *Archives of Pathology & Laboratory Medicine, 134*(4), 545–551.
44. Gertz, M. A. (2016). Immunoglobulin light chain amyloidosis: 2016 update on diagnosis, prognosis, and treatment. *American Journal of Hematology, 91*(9), 947–956.
45. Weiss, B. M., et al. (2014). Increased serum free light chains precede the presentation of immunoglobulin light chain amyloidosis. *Journal of Clinical Oncology, 32*(25), 2699.
46. Benner, R., Hijmans, W., & Haaijman, J. (1981). The bone marrow: The major source of serum immunoglobulins, but still a neglected site of antibody formation. *Clinical and Experimental Immunology, 46*(1), 1.
47. Asofsky, R., & Thorbecke, G. (1961). Sites of formation of immune globulins and of a component of C'3: II. Production of immunoelectrophoretically identified serum proteins by human and monkey tissues in vitro. *The Journal of Experimental Medicine, 114*(4), 471–483.

48. Askonas, B. A., & Humphrey, J. (1958). Formation of specific antibodies and γ-globulin in vitro. A study of the synthetic ability of various tissues from rabbits immunized by different methods. *Biochemical Journal, 68*(2), 252–261.
49. Thorbecke, G., & Keuning, F. (1956). Antibody and gamma globulin formation in vitro in hemopoietic organs. *The Journal of Infectious Diseases*, 157–171.
50. Askonas, B. A., & White, R. (1956). Sites of antibody production in the guinea-pig. The relation between in vitro synthesis of anti-ovalbumin and γ-globulin and distribution of antibody-containing plasma cells. *British Journal of Experimental Pathology, 37*(1), 61.
51. Askonas, B., White, R., & Wilkinson, P. (1965). Production of γ1-and γ2-antiovalbumin by various lymphoid tissues of the guinea pig. *Immunochemistry, 2*(4), 329–336.
52. McMillan, R., et al. (1972). Immunoglobulin synthesis by human lymphoid tissues: Normal bone marrow as a major site of IgG production1. *The Journal of Immunology, 109*(6), 1386–1394.
53. Van Furth, R., Schuit, H. R., & Hijmans, W. (1966). The formation of immunoglobulins by human tissues in vitro: III. Spleen, lymph nodes, bone marrow and thymus. *Immunology, 11*(1), 19.
54. Benner, R., et al. (1981). Frequencies of background immunoglobulin-secreting cells in mice as a function of organ, age, and immune status. *Immunobiology, 158*(3), 225–238.
55. Haaijman, J., Schuit, H., & Hijmans, W. (1977). Immunoglobulin-containing cells in different lymphoid organs of the CBA mouse during its life-span. *Immunology, 32*(4), 427.
56. Haaijman, J., & Hijmans, W. (1978). Influence of age on the immunological activity and capacity of the CBA mouse. *Mechanisms of Ageing and Development, 7*, 375–398.
57. Haaijman, J., et al. (1979). The distribution of cytoplasmic immunoglobulin containing cells over various lymphoid organs of congenitally athymic (nude) mice as a function of age. *Immunology, 36*(2), 271.
58. Radl, J., et al. (1974). The origin of monomeric and polymeric forms of IgA in man. *The immunoglobulin A system* (pp. 57–65). Springer.
59. Vossen, J. (1974). Age related development of humoral immunity in normal children and in children with immunodeficiencies. *Bollettino dell'Istituto sieroterapico milanese, 53*(1), suppl: 152.
60. Skvaril, F., & Morell, A. (1974). Distribution of IgA subclasses in sera and bone marrow plasma cells of 21 normal individuals. *The immunoglobulin a system* (pp. 433–435). Springer.
61. Morell, A., et al. (1975). Cytoplasmic immunofluorescence of bone marrow plasma cells producing immunoglobulins of the four IgG subclasses. *The Journal of Immunology, 115*(2), 579–583.
62. Hijmans, W., Schuit, H. R., & Hulsing-Hesselink, E. (1971). An immunofluorescence study on intracellular immunoglobulins in human bone marrow cells. *Annals of the New York Academy of Sciences, 177*(1), 290–305.
63. Turesson, I. (1976). Distribution of immunoglobulin-containing cells in human bone marrow and lymphoid tissues. *Acta Medica Scandinavica, 199*(1–6), 293–304.
64. van Camp, B. G., et al. (1978). The cellular basis of double paraproteinemia in man. *Clinical Immunology and Immunopathology, 9*(1), 111–119.
65. Deutsch, L. (1899). Contribution a l'etude de l'origine des anticorps typhiques. *Annales de l'Institut Pasteur, 13*, 689.
66. Pfeiffer, R. (1898). Die Bildungsstätte der Choleraschutzstoffe. *Zeitschrift für Hygiene und Infektionskrankheiten, 27*(1), 272–297.
67. Reiter, H. (1913). Studien über Antikörperbildung in vivo und in Gewebskulturen. *Z. Immun.-forschg, 18*(5).
68. Lüdke, H. (1912). Über Antikörperbildung in Kulturen lebender Körperzellen. *Berl. klin. Wschr, 1034.*
69. Thorbecke, G., & Keuning, F. (1953). Antibody formation in vitro by haemopoietic organs after subcutaneous and intravenous immunization. *Journal of Immunology, 70*(2), 129–134.
70. McMillan, R., Yelenosky, R., & Longmire, R. (1976). Antiplatelet antibody production by the spleen and bone marrow in immune thrombocytopenic purpura. *Immunoaspects of the Spleen*, 227–237.

71. Vaughan, J., et al. (1976). Rheumatoid factor-producing cells detected by direct hemolytic plaque assay. *The Journal of clinical investigation, 58*(4), 933–941.
72. Thorbecke, G., et al. (1962). Gamma globulin and antibody formation in vitro: III. Induction of secondary response at different intervals after the primary; the role of secondary nodules in the preparation for the secondary response. *The Journal of Experimental Medicine, 116*(3), 295–310.
73. Janković, B., & Paunović, V. (1973). Immunological responses in the mole rat (Spalax leucodon). I. Antibody production, delayed hypersensitivity and lymphatic tissue. In *Annales d'immunologie*.
74. Knothe, R., Herrlinger, J., & Müller-Ruchholtz, W. (1979). B cell activity in rat bone marrow and spleen during primary and secondary response: Dependence on strength of antigen. *International Archives of Allergy and Immunology, 59*(1), 99–103.
75. Benner, R., et al. (1974). Antibody formation in mouse bone marrow: I. Evidence for the development of plaque-forming cells in situ. *Immunology, 26*(2), 247.
76. Hill, S. (1976). Distribution of plaque-forming cells in the mouse for a protein antigen. Evidence for highly active parathymic lymph nodes following intraperitoneal injection of hen lysozyme. *Immunology, 30*(6), 895.
77. Janković, B., Isaković, K., & Petrović, S. (1973). Direct stimulation of lymphoid tissue of the chicken: III. Haemagglutinin production, haemolysin-forming cells and changes in lymphoid tissues following injection of guinea-pig erythrocytes into the bone marrow. *Immunology, 25*(4), 663.
78. Eipert, E. F., et al. (1979). Bone marrow as a major lymphoid organ in Rana. *Cellular Immunology, 46*(2), 275–280.
79. Gollapudi, V. S., & Kind, L. S. (1975). Phenotypic correction of low reagin production: A genetic defect in the SJL mouse. *The Journal of Immunology, 114*(2 Part 2), 906–907.
80. Kind, L. S., & Malloy, W. F. (1974). Development of reaginic antibody-forming cells in the spleen and bone marrow of immunized mice. *The Journal of Immunology, 112*(4), 1609–1612.
81. Owen, M., & Triffitt, J. (1976). Extravascular albumin in bone tissue. *The Journal of Physiology, 257*(2), 293–307.
82. Studer, R., & Potchen, J. (1971). The radioisotopic assessment of regional microvascular permeability to macromolecules. *Microvascular Research, 3*(1), 35–48.
83. Owen, M., Triffitt, J., & Melick, R. (1973). Albumin in bone. In *Ciba Found Symposium*. Wiley Online Library.
84. Matthews, J., et al. (1973). An ultrastructural study of calcium and phosphate deposition and exchange in tissues. *Hard Tissue Growth Repair and Remineralization*, 187–211.
85. Cooper, R. R., Milgram, J. W., & Robinson, R. A. (1966). Morphology of the osteon: An electron microscopic study. *JBJS, 48*(7), 1239–1271.
86. Neuman, W., & Bareham, B. (1975). Evidence for the presence of secondary calcium phosphate in bone and its stabilization by acid production. *Calcified tissue research, 18*(1), 161–172.
87. Kobayashi, K., et al. (2000). High-level expression of recombinant human serum albumin from the methylotrophic yeast Pichia pastoris with minimal protease production and activation. *Journal of bioscience and bioengineering, 89*(1), 55–61.
88. Fleer, R., et al. (1991). Stable multicopy vectors for high–level secretion of recombinant human serum albumin by Kluyveromyces Yeasts. *Bio/Technology, 9*(10), 968–975.
89. Sleep, D., Belfield, G., & Goodey, A. (1990). The secretion of human serum albumin from the yeast Saccharomyces cerevisiae using five different leader sequences. *Bio/Technology, 8*(1), 42–46.
90. Latta, M., et al. (1987) Synthesis and purification of mature human serum albumin from *E. coli*. *Bio/Technology, 5*(12), 1309–1314.
91. Barash, I., et al. (1993). Synthesis and secretion of human serum albumin by mammary gland explants of virgin and lactating transgenic mice. *Transgenic Research, 2*(5), 266–276.
92. Rabbani, G., & Ahn, S. N. (2019). Structure, enzymatic activities, glycation and therapeutic potential of human serum albumin: A natural cargo. *International Journal of Biological Macromolecules, 123*, 979–990.

93. Fernández-San Millán, A., et al. (2003). A chloroplast transgenic approach to hyper-express and purify human serum albumin, a protein highly susceptible to proteolytic degradation. *Plant Biotechnology Journal, 1*(2), 71–79.
94. Sijmons, P. C., et al. (1990). Production of correctly processed human serum albumin in transgenic plants. *Bio/Technology, 8*(3), 217–221.
95. Huang, L.-F., et al. (2005). Production of human serum albumin by sugar starvation induced promoter and rice cell culture. *Transgenic Research, 14*(5), 569–581.
96. Farran, I., et al. (2002). Targeted expression of human serum albumin to potato tubers. *Transgenic Research, 11*(4), 337–346.
97. Horvathy, D. B., et al. (2016). Serum albumin coating of demineralized bone matrix results in stronger new bone formation. *Journal of Biomedical Materials Research Part B: Applied Biomaterials, 104*(1), 126–132.
98. Skaliczki, G., et al. (2013). Serum albumin enhances bone healing in a nonunion femoral defect model in rats: A computer tomography micromorphometry study. *International orthopaedics, 37*(4), 741–745.
99. Weszl, M., et al. (2012). Freeze-dried human serum albumin improves the adherence and proliferation of mesenchymal stem cells on mineralized human bone allografts. *Journal of Orthopaedic Research, 30*(3), 489–496.
100. Bernards, M. T., Qin, C., & Jiang, S. (2008). MC3T3-E1 cell adhesion to hydroxyapatite with adsorbed bone sialoprotein, bone osteopontin, and bovine serum albumin. *Colloids and Surfaces B: Biointerfaces, 64*(2), 236–247.
101. Horváthy, D. B., et al. (2013). Albumin-coated bioactive suture for cell transplantation. *Surgical innovation, 20*(3), 249–255.
102. Liu, X., et al. (2014). Effects of titania nanotubes with or without bovine serum albumin loaded on human gingival fibroblasts. *International Journal of Nanomedicine, 9*, 1185.
103. Horváthy, D. B., et al. (2019). Serum albumin-coated bone allograft (BoneAlbumin) results in faster bone formation and mechanically stronger bone in aging rats. *Journal of Tissue Engineering and Regenerative Medicine, 13*(3), 416–422.
104. Yim, R. (2005). Administrative and research policies required to bring cellular therapies from the research laboratory to the patient's bedside. *Transfusion, 45*, 144S-158S.
105. Ashton, B., Höhling, H.-J., & Triffitt, J. (1977). Plasma proteins present in human cortical bone: Enrichment of theαHS-glycoprotein. *Calcified Tissue Research, 22*(1), 27–33.
106. Triffitt, J., et al. (1976). Origin of plasma α 2 HS-glycoprotein and its accumulation in bone. *Nature, 262*(5565), 226–227.
107. Dickson, I., Poole, A., & Veis, A. (1975). Localisation of plasma α 2 HS glycoprotein in mineralising human bone. *Nature, 256*(5516), 430–432.
108. Kundranda, M. N., et al. (2004). Annexins expressed on the cell surface serve as receptors for adhesion to immobilized fetuin-A. *Biochimica et Biophysica Acta -Molecular Cell Research, 1693*(2), 111–123.
109. Wang, X. Q., et al. (2010). Fetuin-A promotes primary keratinocyte migration: Independent of epidermal growth factor receptor signalling. *Experimental Dermatology, 19*(8), e289–e292.
110. Makuloluwa, A. K., et al. (2020). Mass spectrometry reveals α-2-HS-glycoprotein as a key early extracellular matrix protein for conjunctival cells. *Investigative Ophthalmology & Visual Science, 61*(3), 44–44.
111. Quelch, K. J., Cole, W. G., & Melick, R. A. (1984). Noncollagenous proteins in normal and pathological human bone. *Calcified Tissue International, 36*(1), 545–549.
112. Wilson, J., Ashton, B., & Triffitt, J. (1976). The interaction of a component of bone organic matrix with the mineral phase. *Calcified Tissue Research, 22*(1), 458–460.
113. Dickson, I. R., & Bagga, M. K. (1985). Changes with age in the non-collagenous proteins of human bone. *Connective Tissue Research, 14*(1), 77–85.
114. Dickson, I., Millar, E., & Veis, A. (1975). Evidence for abnormality of bone-matrix proteins in osteogenesis imperfecta. *The Lancet, 306*(7935), 586–587.
115. Dickson, I. R., Bagga, M., & Paterson, C. R. (1983). Variations in the serum concentration and urine excretion of α 2 HS-glycoprotein, a bone-related protein, in normal individuals and in patients with osteogenesis imperfecta. *Calcified Tissue International, 35*(1), 16–20.

116. Albers, J. J., & Cheung, M. C. (2004). Emerging roles for phospholipid transfer protein in lipid and lipoprotein metabolism. *Current Opinion in Lipidology, 15*(3), 255–260.
117. Damen, J., Regts, J., & Scherphof, G. (1982). Transfer of [14C] phosphatidylcholine between liposomes and human plasma high density lipoprotein partial purification of a transfer-stimulating plasma factor using a rapid transfer assay. *Biochimica et Biophysica Acta -Lipids Lipid Metabolism, 712*(3), 444–452.
118. Tall, A. (1980). Studies on the transfer of phosphatidylcholine from unilamellar vesicles into plasma high density lipoproteins in the rat. *Journal of Lipid Research, 21*(3), 354–363.
119. Tall, A., Abreu, E., & Shuman, J. (1983). Separation of a plasma phospholipid transfer protein from cholesterol ester/phospholipid exchange protein. *Journal of Biological Chemistry, 258*(4), 2174–2180.
120. Jauhiainen, M., et al. (1993). Human plasma phospholipid transfer protein causes high density lipoprotein conversion. *Journal of Biological Chemistry, 268*(6), 4032–4036.
121. Tu, A.-Y., Nishida, H. I., & Nishida, T. (1993). High density lipoprotein conversion mediated by human plasma phospholipid transfer protein. *Journal of Biological Chemistry, 268*(31), 23098–23105.
122. Kostner, G., et al. (1995). Human plasma phospholipid transfer protein accelerates exchange/transfer of α-tocopherol between lipoproteins and cells. *Biochemical Journal, 305*(2), 659–667.
123. Assmann, G., & Gotto Jr, A. M. (2004). HDL cholesterol and protective factors in atherosclerosis. *Circulation, 109*(23_suppl_1), III-8-III-14.
124. Barter, P. J., et al. (2004). Antiinflammatory properties of HDL. *Circulation Research, 95*(8), 764–772.
125. Van Tol, A. (2002). Phospholipid transfer protein. *Current Opinion in Lipidology, 13*(2), 135–139.
126. Huuskonen, J., et al. (2001). The impact of phospholipid transfer protein (PLTP) on HDL metabolism. *Atherosclerosis, 155*(2), 269–281.
127. Van Haperen, R., et al. (2008). Elevated expression of phospholipid transfer protein in bone marrow derived cells causes atherosclerosis. *PLoS One, 3*(5).
128. Jiang, X.-C., et al. (1999). Targeted mutation of plasma phospholipid transfer protein gene markedly reduces high-density lipoprotein levels. *The Journal of Clinical Investigation, 103*(6), 907–914.
129. Hynes, R. O. (1992). Integrins: Versatility, modulation, and signaling in cell adhesion. *Cell, 69*(1), 11–25.
130. Chen, H., et al. (2000) Molecular characterization of human tensin. *Biochemical Journal, 351*(2), 403–411.
131. Miyamoto, S., Akiyama, S. K., & Yamada, K. M. (1995). Synergistic roles for receptor occupancy and aggregation in integrin transmembrane function. *Science, 267*(5199), 883–885.
132. Davis, S., et al. (1991). Presence of an SH2 domain in the actin-binding protein tensin. *Science, 252*(5006), 712–715.
133. Lo, S. H. (2004). Tensin. *The International Journal of Biochemistry & Cell Biology, 36*(1), 31–34.
134. Chen, H., et al. (2002). Tensin1 and a previously undocumented family member, tensin2, positively regulate cell migration. *Proceedings of the National Academy of Sciences, 99*(2), 733–738.
135. Lo, S. H., et al. (1994). Interactions of tensin with actin and identification of its three distinct actin-binding domains. *The Journal of Cell Biology, 125*(5), 1067–1075.
136. Lo, S. H., Weisberg, E., & Chen, L. B. (1994). Tensin: A potential link between the cytoskeleton and signal transduction. *BioEssays, 16*(11), 817–823.
137. Geiger, B. (1979). A 130K protein from chicken gizzard: Its localization at the termini of microfilament bundles in cultured chicken cells. *Cell, 18*(1), 193–205.
138. Burridge, K., & Feramisco, J. R. (1980). Microinjection and localization of a 130K protein in living fibroblasts: A relationship to actin and fibronectin. *Cell, 19*(3), 587–595.

139. Bakolitsa, C., et al. (2004). Structural basis for vinculin activation at sites of cell adhesion. *Nature, 430*(6999), 583–586.
140. Gimona, M., et al. (1988). Porcine vinculin and metavinculin differ by a 68-residue insert located close to the carboxy-terminal part of the molecule. *The EMBO Journal, 7*(8), 2329–2334.
141. Molony, L., & Burridge, K. (1985). Molecular shape and self-association of vinculin and metavinculin. *Journal of Cellular Biochemistry, 29*(1), 31–36.
142. Bass, M. D., et al. (1999). Talin contains three similar vinculin-binding sites predicted to form an amphipathic helix. *Biochemical Journal, 341*(2), 257–263.
143. Bois, P. R., et al. (2005). Structural dynamics of α-actinin-vinculin interactions. *Molecular and Cellular Biology, 25*(14), 6112–6122.
144. Groesch, M. E., & Otto, J. J. (1990). Purification and characterization of an 85 kDa talin-binding fragment of vinculin. *Cell Motility and the Cytoskeleton, 15*(1), 41–50.
145. Brindle, N. P., et al. (1996). The focal-adhesion vasodilator-stimulated phosphoprotein (VASP) binds to the proline-rich domain in vinculin. *Biochemical Journal, 318*(3), 753–757.
146. Bays, J. L., & DeMali, K. A. (2017). Vinculin in cell–cell and cell–matrix adhesions. *Cellular and Molecular Life Sciences, 74*(16), 2999–3009.
147. Carisey, A., & Ballestrem, C. (2011). Vinculin, an adapter protein in control of cell adhesion signalling. *European Journal of Cell Biology, 90*(2–3), 157–163.
148. DeMali, K. A., Barlow, C. A., & Burridge, K. (2002). Recruitment of the Arp2/3 complex to vinculin: Coupling membrane protrusion to matrix adhesion. *The Journal of Cell Biology, 159*(5), 881–891.
149. Mandai, K., et al. (1999). Ponsin/SH3P12: An l-afadin–and vinculin-binding protein localized at cell–cell and cell–matrix adherens junctions. *The Journal of Cell Biology, 144*(5), 1001–1018.
150. Kioka, N., et al. (1999). Vinexin: A novel vinculin-binding protein with multiple SH3 domains enhances actin cytoskeletal organization. *The Journal of Cell Biology, 144*(1), 59–69.
151. DeMali, K. A. (2004). Vinculin–a dynamic regulator of cell adhesion. *Trends in Biochemical Sciences, 29*(11), 565–567.
152. Ziegler, W. H., Liddington, R. C., & Critchley, D. R. (2006). The structure and regulation of vinculin. *Trends in Cell Biology, 16*(9), 453–460.
153. Chen, H., et al. (2005). Spatial distribution and functional significance of activated vinculin in living cells. *The Journal of Cell Biology, 169*(3), 459–470.
154. Johnson, R. P., & Craig, S. W. (1995). F-actin binding site masked by the intramolecular association of vinculin head and tail domains. *Nature, 373*(6511), 261–264.
155. Johnson, R. P., & Craig, S. W. (1994). An intramolecular association between the head and tail domains of vinculin modulates talin binding. *Journal of Biological Chemistry, 269*(17), 12611–12619.
156. Zamir, E., & Geiger, B. (2001). Molecular complexity and dynamics of cell-matrix adhesions. *Journal of Cell Science, 114*(20), 3583–3590.
157. Kanchanawong, P., et al. (2010). Nanoscale architecture of integrin-based cell adhesions. *Nature, 468*(7323), 580–584.
158. Thievessen, I., et al. (2013). Vinculin–actin interaction couples actin retrograde flow to focal adhesions, but is dispensable for focal adhesion growth. *Journal of Cell Biology, 202*(1), 163–177.
159. Mierke, C. T., et al. (2008). Mechano-coupling and regulation of contractility by the vinculin tail domain. *Biophysical Journal, 94*(2), 661–670.
160. Thompson, P. M., et al. (2014). Identification of an actin binding surface on vinculin that mediates mechanical cell and focal adhesion properties. *Structure, 22*(5), 697–706.
161. Ezzell, R. M., et al. (1997). Vinculin promotes cell spreading by mechanically coupling integrins to the cytoskeleton. *Experimental Cell Research, 231*(1), 14–26.
162. Critchley, D. R. (2000). Focal adhesions–the cytoskeletal connection. *Current Opinion in Cell Biology, 12*(1), 133–139.

163. Subauste, M. C., et al. (2004). Vinculin modulation of paxillin–FAK interactions regulates ERK to control survival and motility. *The Journal of Cell Biology, 165*(3), 371–381.
164. Coll, J., et al. (1995). Targeted disruption of vinculin genes in F9 and embryonic stem cells changes cell morphology, adhesion, and locomotion. *Proceedings of the National Academy of Sciences, 92*(20), 9161–9165.
165. Turner, C. E. (2000). Paxillin interactions. *Journal of Cell Science, 113*(23), 4139–4140.
166. Cohen, D. M., et al. (2005). Two distinct head-tail interfaces cooperate to suppress activation of vinculin by talin. *Journal of Biological Chemistry, 280*(17), 17109–17117.
167. Hüttelmaier, S., et al. (1998). The interaction of the cell-contact proteins VASP and vinculin is regulated by phosphatidylinositol-4, 5-bisphosphate. *Current Biology, 8*(9), 479–488.
168. Rottner, K., Hall, A., & Small, J. (1999). Interplay between Rac and Rho in the control of substrate contact dynamics. *Current Biology, 9*(12), 640-S1.
169. Delanoe-Ayari, H., et al. (2004). Membrane and acto-myosin tension promote clustering of adhesion proteins. *Proceedings of the National Academy of Sciences, 101*(8), 2229–2234.
170. Balaban, N. Q., et al. (2001). Force and focal adhesion assembly: A close relationship studied using elastic micropatterned substrates. *Nature cell Biology, 3*(5), 466–472.
171. Giannone, G., et al. (2003). Talin1 is critical for force-dependent reinforcement of initial integrin–cytoskeleton bonds but not tyrosine kinase activation. *The Journal of Cell Biology, 163*(2), 409–419.
172. Zaidel-Bar, R., et al. (2003). Early molecular events in the assembly of matrix adhesions at the leading edge of migrating cells. *Journal of Cell Science, 116*(22), 4605–4613.
173. Munevar, S., Wang, Y.-L., & Dembo, M. (2004). Regulation of mechanical interactions between fibroblasts and the substratum by stretch-activated Ca2+ entry. *Journal of Cell Science, 117*(1), 85–92.
174. Tumbarello, D. A., Brown, M. C., & Turner, C. E. (2002). The paxillin LD motifs. *FEBS Letters, 513*(1), 114–118.
175. López-Colomé, A. M., et al. (2017). Paxillin: A crossroad in pathological cell migration. *Journal of Hematology & Oncology, 10*(1), 1–15.
176. Critchley, D. (2004). Cytoskeletal proteins talin and vinculin in integrin-mediated adhesion. *Biochemical Society Transactions, 32*(5), 831–836.
177. Nayal, A., Webb, D. J., & Horwitz, A. F. (2004). Talin: An emerging focal point of adhesion dynamics. *Current Opinion in Cell Biology, 16*(1), 94–98.
178. Calderwood, D. A. J. J. o. c. s. (2004). Integrin activation. *Journal of Cell Science, 117*(5), 657–666.
179. Devenport, D., et al. (2007). Mutations in the drosophila αPS2 integrin subunit uncover new features of adhesion site assembly. *Developmental Biology, 308*(2), 294–308.
180. Critchley, D. R., & Gingras, A. R. (2008). Talin at a glance. *Journal of Cell Science, 121*(9), 1345–1347.
181. Zaidel-Bar, R., et al. (2007). Functional atlas of the integrin adhesome. *Nature Cell Biology, 9*(8), 858–867.
182. Ratnikov, B., et al. (2005). Talin phosphorylation sites mapped by mass spectrometry. *Journal of Cell Science, 118*(21), 4921–4923.
183. Gingras, A. R., et al. (2005). Mapping and consensus sequence identification for multiple vinculin binding sites within the talin rod. *Journal of Biological Chemistry, 280*(44), 37217–37224.
184. Gingras, A. R., et al. (2008). The structure of the C-terminal actin-binding domain of talin. *The EMBO Journal, 27*(2), 458–469.
185. Senetar, M. A., Moncman, C. L., & McCann, R. O. (2007). Talin2 is induced during striated muscle differentiation and is targeted to stable adhesion complexes in mature muscle. *Cell motility and the Cytoskeleton, 64*(3), 157–173.
186. Humphries, J. D., et al. (2007). Vinculin controls focal adhesion formation by direct interactions with talin and actin. *The Journal of Cell Biology, 179*(5), 1043–1057.

Chapter 7
Bioactivity and Osteogenic Features

Abstract By obtaining detailed and comprehensive knowledge about the internal processes of the body, it becomes possible to utilize bone tissue engineering to create implants and biological materials that possess distinct and suitable characteristics. The crucial characteristics for developing a bone tissue engineering device using bioceramics include the capacity to integrate with bone tissue (osteointegration), stimulate the growth of new bone cells (osteoinductivity), osteoconductivity, be compatible with living tissue (biocompatibility), break down naturally over time (biodegradability), and promote the formation of new blood vessels (angiogenesis). Osteoconduction is the process when an extracellular response is specifically activated on the surface of a material. On the other hand, osteoinduction can be initiated by a chemical that induces reactions both outside and inside the cells at the interface between the tissue and the cell. Biodegradable and porous scaffolds play a crucial role in tissue engineering. Furthermore, it is imperative to employ supplementary methodologies to promote the development of fresh blood vessels to ensure the long-lasting nature of substantial tissue-engineered constructs. Incorporating antibacterial properties into the material might be a highly advantageous characteristic in the design of these materials. Various strategies employed to prevent infections associated with implants included the administration of perioperative systemic antibiotics, implementation of sterile procedures, and conducting surgeries in a controlled and antiseptic environment. The deposition of an apatite layer on the active surface of certain ceramics improves the chemical interaction with adjacent tissues. Various techniques can be employed to generate these coatings. Furthermore, the significance of nanotechnology in the advancement of these materials must not be overlooked. These features are elucidated in this chapter.

Abbreviations

5-ASA	5-aminosalicylic acid
Ake	Akermanite
AMR	Antimicrobial resistance

R. Gholami and S. M. Naghib, *Nanobioceramics for Bone Tissue Engineering and Regenerative Biomedicine*, Biological and Medical Physics, Biomedical Engineering,
https://doi.org/10.1007/978-981-96-0041-0_7

ASCs	Adipose-derived stem cells
bFGF	Basic fibroblast growth factor
Bre	Bredigite
BTE	Bone tissue engineering
CaP	Calcium phosphate
CMCs	Ceramic matrix nanocomposites
CNTs	Carbon nanotubes
Dio	Diopside
GLAD	Glancing angle deposition
HA	Hydroxyapatite
hBMSCs	Human bone marrow mesenchymal stem cells
HCA	Hydroxycarbonate apatite
hFOB	Human fetal osteoblastic cells
MRSA	Methicillin-resistant *S. aureus*
MSCs	Mesenchymal stem cells
MSN	Mesoporous silica nanoparticle
nHA	Nanohydroxyapatite
NO	Nitric oxide
NOS	Nitric oxide synthase
PCL	Polycaprolactone
PLA	Polylactic acid
PLGA	Poly(lactic-co-glycolic acid)
PMCs	Polymer matrix ceramic nanocomposites
PMMA	Poly(methyl methacrylate)
PVD	Photovoltaics
RF	Radiofrequency
ROS	Reactive oxygen species
SBF	Simulated body fluid
SWCNTs	Single-walled carbon nanotubes
TCP	Tricalcium phosphate
TGs	Templated glasses
THR	Total hip replacement
VEGF	Vascular endothelial growth factor
β-TCP	β-tricalcium phosphate

7.1 Introduction

It is necessary for biomaterials to contain certain key properties in order for them to properly fulfill their function in bone tissue engineering. These characteristics include whether the biomaterials are permanent or biodegradable, natural or manufactured. These features include biocompatibility, the capacity to induce bone formation (osteoinductivity), the ability to stimulate bone development (osteoconductivity), the ability to integrate smoothly with existing bone, the presence of a porous structure, and the ability to demonstrate mechanical compatibility with natural bone. The extracellular matrix undergoes modifications to enhance its integration with the surrounding native tissue. These materials serve as the interface for responding to physiological changes in the living organism. In addition, they provide structural guidance, mechanical stability, and locations for anchoring cells. Calcium phosphate ceramics and bioactive glasses have a drawback in that for osteoconduction to occur, the materials must be near the host bone. Nanotechnology may be used as an alternate method for creating porous bioceramics that have improved bioactivity, resorbability, and mechanical strength [1–6].

It is essential to incorporate biomimetic qualities to guarantee that the newly implanted material elicits the necessary biological response and enables the cells adhering to the scaffolds to function in a manner that closely resembles physiological settings. As a result, the development of a coating that exhibits a diverse array of surface gradient patterns and varying degrees of submicron and nanoscale roughness is highly intriguing for biomimetic applications [7, 8]. There has been a substantial amount of interest in the modification of nanotopography over the course of the last several years [9–12]. As a direct result of this, techniques for film deposition have been devised with the purpose of enhancing the inherent morphological characteristics of the material [13]. Tissue regeneration and bioactivity reactions are contingent upon the controlled dissolution of bioceramics. In this regard, TCP is distinguished by a high rate of degradation, which accelerates the substance's absorption and induces an immune response [14, 15]. The chemical stability of stoichiometric HA is greater than that of TCP, which restricts its ability to dissolve in living organisms [16–18].

The development of a biomaterial that replicates the functionality of an autograft is a significant undertaking. Bone is synthesized through a multiphase process by osteoblasts, which are generated from stem and progenitor cells. The proteins found in the natural bone matrix can draw mesenchymal stem cells (MSCs) from the bone marrow and periosteum to the site of repair. The growth and multiplication of MSCs are facilitated by this attraction, which results in the production of osteoblasts [1]. Porous hydroxyapatite (HA) implants are intended to address minor bone defects and are incapable of supporting significant burdens due to a substantial reduction in strength as the number of openings increases. Consequently, the development of HA-based bone and organ surrogate materials is contingent upon the development of scaffolds or matrices that exhibit a suitable macro- and microstructural morphology. The following are included in this category: biocompatibility, mechanical strength, biodegradability, osteoconductivity, pore size, and pore interconnectivity [19].

Nanotechnology is distinguished by its focus on nano-bioceramics and its ability to modify the structure and chemistry of scaffolds. The development of innovative bioceramics systems has broadened the study field, leading to enhanced tissue engineering implants and biosensors. The design, development, and utilization of new bioceramic materials with specific desired properties (such as bioactivity, strength, and toughness) pose a significant challenge for material scientists and medical professionals working on the next generation of biomaterials for complex tissue engineering

[19]. The existing techniques used for the fabrication of biomimetic nanofibrous scaffolds include many drawbacks. Printing nanofibrous structures with high precision is a significant difficulty for most 3D printers, leading to longer processing times and the need for costly equipment [20].

7.2 Osteointegration

Polymers and ceramics have been investigated and used as substitutes for bone [21]. Glass–ceramics may be valuable bioactive materials for both soft and hard tissue engineering if the appropriate combination is chosen, due to their ideal chemical reactivity. Consequently, they may be divided into two types (A and B) depending on their bioactivity. Ceramics that promote the process of "osteoinduction," often referred to as osteoproduction, are classified as group A. Ceramics that enable "osteoconduction" are categorized as group B. Osteoconduction refers to the activation of an extracellular response only at the surface of a substance [22]. According to research, osteoinduction may be triggered by a chemical that causes reactions both outside and within the cells at the boundary between the tissue and the cell [23].

7.2.1 Osteoconductivity

The substrate scaffold in this process facilitates the rapid dispersion of osteogenic cells and the formation of blood vessels [24]. Advancements in nanotechnology have enabled the creation and manufacture of innovative materials on a tiny level. Due to their structural and biochemical similarities to physiological tissues, these nanoparticles are believed to constitute the next generation of advanced materials. The nanotopological properties of the material also promote favorable biological reactions, hence enhancing the efficiency of tissue regeneration. Nano-bioceramics are very pertinent in bone tissue engineering (BTE) because of their chemical similarity to physiological bone and excellent biocompatibility. The mechanical and biological properties of these materials may be enhanced by nanoscale engineering [25].

Upon introduction into a living environment, serum proteins readily adhere to the surface of a scaffold. Cells interact with the protein layer rather than the scaffold itself. If the physical and chemical properties of a surface are appropriate, cells will adhere and function as planned. However, if these properties are unsuitable, the cells will remain spherical and show little or no adhesion. This phenomenon is caused by the selective adsorption of certain extracellular matrix proteins, including laminin, COL-I, bone sialoprotein, vitronectin, and fibronectin, to the surface of the biomaterial [26]. In order to create and develop BTE structures, it is crucial to understand the hierarchical organization of natural bone. The combination of these characteristics results in the formation of a unique and clearly defined tissue that provides bone with its mechanical properties and promotes the continuous movement, specialization, and growth of cells within a supportive microenvironment [25]. BTE nanoparticles must enhance osteointegration while also providing mechanical and structural integrity. Hematoma forms in the area after a bone fracture, and an immediate inflammatory reaction takes place to prevent infection. Concurrently, the release of osteoinductive growth factors stimulates the differentiation, migration, and proliferation of mesenchymal stem cells (MSCs) and fibroblasts [27].

In order to function as intended, a BTE scaffold requires certain characteristics. First, it should promote the process of osteoinduction by facilitating the movement, growth, and specialization of fibroblasts and MSCs. To facilitate the growth of new blood vessels and promote the formation of bone-like tissue, it is important to stimulate angiogenesis, vascularization, and fibrocartilage mineralization. Additionally, it should support the process of bone formation (osteoconduction). Ultimately, it must guarantee the fusion of healthy bone with the scaffold by promoting the formation of new bone tissue in a controlled manner. To effectively reconstruct, it is necessary to pass mechanical stresses through the scaffold, establish enough strength between the scaffold and bone surfaces, and prevent stress shielding effects that might lead to implant loosening [28]. In order to meet these requirements, many approaches have been devised, such as using bioceramics to stimulate the formation of new bone tissue [29], and utilizing porous scaffolds that include growth factors to facilitate the development of blood vessels and new bone tissue [30, 31].

Nano-bioceramics address the limitations of traditional biomaterials by enabling precise control of cellular activity at the nanoscale. This enables the exploration of innovative concepts and hypotheses, hence enhancing our comprehension of bone regeneration. Enhanced stability and desired results may be achieved by including factors that affect the physiological response at each step of bone regeneration via the use of BTE scaffolds made with nano-bioceramics [25]. Hydroxyapatite may be replaced with anionic or cationic materials to enhance osteointegration and bioactivity because of its unique properties. For instance, the presence of Zn-substituted nano-HA inhibits the activity of Staphylococcus aureus, enhances the process of bone formation, and promotes the growth of mesenchymal stem cells derived from adipose tissue [32]. Replacing HA thin films with nanocrystalline silicon enhances the production of extracellular matrix, cell adhesion, and proliferation [33]. The study on nano-HA with carbonate and zinc substitution reveals that the equilibrium of substituted ions has an impact on the in vitro apatite formation process [34].

Crystallinity is a crucial element of nano-bioceramics. For example, as compared to amorphous calcium phosphate, extremely crystalline nano-HA thin films have been shown to enhance cell proliferation [35]. Ensuring the optimal cellular responses requires precise control over the size and phase of nano-bioceramics, however further study is necessary [25]. β-Tricalcium phosphate (β-TCP) may be used to fabricate nanostructured scaffolds that are biodegradable and are absorbed by the body after healing. This promotes osteointegration and bone remodeling [36]. Furthermore, β-TCP nanoparticles have the potential to serve as carriers for delivering drugs in medical treatments such as antibacterial, anti-tumor, and anti-inflammatory therapies [37].

In future, orthopedic implants might be enhanced with nano-calcium phosphate coatings that provide osteoconductivity and the ability to transport drugs. This would increase the lifespan of the implant and speed up the healing process [25]. A notable advancement is the fabrication of multi-material nanocomposite scaffolds. These nanocomposites provide improved characteristics. Some examples of these materials include ceramic matrix nanocomposites (CMCs) and polymer matrix ceramic nanocomposites (PMCs). PMCs' nanobioceramic-reinforced polymer matrices provide exceptional rigidity, resistance to fractures, and ability to deform like bone, while CMCs' nano-bioceramic matrices are strengthened with other materials [25].

SYNTHETBONE is a notable advancement in the field of bone fillers. It is composed of bioactive glass–ceramics, namely bioglass (45S5), biphasic HA-TCP, TCP, and HA. A clinical investigation has shown the effectiveness of this composite material in repairing bone abnormalities, even the most severe skull deformities, without any instances of implant rejection. Experiments conducted in living organisms demonstrate that it stimulates the formation of new bone tissue and the breakdown of implants [25]. To achieve long-term implant success and osteointegration, the objective is to combine innovative nano-bioceramics with pre-existing biomaterials in order to recreate a biomimetic environment. Prior to the implementation of nanocomposites in clinical environments, several unresolved issues need to be addressed. These include enhancing the strength of the implant and ensuring that it maintains its phenotypic expression. Additionally, there is a need to establish vascularization and angiogenesis throughout the implant to facilitate nutrient exchange and the delivery of biomolecular cues in a spatial–temporal manner, which is crucial for proper mechanotransduction [25]. Nano-bioceramics show considerable potential for drug delivery due to their capacity to release therapeutic chemicals in a controlled manner via mesoporous nanostructures and functionalized pore walls [25]. Advancements in sol–gel synthesis methods have made it possible to manufacture nano-bioceramics with various compositions, crystallinity, nanostructures, and particle sizes at low temperatures while maintaining control over the nanostructures [38]. Nanostructured surfaces and smaller crystals have a higher specific surface area, leading to increased bioactivity due to their faster disintegration compared to micro-sized crystals. Nanosized bioceramics boost chemical interaction with newly produced bones, facilitate cell adhesion and differentiation, and accelerate the formation of the physiologically active apatite layer [39].

Like the hydroxyapatite crystals found in human tissues, nanoparticles of nanohydroxyapatite (nHA) measuring 30 nm in size enhance the performance of osteoblasts and the process of biomineralization [40–43]. When bioactive glass nanoparticles are compared to micrometric equivalents, they promote higher rates of mineralization and improve the mechanical and bioactive properties in composite tissue engineering [44–46]. A novel cohort of bioactive glasses with well-organized mesoporosity has been synthesized by the integration of supramolecular chemistry and sol–gel methods, resulting in enhanced effectiveness [47, 48]. In summary, nanobioceramics demonstrate greater bioactivity compared to their counterparts in the micro size range. Their bioactive behavior is influenced by factors such as their size, shape, internal nanostructure, composition, crystallinity, surface area, nanoporosity, and internal nanostructure. The precise nanoscale features that govern bioactivity and their impact on the bioactive response remain unclear, as stated by Covarrubias et al. [48]. Recent advancements in improving the osteoconductivity of titanium implants have led to significant improvements in their ability to integrate with bone tissue, known as osseointegration. Acid-etching and sandblasting are two techniques that create small-scale changes on the surface, enhancing its ability to support bone growth and eliminating chemical impurities. In order to preserve the osteoconductive properties of the surface, it is essential to provide sufficient care [49–54]. To enhance osteoconductivity, implant surfaces might be coated with substances that mimic bone minerals, such as calcium phosphate [51, 55]. Most biocompatible ceramics have the property of being osteoconductive, whereas absorbable ceramics have the property of being osteoinductive [21].

7.2.2 *Osteoinductivity*

Osteoinduction refers to the recruitment of osteoprogenitor cells to the location where bone grafting is performed, stimulating the transformation of undifferentiated and mesenchymal cells into osteoblasts [24]. Studies have shown that the combination of collagen mineral composites in porous ceramics with autologous bone marrow or marrow cells may stimulate osteogenesis [21, 56, 57]. Polymers and ceramics may be classified into three categories based on their degradation rate: bioabsorbable (also called biodegradable), semi-absorbable, or non-absorbable (as shown in Table 7.1). Several factors, including porosity, surface quality, size, and size distribution, influence the rate at which bone materials absorb information. When non-absorbable substances are broken down into small particles that may be consumed by macrophages, they can be absorbed by the body. In order to ascertain the absorption capacity and rate of a material, it is crucial to offer a precise description of its state [21].

It is crucial to conduct a comprehensive investigation of degradation byproducts since they might potentially affect cellular functioning [21]. Until the implants are modified with a clinically approved osteogenic agent, such as one of the bone morphogenetic proteins (BMPs), which provide the implants the ability to stimulate bone growth, it is improbable that any progress in osteogenic activity will occur. BMP-2 and BMP-7 have clinical uses in promoting bone non-union healing (BMP-7) [58, 59] and enhancing the speed and effectiveness of spinal fusion (BMP-2)

Table 7.1 Classification of ceramics based on their absorption properties [21]

Materials	Duration	Classification
$CaSO_4$, CaP compounds	< Months	Absorbable
Glass–ceramic (Bioglass®)	> Years	Semi-absorbable
Diamond, ZrO_2, Al_2O_3	> Many years or permanent	Non-absorbable

[60, 61]. Currently, the agents are either coupled with a carrier that they gradually absorb, or they are provided in a freely soluble form. For optimal effectiveness, it is recommended to use high doses, often in the milligram range [62, 63]. Consequently, several medication delivery methods are deficient in terms of efficiency and are not economically viable [64, 65]. In addition, there have been efforts to directly bind BMPs to the surface of the implant. However, this has not led to an enhancement in osteoinductivity. Indeed, no progress is expected. Due to the burst-release kinetic profile of an adsorbed protein depot, an osteogenic substance is rapidly released into the peri-implant region to induce a long-lasting osteoinductive response [55, 66, 67]. The only technique to evaluate osteointegration is by measuring the neoformation of bone in the "osteoinductive space," which refers to the region between the implant's surface and the parent bone bed [55].

In order to enhance the structural durability and integration of implants soon after surgery, it is essential to assess the osteoinductive reaction to various methods of delivering BMP-2. To properly analyze the effects of osteoinduction or osteoconduction on the peri-implant regions, it is necessary to establish a clear separation in both time and space. Typically, individual growth factors (BMPs) are included in the existing acellular osteoinductive materials [68]. These procedures need higher than normal concentrations of the morphogen due to the lack of necessary supplementary signals and ECM ligands that enhance their efficiency. This raises questions regarding the safety and expense of these techniques. On the other hand, the ability of the devitalized hypertrophic pattern to induce bone formation depends on a combination of factors that are delivered by an ECM molecular structure and accumulate in appropriate amounts within normal physiological levels. Therefore, the use of devitalized hypertrophic cartilage from apoptosis could be a promising substitute for existing osteoinductive materials. It has the potential to be as effective as cell-based grafts, without the need for controversial allogeneic MSCs or the logistical challenges and expenses associated with using autologous cells. Unlike naturally occurring tissue-derived ECMs, which have attracted considerable therapeutic attention [69], this technique has the capability to augment the engineered patterns in specific proteins while also emulating a systematic procedure for efficient bone tissue formation. This might be achieved by inducing the overexpression of certain factors, such as BMP-2 and VEGF-α, in transformed cells during their cultivation in a laboratory setting. This would result in the preservation of these factors via the initiation of apoptosis and their incorporation into the extracellular matrix [68]. In order to address the needs of particular patients or challenging environmental conditions at the repair site (such as cases involving minor bone losses that only require osteoinduction, or atrophic nonunions that necessitate extensive vascularization), it is necessary to develop modified grafts that have enhanced osteoinductive or angiogenic capabilities [70, 71]. The suggested technique is clearly different from the "smart" ceramic materials that have shown osteoinductive properties in large animal models. This strategy includes embedding different growth factors in the extracellular matrix in a manner that can be adjusted and managed to meet particular therapeutic requirements [68, 72].

7.3 Biodegradability

In recent years, the idea of regeneration has superseded the trend from replacement to repair. The approach may be shown by examining third-generation bioceramics, which are specifically engineered to give cells the necessary framework to facilitate their bone regeneration role. The powerful tool of bone tissue creation is backed by three crucial components: scaffolds, signals (biochemical factors), and cells [73–75]. Mesoporous bioactive glasses, also known as templated glasses (TGs), have a composition comparable to regular sol–gel glasses ($SiO_2 - CaO - P_2O_5$) and possess well-organized mesoporous structures. In addition, they exhibit rapid responses to bioactive stimuli [47, 76–81]. The efficiency of these nanostructured materials as controlled delivery systems is due to their well-organized mesoporous cavity architecture, which allows for the precise release of various therapeutic compounds after containment. The chemistry of silica is mostly influenced by the quantity of silanol groups present on its surface [82]. Thus, to enhance the regulation of the drug loading and release mechanisms, it is feasible to modify the silanol-containing mesoporous walls with various organic groups [83–85]. Mesoporous materials are excellent choices for creating 3D scaffolds for BTE because they have the unique potential to serve a twofold role, promote bioactive reactions, and release drugs in a regulated manner [83]. Additionally, it is possible to affix osteoinductive chemicals onto the surface of scaffolds, acting as signals to promote or enhance the generation of new bone. These substances include proteins, peptides, and growth factors [86].

Mesostructured materials have developed into a broad classification of bioceramics that may meet the needs for bone tissue regeneration and facilitate the regulated delivery of medications at the local level. TGs are an innovative group of nanostructured materials that exhibit faster bioactive properties compared to conventional bioactive glassware. The improved bioactive features have been linked to the rigorous structuring of mesopores that are consistently sized [87]. Mesostructured bioceramics are a very suitable option for initiating the production of 3D macroporous scaffolds for bone tissue assembly. This is because they provide the combined benefits of controlled drug release and enhanced bioactive properties. Special emphasis should be placed on the possibility of chemically bonding osteoinductive substances such as growth factors, proteins, and peptides, to the surface of the 3D scaffolds in this context. This would serve as an attractive stimulus to bone cells, therefore accelerating the process of bone regeneration [87]. The calcium phosphate included in calcium phosphate ceramics is often subjected to sintering at temperatures ranging from 1000 to 1500 °C. The final firing conditions play a crucial role in determining whether the resulting calcium phosphate is in the form of α-TCP or HA. Nevertheless, it is common for both stages to be integrated into the same end product. The phases formed at high temperatures during sintering are not only affected by the sintering temperature, but also by the partial pressure of water in the sintering atmosphere. Hydroxyapatite is a thermally stable phase that may exist at temperatures up to 1360 °C and can form in the presence of water. As the water partial pressure rises, the rate at which tricalcium phosphate or tetracalcium phosphate transforms into HA and the temperature range in which HA stays stable both also increase [1].

However, studies with biodegradable porous TCP implanted at places outside of the skeleton have shown that the material degrades quickly and does not promote bone formation. Extensive study has been carried out on mixed calcium phosphates, which consist of tricalcium phosphate (TCP) combined with either a polymer or hydroxyapatite. The main reason for this is the quick breakdown and inadequate mechanical characteristics of TCP. However, the biocompatibility and bone growth are also influenced by the purity of TCP [88]. Extended durations of absorption or inflammatory responses might lead to design challenges in polymeric materials derived from PLA. These polymer varieties often experience hydrolysis in acidic conditions and degrade more quickly than ceramic materials. During the process of breaking down substances, acidic metabolites may be produced. These acidic metabolites can create a localized acidic environment that hinders the healing of wounds. Moreover, foreign body giant cell reactions are often linked to carriers made from polylactic acid (PLA). In a highly inflammatory setting, there is an increased probability of infection and bone resorption [89].

Subsequently, the construct would be surgically re-implanted in the patient. The scaffold should ultimately disintegrate inside the body into benign degradation byproducts, synchronously with the cells' production of extracellular matrix. Ceramics have intrinsic strength, despite their fragility, and functional scaffolding systems must withstand physiological pressures. Nevertheless, the material's strength significantly diminishes with an increase in porosity, therefore necessitating the advancement of reinforcing technologies. In order to preserve the crucial mechanical qualities of the structure, it is essential that the pace at which the scaffold degrades aligns with the rate at which bone develops via the interconnected capillaries of the ceramic material. Altering the phase and chemical composition of bioactive ceramics has been the subject of a number of investigations, with the goal of regulating the release of certain ions from scaffolds into the tissues that surround them. These alterations have the potential to have a major impact on the osteointegration qualities of the structure. Producing bioceramics that have improved bioactivity, resorbability, and mechanical strength may be accomplished via the use of nanotechnology, which provides an alternate method. The manipulation of materials at the atomic, molecular, and supramolecular levels is what can be accomplished using nanobiotechnology. The compressive strength of the biomaterial is shown to be increased when the particle size is reduced to the nanoscale. In addition, bioceramics with nanograin microstructures exhibit higher levels of bioactivity compared to traditional bioceramics. Decreasing the size of particles leads to a notable rise in the number of grain boundaries, which greatly affects the overall physical and chemical characteristics and especially the qualities connected to the surface. Therefore, it is possible to improve the properties related to surface chemistry, energy, and roughness, such as protein adsorption and dissolution, in order to increase cell adhesion, proliferation, and differentiation [1].

The study in the area of biomaterials for the replacement or repair of stiff tissue is both demanding and captivating. There are two approaches to manipulate cell and tissue activity on biomaterial surfaces: chemically modifying surfaces with biologically active macromolecules or physically changing the surface topography at the nanoscale [90]. Polymers often have lower mechanical strength compared to bone. However, polymers have considerable promise as biodegradable and mechanically better materials for replacing bones. Polymer matrix composites offer potential benefits for bone replacement, including the ability to prevent stress shielding, eliminate the need for a second surgery to remove the implants, and address the problem of metal implants releasing ions, provided that the implants can be made biodegradable [90]. Enhancing the mechanical and bone bonding properties of a polymer may be achieved by incorporating a bioactive component, such as HAP or bioglass particles, into it. The implantation of these composites in vivo stimulates the formation or synthesis of new bone. Ultimately, the implant is replaced by bone tissue after the biodegradable polymer matrix decomposes. Consequently, the weight may progressively shift to the growing bone. The notion of using hydroxyapatite as a reinforcement has been instrumental in the creation of many biodegradable polymer composites, including HA/polyhydroxybutyrate and HA/polylactide [91–93]. Biodegradable and porous scaffolds are essential in tissue engineering. Scaffolds provide cells with physical structures to cling to, multiply, and guide the development of new tissues by imitating the natural extracellular matrix [20].

7.4 Antimicrobial Characteristics of Materials

The use of biomaterials in the treatment of bone defects may increase the susceptibility to infection associated with implants and surgical methods. These ailments are caused by the formation of biofilms, which happen when bacteria attach to the implant site. A fundamental obstacle that must be overcome is the elimination of post-surgical infections. This is because bacterial biofilm has an intrinsic resistance to the majority of treatments. After the biomaterials have been successfully implanted, it is especially important for the tissue to integrate as rapidly as possible in order to reduce the amount of bacterial colonization that occurs on the implant [94]. In every single case of bone infection, antibiotics are used for the objectives of both preventing and treating the infection [95].

Several measures used to avoid infections linked with implants included administering perioperative systemic antibiotics, using sterile processes, and performing operations in a controlled, antiseptic atmosphere. In addition, some surface modification methods have been developed that demonstrate enhanced characteristics, despite their decreased relief. The approaches include plasma chemical oxidation [96], nanostructure coatings [97], anti-adhesive polymeric coatings [98], ceramic coatings [99], and ion insertion [100]. Resolving this problem in a smooth way remains quite tough. Since the inception of nanomedicine, scientists have been increasingly concentrated on devising antibiotic administration and release kinetics tactics that are both specific and controlled. Localized medications may overcome limitations by delivering a concentrated dose of treatment directly to or near the desired target area, achieving optimum effectiveness while minimizing systemic side effects. Specifically, the usage of biomaterials infused with antibacterial drugs has been strongly suggested for directly combating some diseases associated with implants. In the field of orthopedic surgery, antibiotic- or drug-loaded nanofibers are commonly used materials that have been shown to be useful in the treatment of osteomyelitis [101–106]. Any alteration made only to solve or develop an antibacterial problem compromises the bioactive characteristics of the material on numerous levels. An important obstacle in this sector is developing a bio-interactive surface for orthopedic implants that improves biological activity and also inhibits or eliminates bacterial attachment [101].

It is commonly acknowledged that bacteria may attach and multiply on the surface of TCP [107], leading to serious illnesses related to implants. Administering antibiotics either topically or systemically after surgery is one approach to tackling this problem. Calcium phosphate cements have the capability to bind antibiotics and gradually release them, which enables the latter process [108]. Nevertheless, the use of antibiotics is expensive and is associated with the worldwide issue of the rise of bacteria that are more resistant to traditional antibiotic treatments [109]. One novel strategy for addressing the limits of conventional antibiotics is to replace calcium ions in the structure of the TCP with antibacterial cations such as silver ions, copper ions, and zinc ions [110–113]. Copper, when added to cement, has a dual effect: It inhibits the development of Gram-negative bacteria while simultaneously promoting the growth of primary lung fibroblasts, K7M2 bone cells, and E297 glial cells. Copper also enhances the formation of bone cells. On the other hand, it does not have any impact on U87 glial cells and it only demonstrates a modest increase in the proliferation of Gram-positive S. aureus. The cements containing copper showed a far higher antibacterial effect against Gram-negative bacteria and a greater ability to stimulate the growth of mammalian cells, even though they contained less copper than the original granules [112]. The advancement of nanotechnology has enabled the production of HA particles that are smaller than 100 nm, include additional functionalities, and are biodegradable. The intrinsic antibacterial properties of HA nanoparticles have been shown to be effective against Gram-positive and Gram-negative bacterial strains and have been demonstrated by Wu and colleagues [114, 115].

Bacterial colonization during the process of bone tissue regeneration is a significant obstacle that may result in implant failure and tissue infection. Antibiotic treatment is a method used to address this problem. However, there are still difficulties in this area, such as the rise in bacterial resistance caused by the improper use of antibiotics. According to the paper, the introduction of magnesium oxide (MgO) to the HA resulted in a reduction in biofilm formation and bacterial growth. This reduction occurred in a way that was dependent on the concentration of MgO. Compared to regular hydroxyapatite, hydroxyapatite doped with magnesium (Mg-doped HA) showed better antibacterial properties and inhibited the formation of biofilm when the concentration of MgO was equal to or more than 3 weight percent. This Mg-doped HA also showed nontoxic behavior and did not stimulate the creation of reactive oxygen species by the cells [116]. Aside from Mg^{2+} ions, several antimicrobial metal ions may be included into the Hap structure to mitigate the likelihood of bone infections and enhance bone regeneration. Several types of ions, including monovalent ions like Ag^+ and F^-, divalent ions like Zn^{2+}, Cu^{2+}, and Sr^{2+}, and trivalent ions like Ce^{3+} and Ga^{3+}, have been shown to exhibit antibacterial properties against a diverse spectrum of Gram-positive and Gram-negative organisms [117–123]. The latest evaluation of the significant impact of dopant site specificity on the antibacterial efficacy of Zn^{2+}-doped HA against Bhattacharjee et al. [124] did a research on the bacteria Staphylococcus aureus and Escherichia coli. Aureus. The findings indicated that the artificial samples were ineffective in functioning as an antibacterial agent when Zn^{2+} ions were integrated into the OH^- channels of HA. This is a result of the limited leaching of dopants, with a concentration of around 5.28 ppb. However, if Zn^{2+} ions are replaced in the Ca^{2+} sites of HA, they may exhibit antibacterial activities. The dopant is leached at a rate of roughly 413 ppb. Scientists have shown that under some situations, antimicrobial ions may be coupled with other osteogenic ions, such as Sr^{2+} ions, to expedite the bone regeneration process. For instance, the findings indicated that the HA doped with Ce^{3+} showed enhanced antibacterial efficacy, while its biocompatibility and ability to generate apatite were diminished [125]. The samples that were co-doped with both Sr^{2+} and Ce^{3+} demonstrated simultaneous antibacterial activity against Pseudomonas aeruginosa and S. aureus strains, cytocompatibility, the capacity to form apatite, and high thermal stability up to 700 °C. These characteristics are highly desirable for bone implant applications. Although the Sr^{2+}-doped samples exhibited good cytocompatibility, the dual-doped samples showed even better cytocompatibility [114].

Administration of antibiotics directly at the site of infection by healthcare professionals has been proposed as a viable method for effectively treating bone infections [126]. Peppermint essential oil and vancomycin, along with other natural ingredients and pharmaceuticals, were added to both bare and metal-doped HA on this item. By chemically modifying HA with different agents, it leads to a sustained and continuous release of antibiotics in a specific area [127]. Lazic et al. chemically modified rod-shaped HA particles by attaching 5-aminosalicylic acid (5-ASA) to them. This led to the chemical bond formation between the surface of HA and the nearby carboxyl and hydroxyl groups of 5-ASA [128]. Because of the bonding that occurs between the amino groups in 5-ASA, it is possible for the amino groups to convert Ag^+ ions into metallic silver. At a concentration of 1 mg L^{-1} in the environment, the samples demonstrate a level of antibacterial activity that is considered to be suitable for human consumption [114]. Studies have indicated that incorporating inorganic antibacterial metallic ions, such as Sr, Zn, Cu, and Ag, into the HA lattice may enhance its antibacterial characteristics [129–136]. Nevertheless, there exists a concrete approach for developing antimicrobial surfaces. Research has shown that the use of nanopatterns is an effective approach to controlling the destiny of stem cells. High aspect ratio nanopatterns have the capability to eradicate bacteria and prevent the formation of biofilms, which is an additional benefit [137].

Making nanostructured thin coatings using the glancing angle deposition (GLAD) arrangement is a potentially beneficial method. The application of dielectric or metallic coatings for optical systems has been the main usage of this technique.

The field of biomaterials has just lately begun to use this technique [138]. The deposition of patterned structures in a variety of sizes and forms, including inclined columnar-like formations with modifications at the nanoroughness level, is often made possible by the growth direction of these coatings [13]. Most studies have used the GLAD strategy, using line-of-sight deposition techniques such as thermal evaporation. However, the self-shadowing phenomena may be beneficial for other PVD processes, such as magnetron sputtering [13]. When it comes to the application of bioactive coatings, the photovoltaics (PVD) sector has recently introduced a unique approach known as radiofrequency (RF) magnetron sputtering. Magnetron sputtering is often used in the production of coatings that are suitable for a variety of applications. Scientists are fascinated by this technology because it has the ability to alter the physicochemical properties and coating structure by modifying the circumstances under which the deposition occurs [139]. The use of radiofrequency magnetrons to deposit bioactive calcium phosphate thin films has garnered considerable interest [140, 141]. This technique has the capability to provide CaP coatings that demonstrate strong adherence to the substrate. Additionally, RF magnetron sputtering is generally acknowledged as a precise and regulated method for depositing materials [13].

Despite the remarkable clinical success, there is a significant need for prompt attention to the issue of revision surgery. An important problem that arises is the occurrence of inflammatory osteolysis caused by the debris from wear and tear around hip implants. This often results in the loosening of the acetabular and femoral components of the implants. In addition to osteonecrosis, fractures, and hip dysplasia, this condition is among the several factors that might lead to the need for revision of a total hip replacement (THR) [142]. The occurrence of aseptic loosening is greatly affected by an inadequate level of osseointegration. Therefore, it is crucial to use bioceramic coatings that are properly formulated and applied [143] to avoid premature implant failure [144]. Furthermore, the coatings must possess sufficient mechanical stability to avoid early separation from the implant surface when exposed to physiological stresses related to movement. In order to reduce the danger of postimplantation periprosthetic osteomyelitis, a disease caused by S. In addition to being aureus-resistant, the implant coating should also possess antibacterial capabilities. Aureus. It has been found that none of the coatings now available on the market can completely meet all of the previously listed parameters. This highlights the need of carrying out more research and development to provide cutting-edge and better biomedical coatings for orthopedic implants [144]. Systemic antibiotic therapy often leads to toxicological and unpleasant consequences. Additionally, it is impacted by factors such as the amount, timing, and duration of the medicine. Before being administered systemically, drugs undergo biotransformation and elimination from the body. As a result, they cannot adequately deliver a concentrated amount of medication to the affected area in instances of bone infections with limited blood supply [145].

Targeted use of antibiotics may effectively address the described problems. The local delivery of medicine helps maintain its therapeutic concentration at the site of damage by minimizing the occurrence of undesirable effects [146]. Biofilms become more vulnerable to antibiotics because of their heightened local concentration, allowing them to spread more broadly into poorly vascularized areas of the bone. Additional benefits of the local route include a decreased antibiotic dose, improved absorption, and an extended duration of release. Therefore, it is more beneficial to locally give antibiotics to treat infections related to bones and devices [95]. Currently, in the United States, the principal options for locally administering antibiotics are the polymethylmethacrylate (PMMA) polymeric cement approved by the US FDA and the PMMA beads utilized in Europe [147]. PMMA's main disadvantage is its lack of biodegradability, which requires surgical removal after antibiotic release is finished. A comprehensive investigation has been carried out on biodegradable ways of delivering medicine in order to avoid the need of further interventions. The many classifications of biodegradable antibiotic carriers include bone grafts and replacements, natural and synthetic polymers, and further biodegradable substances [145].

Bioceramics, which are bone substitutes, are a kind of biodegradable material that may be used for bone restoration. They have the potential to release antibiotics at a customizable pace, which can be adjusted to match the rate at which the bone regenerates [95]. Nanoparticles possess distinctive features such as a significant surface area, adjustable porosity, controllable size and shape, and adaptable functioning. These drug delivery systems have the potential to offer sustained, targeted, and stimuli-sensitive release of drugs. They also have a greater drug payload, longer drug half-life, higher drug solubility, and enhanced cellular absorption [148]. As a result of their slower rate of degradation compared to polymers, bioceramic nanoparticles are advantageous for the long-term preservation of pharmaceuticals. Additionally, they have remarkable resilience in situations that include water. In addition, bioceramic nanoparticles may be designed to have crystalline structures and a chemistry that is similar to that of bone [148, 149]. It has been shown that bioceramic nanocoatings improve the differentiation, proliferation, and adhesion of osteoblastic cells [150]. This is because they increase the bioactivity of orthopedic implant surfaces. Electron and magnetron sputtering, electrophoretic, pulsed laser, and ion beam deposition, plasma spraying, and biomimetic precipitation from simulated body fluid (SBF) are some of the techniques that are used in the process of coating metallic implants with bioceramics [95].

Furthermore, nanoscaffolds have the potential to function as biomimetic frameworks for bone regeneration or repair, as well as vehicles for the controlled delivery of drugs. The enormous surface area, moderate degradation rate, and porous structure of 3D nano-bioceramic scaffolds make them extremely well-suited for controlled drug release. This specific use is very advantageous. Electrospinning and anodization are two processes that are often used in the process of generating nanoscaffolds [148]. Table 7.2 enumerates the various forms of calcium phosphates, including cements, scaffolds, particulates, and composite forms the various kinds of surfaces modified nanosilica (with a focus on pH-controlled release of drugs), other BGN and TiO_2 nanoparticles studied as antibiotic carrier and controlled antibiotic delivery for bone infections [95].

Table 7.2 Different types of nano-bioceramics as antibiotic carriers for bone

	Antibiotics	Carriers	Applications
Calcium phosphate nano-bioceramics	Minocycline	Nano-HA/collagen/α-calcium sulfate composite	Antibacterial biomaterial for bone regeneration
	Gentamicin	HA/gelatin nanocomposite	Bone cements with improved drug release
		CaP/PLGA nanoparticles	Osteomyelitis
	Vancomycin	Nano-HA incorporated acrylic cement	MRSA-induced chronic osteomyelitis
		Nano-HA/collagen granules	Controlled DDS
		Nano-HA pellets	MRSA-induced osteomyelitis
	Ceftazidime	Nano-HA/β-TCP ceramic scaffolds	Bacterial biofilm inhibition
	Clindamycin	Nano-HA/PLGA core–shell nanospheres	Controlled DDS
	Fosfomycin, gentamicin, and vancomycin	Microporous HA nanoparticles	Osteomyelitis
	Ciprofloxacin	HA nanoparticles	Osteomyelitis
	Tigecycline	HA nanogranules	Osteomyelitis
Silica nano-bioceramics	Gentamicin sulfate	MSNs/chitosan composite hydrogels	Cartilage regeneration with infection control
		MSNs/chitosan composite hydrogels	Cartilage regeneration with infection control
	Tetracycline	MSNs in carboxymethyl chitosan scaffold	Wound dressing
	Erythromycin	PEG/Chitosan-coated MSNs with core–shell structure	pH-sensitive drug delivery
	Gentamicin	Polyelectrolyte-coated MSNs	pH-sensitive drug delivery
	Amoxicillin	Self-assembled biohybrid layers coated MSNs	Bacterial infections
	Penicillin G	Carboxylic acid grafted silica nanoparticles	MRSA and other drug-resistant infections
Bioactive glass nano-bioceramics	Ampicillin	Chitosan-nano bioactive glass coatings	Long-term drug delivery
		Chitosan-nano bioactive glass coatings	Long-term drug delivery

(continued)

Table 7.2 (continued)

	Antibiotics	Carriers	Applications
	Tetracycline	Mesoporous bioactive glass nanoparticles in polymer scaffolds	Sequential drug delivery
		Nanobioactive glass/ collagen composite	Biomaterial-related infections
	Gentamicin	Nanoporous bioactive glass ultrathin fibers	Controlled drug release
	Sodium ampicillin	Amino functionalized mesoporous bioactive glass nanoparticles	Delivery of therapeutic molecules
Titania nano-bioceramics	Gentamicin	Poly(lactic-co-glycolic acid)/chitosan-coated titania nanotubes	Long-term antibacterial activity
		Chitosan film covered titania nanotubes	Controlled release
		Titania nanotubes	Implant-associated infections
	Enrofloxacin	Silane modified titania nanotubes	Long-acting drug delivery systems
	Cefuroxime	Titania nanotubes	Controlled drug delivery

Adapted with permission from [95]
PLGA poly(lactic-co-glycolic acid), *MSN* mesoporous silica nanoparticle, *MRSA* methicillin-resistant *S. aureus*, *PLGA* poly(lactic-co-glycolic acid), *MRSA* methicillin-resistant *S. aureus*, *HA* hydroxyapatites, *CaP* calcium phosphate, *β-TCP* β-Tricalcium phosphate

The emergence of antimicrobial resistance (AMR) in bacteria presents a grave danger to the substantial progress achieved in healthcare over the last two centuries. The effectiveness of antimicrobial, antiviral, and antifungal medications is reduced by AMR, leading to higher patient death rates and longer hospital admissions for patients. Bacteria, such as S, that are linked to bone infections, often display antimicrobial resistance. Aureus, E. Klebsiella pneumoniae linked to *E. coli* is referenced in source [95]. A strategy to address AMR involves the research and production of more efficacious antibiotics, however this approach is unfortunately both resource-intensive and time-consuming. Nanoantibiotics, also known as antimicrobial nanomaterials, provide a more efficient strategy for tackling antibiotic resistance [95]. The bactericidal effect of Titanium dioxide (TiO_2) nanoparticles has been ascribed to the impairment of the bacterial cell wall and membrane caused by reactive oxygen species (ROS) [151, 152]. TiO_2 has photocatalytic properties and becomes bactericidal when exposed to UV light. The composition of the bacterial cell membrane has an impact on the photocatalytic antibacterial activity. Reactive oxygen species may have a more substantial effect on Gram-negative bacteria than they do on Gram-positive bacteria, according to the findings of several studies [153].

7.5 Angiogenesis in Bioceramic-Based Nanomaterials

To ensure its survival after being implanted in a living organism, a tissue-engineered construct that is larger than what can be sustained by oxygen diffusion alone must undergo vascularization. This process entails the formation of a network of small blood vessels inside the construct, which may provide nutrients to the cells. Angiogenesis has lately been recognized as a major obstacle in tissue engineering [154]. The rate of spontaneous vascular ingrowth from pre-existing blood vessels in the tissue, induced by in vivo signals, is usually a few tenths of micrometers each day. However, this pace is inadequate to provide the cells inside the construct with enough nourishment [155]. Therefore, it is crucial to use additional techniques to stimulate the growth of new blood vessels (angiogenesis) in order to assure the durability of significant tissue-engineered structures [156]. Vascularization is indispensable for the process of bone regeneration. Implants that contain DNA and cells that promote the growth of blood vessels in bones have been demonstrated to stimulate the formation of bone vessels. The process of osteogenesis and osteoinductivity, as well as the formation of Haversian canals, have been observed in calcium phosphate (β-TCP) implants. During bone restoration, bioceramics may be essential for the process of angiogenesis [157]. It has been established via further study on bioactive glasses that there are ionic dissolution products that promote the gene synthesis of basic fibroblast growth factor (bFGF) and vascular endothelial growth factor (VEGF) in fibroblasts, respectively. These results suggest that this specific silicate substance has the capacity to promote angiogenesis [158–161]. There are several angiogenic growth factors that govern the process of angiogenesis, which is the process of generating new blood vessels. These growth factors include VEGF, bFGF, and TGF-β. Endothelial cells are necessary for angiogenesis because they are responsible for the proliferation, migration, and tube formation. These factors have an effect on these processes. Similarly, it is hypothesized that the receptors (R) present on the outer layer of endothelial cells, such as VEGFR, FGFR, and TGFβR, have a role in regulating angiogenesis in living organisms [162–164]. Activation of nitric oxide synthase (NOS) occurs and leads to an increase in nitric oxide (NO) levels upon binding of proangiogenic substances to their receptors. The proliferation, migration, and creation of arteries in endothelial cells are induced by heightened nitric oxide signals in the nucleoli [165]. VEGFR, FGFR, TGFβR, NOS, and NO play a crucial role in angiogenesis [158].

In cases of osteoporotic pathological situations, the process of bone regeneration is greatly hindered due to a deficiency in angiogenesis, osteogenesis, and remodeling [166]. The existing treatments have specific restrictions, including the positioning of vascularized grafts from the iliac crest and fibula, the use of autograft and allograft cancellous bone, and the application of additional bone transport technologies [167]. Autograft harvesting, often done from the iliac crest, is linked to complications at the donor site such as infection, hemorrhage, and anatomical limitations. Furthermore, it is expensive. Genetic differences limit the use of allografts due to the risk of transmitting diseases and triggering an immune response, whereas vascularized grafts are excessively costly [168, 169]. Vascularization is a crucial factor that influences the amount and growth of new bone formation at areas where there are bone defects [170]. Therefore, the process of developing blood vessels inside large bone grafts continues to be a major obstacle in the field of bone tissue engineering. Several techniques have been suggested to improve the development of blood vessels. For instance, one method involves using endothelial progenitor cells or co-cultivating endothelial cells and osteogenic cells to pre-vascularized a scaffold. Another approach is to design a porous nanocomposite scaffold that promotes the growth of blood vessels. Additionally, adipose-derived stromal cells can be used to release vascular endothelial growth factor [30]. Ensuring the success of any transplant requires achieving perfect integration of grafts at several hierarchical levels, including between individual graft components and the surrounding host tissue [171]. According to Mercado-Pagan et al. [171], a practical method for creating vascularized bone tissue structures involves progressively integrating synthetic tissue vascular grafts and cell-rich hydrogels into a porous, stiff frame that promotes bone growth. Because of its inflexible structure, the hydrogel is able to transfer signals, enclose, and sustain cells that are necessary for the production of bone tissue and the growth of tiny blood arteries [172]. Additionally, the hydrogel is able to give increased structural direction and mechanical strength.

Creating a range of bioactive materials with enhanced mechanical strength and the capability to promote osteogenesis and angiogenesis via the controlled release of certain bioactive ions is very desired. Wu et al. produced and analyzed a range of bioceramics including calcium, magnesium, and silicon. These bioceramics, namely diopside (Dio) $\left(CaMgSi_2O_6\right)$, akermanite (Ake) $\left(Ca_2MgSi_2O_7\right)$, and bredigite (Bre) $\left(Ca_7MgSi_4O_{16}\right)$, showed superior mechanical strength in laboratory conditions compared to bioglass [173–176]. In addition, it has been shown that akermanite dissolution extracts have the capacity to induce the differentiation of osteoblasts, human bone marrow mesenchymal stem cells (hBMSCs), and human adipose stem cells into osteoblasts, even without the presence of additional cytokines. Akermanite ceramics have a higher ability to stimulate the growth of new blood vessels and mend bones compared to β-TCP bioceramics. Based on these findings, bioceramics containing calcium, magnesium, and silicon have the potential to be beneficial biomaterials for the regeneration of bone tissue. Bredigite, akermanite, and diopside are bioceramics that consist of calcium, magnesium, and silicon, despite their variations in composition, degradation rate, and ionic concentration when extracted [176]. To create the most efficient bioceramics for bone regeneration and tissue engineering, it is crucial to analyze how varied levels of calcium, magnesium, and silicon ions affect the development of bone cells and the formation of new blood vessels. Additionally, it would be interesting to explore if this particular group of bioceramics, as opposed to just one bioceramic, has the ability to promote both bone formation and blood vessel growth at the same time [177].

The silicate bioceramic degradation's ionic chemicals promote angiogenesis, osteogenic differentiation, and cell proliferation. The concentration of silicon (Si) in the extracts of silicate bioceramic differed considerably from that in the control medium and TCP extract. Upon comparing the three silicate bioceramics, it is clear that Ake and Bre release a higher amount of Si ions and have a stronger stimulating impact on angiogenesis and osteogenesis compared to Dio. The Ca–Mg–Si family of bioceramics has a unique ability to stimulate both bone formation and blood vessel growth, as shown by the aforementioned research results. As a result, they can be utilized as biomaterials for the purpose of tissue engineering and bone regeneration [177]. The akermanite bioceramic is the first silicate bioceramic to exhibit the ability to concurrently promote both bone formation and blood vessel growth. Akermanite bioceramics promote bone regeneration by releasing a suitable concentration of silicon ions when they break down. This mechanism stimulates the growth and formation of new blood vessels by HAECs. Akermanite bioceramics showed a greater increase in neovascularization compared to β-TCP bioceramics after being implanted in vivo. The results indicate that bioceramics made of silicate and magnesium, namely akermanite bioceramics, have the ability to promote angiogenesis. This characteristic may play a role in their ability to improve the regeneration of bone tissue [158].

VEGF and bFGF are often used as angiogenic growth factors to induce angiogenesis in tissue-engineered structures. Unfortunately, the distribution of growth factors may be limited in this way, since TE scaffolds or biomaterials are produced under conditions that are typically unfriendly to proteins. Moreover, controlling the release of proteins in living organisms is difficult and may be risky owing to the intricate in vivo surroundings and the short lifespan of proteins [160]. Some silicate bioactive materials have been shown to have the ability to promote the formation of blood vessels, which might be used instead of recombinant inductive growth factors. As a consequence, there has been a recommendation to use biomaterials to promote the growth of new blood vessels [158, 160]. Studies have shown that bioactive glasses promote the proliferation of endothelial cells and the development of endothelial tubules [161, 178, 179].

According to the findings of many studies, the process of angiogenesis involves the participation of fibroblasts in addition to endothelial cells. Consequently, in order to examine the causes of angiogenesis, a cell co-culture model that is established in the laboratory and combines endothelial cells and fibroblasts has been employed. According to the findings [180], adding fibroblasts to the culture of endothelial cells results in a considerable increase in the endothelial cells' ability to generate new blood vessels. When it comes to the intimate connection that occurs between fibroblasts and endothelial cells during the process of angiogenesis, the paracrine impact is an essential component, as shown by these studies [156]. In conjunction with angiogenic medicines, which are substances that accelerate the process of angiogenesis, adherens molecules are important for the creation of tubular-like network structures that are required for angiogenesis. Vascular endothelial cadherin, also known as VE-cad, is the major molecule that is responsible for cell adhesion in endothelial cells. It is found just at the junctions inside blood arteries where endothelial cells interact with one another. The osteoinductive ability and bioactivity of a scaffold may be influenced by its structure, as demonstrated by prior research. The properties of micropore size, porosity, and interconnectivity are significantly correlated with bone mineralization and vascularization, according to reports [181, 182]. In addition, research has demonstrated that the microscopic texture of a metal implant can regulate the controlled release of substances that stimulate the growth of blood vessels by bone-forming cells, thereby facilitating the formation of new blood vessels in living organisms [183].

Adipose-derived stem cells, also known as ASCs, have been the subject of substantial research in the area of bone tissue engineering. These cells have been considered a viable alternative to bone marrow-derived mesenchymal stem cells, also known as bMSCs. Compared to bMSCs, ASCs have a number of useful benefits. These consist of a less invasive harvesting method, the capacity to generate a greater number of stem cell progenitors from the same volume of tissue, improved cell growth and specialization capabilities, and superior characteristics in the areas of bone development and blood vessel formation. Furthermore, osteogenic differentiated ASCs have been demonstrated to facilitate the development of new bone tissue and blood vessel growth when affixed to decellularized bone allografts. This is in contrast to non-differentiated ASCs and non-differentiated/differentiated bMSCs [184]. Recent research has demonstrated that the osteogenic differentiation process in ASCs can be facilitated in a laboratory environment by utilizing nanostructured topography on metal disks [185, 186]. Highly interconnected macroporous hydroxyapatite scaffolds, which featured surface topographies such as nanosheets, nanorods, and a combination of nanorods and microrods, enhanced the attachment, proliferation, spreading, and osteogenic differentiation of rat adipose-derived stem cells (ASCs). This was in addition to the fact that these scaffolds increased the expression of angiogenic factors. Through the activation of the Akt signaling pathway, the micro-nanohybrid surface demonstrated the most substantial improvement. When applied to a mouse model with critical-sized calvarial lesions, these HAp scaffolds, in conjunction with the micro-nanohybrid surface and ASCs, have the ability to stimulate bone growth as well as the creation of blood vessels. According to the findings of a study, HAp bioceramic scaffolds that have surface alterations that are micro-nanohybrid might potentially act as carriers for adipose stem cells, therefore enabling the integration of ASCs with pre-existing cells and ultimately leading to the creation of vascularized tissue-engineered bone [186].

7.6 Apatite-Forming Ability

Some ceramics and other biomaterials exhibit significant bioactivity due to the formation of an apatite layer on their active surface. This layer enhances chemical interaction with surrounding tissues [187]. Various techniques have been devised to apply bioactive ceramics, such as hydroxyapatite, onto metals like titanium alloys and stainless steel. This results in the creation of robust bioactive materials that are suitable for demanding applications like hip joints [188]. In order for the substance to cling to living bone, apatite, which has resemblance to bone, must form on artificial material surfaces. Under normal conditions, the occurrence of this process is possible due to the natural state of apatite ions being supersaturated in human body fluid [189]. A cell-free simulated body fluid (SBF) with ion concentrations similar to those seen in human blood plasma may be used to replicate this phenomenon [188]. If a synthetic material has effective functional groups on its surface, it will spontaneously generate apatite when placed in SBF [189].

Research has shown the efficacy of functional groups including Ta–OH [190], Nb–OH [191], Zr–OH [192], Ti–OH [193], and Si–OH [194] in promoting the formation of apatite on metal oxide gels [189, 195]. By applying hydroxyapatite coatings to their surfaces, bioinert ceramics such as zirconia and alumina, which do not naturally interact with living bone, may be converted into bioactive materials. These surfaces may be subjected to chemical treatments that induce the formation of hydroxyl groups, hence facilitating the nucleation of apatite in SBF [196, 197]. Recent study has shown that heat and NaOH treatments may facilitate the formation of a bond between living bone and metals with high fracture toughness, such as tantalum [198] and titanium [199, 200]. Although treated titanium metal has a higher elastic modulus than human cortical bone, it is considered a viable material for bone replacement owing to its strong capacity to generate apatite and bond with bone [188, 189].

Several methods may be used to produce hydroxyapatite coatings, including plasma spraying [201], electrochemical deposition [202], radio-frequency magnetron sputtering [203], and biomimetic approaches [204–206]. Functional groups, such as carboxylic, hydrogen phosphate, and sulfonic groups, may assist in the nucleation of apatite on substrates [207]. By immersing substrates in highly concentrated SBF or other solutions with an excess of calcium and phosphate, comparable outcomes may be achieved [208]. Due to their favorable properties of elasticity and hardness, polymeric materials are being considered as potential replacements for bones. However, they often lack the ability to establish an inherent connection with living tissues [209]. Despite their limited deformability and fracture toughness, certain ceramics including glass–ceramic A-W [210], sintered hydroxyapatite [211], and Bioglass [212] are now used in clinical settings as bioactive materials that can bond with living bone [189, 213–215]. Diopside $\left(MgCaSi_2O_6\right)$, a ceramic derived from the $SiO_2-CaO-MgO$ system, has exceptional osteoinductivity and fracture toughness. Diopside has the ability to bind to newly produced bone due to its advantageous degradation rate [216, 217]. Sol–gel and co-precipitation are widely used wet-chemical synthesis methods for creating very pure nanostructured bioceramics that exhibit enhanced surface activity [218–221]. Nevertheless, maintaining the nanostructure intact throughout the process of high-temperature calcination remains a challenging endeavor [222, 223].

A recent advancement involves creating copolymer fiber composites by altering polymer fibers with TiO_2 gel [224] or $CaO - SiO_2$ [225] and immersing them in SBF. These composites consist of apatite and ethylene–vinyl alcohol. Mesoporous bioactive ceramics are used for the fabrication of 3D scaffolds to facilitate bone regeneration. Some examples of these materials include glasses with a certain structure and silica with very small pores [226]. Apatite is generated on the surfaces of bioactive glass by a series of chemical reactions in SBF. These processes include the hydrolysis of silica, ion exchange, and the subsequent deposition of calcium phosphate, leading to the formation of a hydroxycarbonate apatite (HCA) layer [227–229].

7.7 Proliferative Materials

In order to create bone tissue scaffolds, it is essential to have a thorough understanding of the role of biomolecules and the processes involved in the natural formation of bone. Tissue engineering aims to enhance bone regeneration by using osteoprogenitor support, osteoinduction via growth factors, osteoconduction, and osteoblast adhesion [226]. Age-related declines in osteoblast activity are linked to heightened levels of oxidative stress and inflammation [226, 230, 231]. Runx2 is triggered to initiate the formation of osteoblasts, whereas osteoprogenitors are promoted by pathways including Wnt/β-catenin and MAPK [232, 233]. These pathways, in turn, control the differentiation and proliferation of osteoblasts. The MAPK signaling pathway may limit the use of therapeutic interventions to the first phases of inflammation, since it plays a critical role in the early stages of bone healing [234–236]. To assess the biosafety of nanoparticles, it is essential to comprehend their chemical and physical properties in the condition in which they are supplied or produced. The behavior of nano-bioceramics and nanocomposite materials is influenced by the surrounding medium [237]. Some characteristics of nano-bioceramics and nanocomposite materials are given in Table 7.3.

Nanoparticles enhance the performance of biomaterials by affecting integrin binding, promoting protein adsorption and surface energy, and releasing medicines or ions to stimulate osteoblast growth, proliferation, adhesion, and recruitment [238, 239]. In addition, they improve mechanical properties including fracture toughness, shear strength, compressive strength, and tensile strength [240, 241]. In addition, nanoparticles serve as radiopacifiers to improve X-ray imaging [242]. Bioactive ions, including strontium (Sr^{2+}), magnesium (Mg^{2+}), silicon (Si^{2+}), and calcium (Ca^{2+}), enhance the activity of osteoblasts on typically unreactive surfaces [238, 239]. The addition of nanobioglass comprising silicon, calcium, and phosphorus to polymers results in superior bioactivity compared to nanohydroxyapatite (nHA) within the same matrix [242–244]. Strontium ions have the potential to modify nanoparticles like TiO_2 on poly(methyl methacrylate) (PMMA) in order to enhance the formation of osteoblasts [245].

nHA is often included into collagen [246, 247] or carbon nanotubes (CNTs) [248], as well as bioinert injectable bone cements, due to its similarity in structure to bone apatite [249–251]. It may be added either as a mineralized component or used independently. To reduce the formation of fibrous tissue and prevent the loosening of the implant, it is crucial to establish a direct bioactive connection with the host bone [242]. Scientists have replicated the structure of bone by mineralizing nanohydroxyapatite onto collagen, creating an organic–inorganic composite. The incorporation of nHA-collagen into injectable bone cements including calcium sulfate and chitosan enhances the growth and specialization of osteoblasts [242].

Table 7.3 Properties of nano-bioceramics and nanocomposite materials [237]

Nano-bioceramics/ nanocomposite materials	Application	Main results or biocompatibility	Evaluated parameters	Cell lines
Selenium-ferroferric oxide nanocomposites	Bone tissue engineering	Inhibited cell proliferation, induced cell apoptosis	MTT assay	Human osteoblast-like MG-63 cells
AgNP soft nanocomposite	As bactericidal agent	Increased cell viability	Cell viability	HepG2 cells
Nanohydroxyapatite/gelatin/ carboxymethyl chitin (n-HA/ gel/CMC)	Bone tissue engineering	Enhancement of network formation, cell proliferation and adhesion	Cell morphology, MTT assay, and hemocompatibility (protein adsorption, platelet adhesion, and hemolysis)	MG63 osteoblast cells
Poly(3-hydroxybutyrate) (PHB) composite scaffolds	Bone tissue engineering	Considerably increased cell attachment, inducing better osteoconductivity, and improved cell proliferation	Cell proliferation and alkaline phosphatase (ALP) assays	MG63 cell line
Nanohydroxyapatite/gelatin	Bone regeneration	Greater osteoblast proliferation and attachment, more osteoblast differentiation on 2D cell culture; did not alter morphology	Histomorphometry, Alizarin red test	Rat somatic stem cells
Halloysite nanotube (HNT)-doped poly(lactic-co-glycolic acid) (PLGA) composite nanofibers	Drug delivery and tissue engineering	Elevation of cell proliferation and attachment of HNT-doped PLGA scaffolds allow more protein adsorption	SEM and MTT assay	Mouse fibroblast cells
Hydrogel nanocomposites	A potential UV/blue light filtering material for ophthalmic lenses	No harmful influence on the biocompatibility	In vitro cell adhesion test	Lens epithelial cells (LECs)

(continued)

Table 7.3 (continued)

Nano-bioceramics/ nanocomposite materials	Application	Main results or biocompatibility	Evaluated parameters	Cell lines
Thermoresponsive nanocomposites generated by poly(*N*-isopropylacrylamide)	As an embolization agent in cancer treatment	Excellent biocompatibility	Cytotoxicity test	Human umbilical vein endothelial cells
Biogenic silica-metal phosphate nanocomposites (BSMPNs)	Bone tissue engineering	Nontoxic and presented excellent biocompatibility	Cell viability assay	Human mesenchymal stem cells
Mg-Zn-Zr nanocomposites	Biodegradable bone fixation material	Better proliferation and adhesion of cells	In vitro corrosion resistance and cytocompatibility	Osteoblasts
Liquid crystalline epoxy nanocomposites	As fabricated aesthetic brackets for orthodontic treatment	Presented greater microhardness and suitable biocompatibility	Thermocycling and MTT	–
Titania-graphene nanocomposites	As biomimicking agent	More favorable materials to biomimetic bone action	FITR, SEM, confocal microscopy, MTS assay	Human cell lines (MG-63 and AGS) and simulated body fluid
Epoxy/clay nanocomposite	As antimicrobial agent	Excellent cytocompatibility	RBC hemolytic and MTT assays	*Candida albicans* (ATCC10231), *Mycobacterium smegmatis* (ATCC14468), *Escherichia coli* (ATCC 10,536), and Staphylococcus aureus (ATCC11632) strains
	Tissue regeneration	Nontoxic, proliferation and adherence of cells	Histopathologic study of the skin, liver, heart, and brain, and proliferation of dermatocytes without induction of any sign of toxicity to the above organs	Wistar rat

Nanohydroxyapatite enhances both the drug encapsulation efficiency and the role as a carrier for hydrophilic drugs, such as alendronate, in injectable polycaprolactone (PCL) microspheres [249]. Alendronate is recognized for its ability to stimulate osteoblast formation and inhibit osteoclast activity. Compared to directly loading into PCL microspheres, alendronate-loaded nHA provides extended drug release and stimulates the formation of new bone cells [242, 249].

Studies have shown that biomaterials with nanoparticles have a higher level of improvement in cellular activity compared to biomaterials without nanoparticles, even when the nanoparticles themselves do not possess inherent bioactivity [242, 252]. When comparing larger particles to titania and alumina nanoparticles, it is seen that osteoblast proliferation and spreading are more pronounced on the latter. The incorporation of ZrO_2 nanoparticles into PMMA enhances both the adhesion and proliferation of osteoblasts compared to microparticles [242, 253]. In addition, researchers have developed graphene nanosheets and carbon nanotubes modified with HA (Gr-HA and CNTs-HA), which have been shown to enhance the differentiation and proliferation of human fetal osteoblastic cells (hFOB 1.19) [254]. These materials have potential when used as scaffolds in bone tissue creation. Chemically modified single-walled carbon nanotubes (SWCNTs) containing hydroxyapatite assist in the mineralization of bone cells. The addition of carbon nanotubes (CNTs) to hydroxyapatite coatings by plasma spraying leads to an improvement in both crystallinity and fracture toughness. The HA-CNT coating demonstrates biocompatibility with living cells by facilitating cell growth and proliferation [255, 256]. Despite their benefits, several studies indicate that low protein affinity on ZrO-rich scaffolds hinders cell adhesion and proliferation [257, 258]. Calcium zirconate is formed when ZrO and HA undergo sintering [259]. This material has exceptional mechanical robustness and serves as a valuable additive in chitosan scaffolds. Osteopontin synthesis may be facilitated by the release of calcium from bio-cements, which enhances cell adhesion and proliferation [260–262].

7.8 Conclusion

To effectively generate and develop BTE structures, a comprehensive understanding of the hierarchical organization of natural bone is essential. The primary objective of BTE nanoparticles is to augment osteointegration while simultaneously ensuring mechanical and structural integrity. Osteoconduction is the process by which an extracellular reaction is specifically initiated solely on the surface of a material. Osteoinduction is the process of attracting osteoprogenitor cells to the site where bone grafting is done, which triggers the conversion of undifferentiated and mesenchymal cells into osteoblasts. Research suggests that osteoinduction can be initiated by a chemical that induces responses in both the extracellular and intracellular regions at the interface between the tissue and the cell. To meet the requirements of specific patients or difficult environmental conditions at the repair site, it is essential to create customized grafts with improved capacities to stimulate bone formation or promote blood vessel growth. Polymers and ceramics can be categorized into three groups according to their rate of degradation: bioabsorbable (also known as biodegradable), semi-absorbable, or non-absorbable. Scaffolds that are both biodegradable and porous play a crucial role in tissue engineering. Polymer matrix composites have the potential to provide advantages for bone replacement, such as the capacity to minimize stress shielding, avoid the necessity of a second surgery to remove the implants, and resolve the issue of metal implants releasing ions, as long as the implants can be rendered biodegradable. In vivo, the introduction of these composites prompts the generation or production of fresh bone. Eventually, the implant is substituted with bone tissue after the biodegradable polymer matrix disintegrates. Tissue engineering seeks to improve the process of bone regeneration by employing osteoprogenitor support, osteoinduction through growth factors, osteoconduction, and osteoblast adhesion. The utilization of biomaterials in the management of bone abnormalities may heighten the vulnerability to infection linked with implants and surgical techniques. Once the biomaterials have been effectively implanted, it is crucial for the tissue to integrate quickly to minimize bacterial colonization on the implant. Various strategies employed to prevent infections associated with implants included the administration of perioperative systemic antibiotics, implementation of sterile procedures, and conducting surgeries in a controlled and antiseptic environment. Furthermore, several surface alteration techniques have been devised that exhibit improved properties, even when their elevation has been reduced. The methods encompass plasma chemical oxidation, nanostructure coatings, anti-adhesive polymeric coatings, ceramic coatings, and ion insertion. An innovative approach to overcome the limitations of traditional antibiotics involves substituting calcium ions in the TCP structure with antibacterial cations such silver ions, copper ions, and zinc ions. The inherent antibacterial capabilities of HA nanoparticles have demonstrated efficacy against both Gram-positive and Gram-negative bacterial pathogens. Local administration of medication ensures that the therapeutic concentration remains concentrated at the site of injury, while reducing the occurrence of unwanted side effects. Vascularization is a critical determinant that impacts the quantity and development of new bone formation in regions with bone defects. In order for a tissue-engineered construct to survive after being implanted in a living creature, it must undergo vascularization if it is larger than what can be sustained by oxygen diffusion alone. This process involves the development of a network of tiny blood veins within the structure, which can supply nutrients to the cells. Recently, angiogenesis has been identified as a significant barrier in tissue engineering. Implants incorporating DNA and cells that facilitate angiogenesis in bones have been shown to increase bone vessel development.

References

1. El-Ghannam, A. (2005). Bone reconstruction: From bioceramics to tissue engineering. *Expert Review of Medical Devices, 2*(1), 87–101.

2. Kalantari, E., et al. (2018). Nanostructured monticellite: An emerging player in tissue engineering. *Materials Today: Proceedings, 5*(7, Part 3), 15744–15753.
3. Kalantari, E., et al. (2019). Biocomposites based on hydroxyapatite matrix reinforced with nanostructured monticellite ($CaMgSiO_4$) for biomedical application: Synthesis, characterization, and biological studies. *Materials Science and Engineering: C, 105*, 109912.
4. Kalantari, E., & Naghib, S. M. (2019). A comparative study on biological properties of novel nanostructured monticellite-based composites with hydroxyapatite bioceramic. *Materials Science and Engineering: C, 98*, 1087–1096.
5. Kalantari, E., et al. (2018). Nanostructured monticellite for tissue engineering applications—Part I: Microstructural and physicochemical characteristics. *Ceramics International, 44*(11), 12731–12738.
6. Kalantari, E., et al. (2018). Nanostructured monticellite for tissue engineering applications—Part II: Molecular and biological characteristics. *Ceramics International, 44*(12), 14704–14711.
7. Rau, J. V., et al. (2009). Physicochemical investigation of pulsed laser deposited carbonated hydroxyapatite films on titanium. *ACS Applied Materials & Interfaces, 1*(8), 1813–1820.
8. Costa, D. O., et al. (2013). The differential regulation of osteoblast and osteoclast activity by surface topography of hydroxyapatite coatings. *Biomaterials, 34*(30), 7215–7226.
9. Karazisis, D., et al. (2017). The influence of controlled surface nanotopography on the early biological events of osseointegration. *Acta Biomaterialia, 53*, 559–571.
10. Lee, L. C., et al. (2017). Nanotopography controls cell cycle changes involved with skeletal stem cell self-renewal and multipotency. *Biomaterials, 116*, 10–20.
11. Chen, Z., et al. (2017). Nanotopography-based strategy for the precise manipulation of osteoimmunomodulation in bone regeneration. *Nanoscale, 9*(46), 18129–18152.
12. Kim, H. N., et al. (2013). Nanotopography-guided tissue engineering and regenerative medicine. *Advanced Drug Delivery Reviews, 65*(4), 536–558.
13. Prosolov, K. A., et al. (2019). Glancing angle deposition of Zn-doped calcium phosphate coatings by RF magnetron sputtering. *Coatings, 9*(4), 220.
14. Eggli, P., Müller, W., & Schenk, R. (1988). Porous hydroxyapatite and tricalcium phosphate cylinders with two different pore size ranges implanted in the cancellous bone of rabbits. A comparative histomorphometric and histologic study of bony ingrowth and implant substitution. *Clinical Orthopaedics and Related Research*, (232), 127–138.
15. Ducheyne, P., & Qiu, Q. (1999). Bioactive ceramics: The effect of surface reactivity on bone formation and bone cell function. *Biomaterials, 20*(23–24), 2287–2303.
16. Daculsi, G., et al. (1990). Formation of carbonate-apatite crystals after implantation of calcium phosphate ceramics. *Calcified Tissue International, 46*(1), 20–27.
17. Nery, E. B., et al. (1992). Tissue response to biphasic calcium phosphate ceramic with different ratios of HA/βTCP in periodontal osseous defects. *Journal of Periodontology, 63*(9), 729–735.
18. Jarcho, M., et al. (1976). Hydroxylapatite synthesis and characterization in dense polycrystalline form. *Journal of Materials Science, 11*(11), 2027–2035.
19. Das, R. (2016). Application of nanobioceramics in bone tissue engineering. *Nanobiomaterials in hard tissue engineering* (pp. 353–379). Elsevier.
20. Zeinali, R., et al. (2021). Recent progress on biodegradable tissue engineering Scaffolds prepared by thermally-induced phase separation (TIPS). *International Journal of Molecular Sciences, 22*(7), 3504.
21. Park, J. (2009). *Bioceramics: Properties, characterizations, and applications* (Vol. 741). Springer Science & Business Media.
22. Hench, L. (1994). Bioactive ceramics: Theory and clinical applications. *Bioceramics* (pp. 3–14). Elsevier.
23. Kokubo, T., et al. (1985). Mechanical properties of a new type of apatite-containing glass-ceramic for prosthetic application. *Journal of Materials Science, 20*(6), 2001–2004.
24. ur Rehman, I. (2004). Nano bioceramics for biomedical and other applications. *Materials Technology, 19*(4), 224–233.

25. Feng, L., & San Thian, E. (2013). Applications of nanobioceramics to healthcare technology. *Nanotechnology Reviews, 2*(6), 679–697.
26. Grzesik, W. J., & Robey, P. G. (1994). Bone matrix RGD glycoproteins: Immunolocalization and interaction with human primary osteoblastic bone cells in vitro. *Journal of Bone and Mineral Research, 9*(4), 487–496.
27. Dimitriou, R., Tsiridis, E., & Giannoudis, P. V. (2005). Current concepts of molecular aspects of bone healing. *Injury, 36*(12), 1392–1404.
28. Lichte, P., et al. (2011). Scaffolds for bone healing: Concepts, materials and evidence. *Injury, 42*(6), 569–573.
29. Thian, E., & Best, S. (2008). Materials viewpoints in bone regenerative medicine: Progress and prospects. *Materials Science and Technology, 24*(9), 1027–1030.
30. Jabbarzadeh, E., et al. (2008). Induction of angiogenesis in tissue-engineered scaffolds designed for bone repair: A combined gene therapy–cell transplantation approach. *Proceedings of the National Academy of Sciences, 105*(32), 11099–11104.
31. Baylink, D. J., Finkelman, R. D., & Mohan, S. (1993). Growth factors to stimulate bone formation. *Journal of Bone and Mineral Research, 8*(S2), S565–S572.
32. Thian, E., et al. (2013). Zinc-substituted hydroxyapatite: A biomaterial with enhanced bioactivity and antibacterial properties. *Journal of Materials Science: Materials in Medicine, 24*(2), 437–445.
33. San Thian, E., et al. (2006). The response of osteoblasts to nanocrystalline silicon-substituted hydroxyapatite thin films. *Biomaterials, 27*(13), 2692–2698.
34. Kumar, G. S., et al. (2012). Synthesis, characterization and in vitro studies of zinc and carbonate co-substituted nano-hydroxyapatite for biomedical applications. *Materials Chemistry and Physics, 134*(2–3), 1127–1135.
35. Hu, Q., et al. (2007). Effect of crystallinity of calcium phosphate nanoparticles on adhesion, proliferation, and differentiation of bone marrow mesenchymal stem cells. *Journal of Materials Chemistry, 17*(44), 4690–4698.
36. de Groot, K. (1993). Clinical applications of calcium phosphate biomaterials: A review. *Ceramics International, 19*(5), 363–366.
37. Yu, D., et al. (1992). Self-setting hydroxyapatite cement: A novel skeletal drug-delivery system for antibiotics. *Journal of Pharmaceutical Sciences, 81*(6), 529–531.
38. Lukowiak, A., et al. (2013). Bioactive glass nanoparticles obtained through sol–gel chemistry. *Chemical Communications, 49*(59), 6620–6622.
39. Caridade, S. G., et al. (2013). Chitosan membranes containing micro or nano-size bioactive glass particles: evolution of biomineralization followed by in situ dynamic mechanical analysis. *Journal of the Mechanical Behavior of Biomedical Materials, 20*, 173–183.
40. Webster, T. J., et al. (2000). Specific proteins mediate enhanced osteoblast adhesion on nanophase ceramics. *Journal of Biomedical Materials Research: An Official Journal of The Society for Biomaterials, The Japanese Society for Biomaterials, and The Australian Society for Biomaterials and the Korean Society for Biomaterials, 51*(3), 475–483.
41. Webster, T. J., et al. (2000). Enhanced functions of osteoblasts on nanophase ceramics. *Biomaterials, 21*(17), 1803–1810.
42. Ferraz, M., Monteiro, F., & Manuel, C. (2004). Hydroxyapatite nanoparticles: A review of preparation methodologies. *Journal of Applied Biomaterials and Biomechanics, 2*(2), 74–80.
43. Sanosh, K., et al. (2009). Synthesis of nano hydroxyapatite powder that simulate teeth particle morphology and composition. *Current Applied Physics, 9*(6), 1459–1462.
44. Misra, S. K., et al. (2008). Comparison of nanoscale and microscale bioactive glass on the properties of P (3HB)/Bioglass® composites. *Biomaterials, 29*(12), 1750–1761.
45. Valenzuela, F., et al. (2012). Preparation and bioactive properties of novel bone-repair bionanocomposites based on hydroxyapatite and bioactive glass nanoparticles. *Journal of Biomedical Materials Research Part B: Applied Biomaterials, 100*(6), 1672–1682.
46. Vollenweider, M., et al. (2007). Remineralization of human dentin using ultrafine bioactive glass particles. *Acta Biomaterialia, 3*(6), 936–943.

47. Yan, X., et al. (2004). Highly ordered mesoporous bioactive glasses with superior in vitro bone-forming bioactivities. *Angewandte Chemie International Edition, 43*(44), 5980–5984.
48. Covarrubias, C., et al. (2015). The effect of the nanoscale structure of nanobioceramics on their in vitro bioactivity and cell differentiation properties. *Journal of Nanomaterials, 2015.*
49. Buser, D., et al. (1991). Influence of surface characteristics on bone integration of titanium implants. A histomorphometric study in miniature pigs. *Journal of Biomedical Materials Research, 25*(7), 889–902.
50. Gotfredsen, K., Berglundh, T., & Lindhe, J. (2000). Anchorage of titanium implants with different surface characteristics: An experimental study in rabbits. *Clinical Implant Dentistry and Related Research, 2*(3), 120–128.
51. Wong, M., et al. (1995). Effect of surface topology on the osseointegration of implant materials in trabecular bone. *Journal of Biomedical Materials Research, 29*(12), 1567–1575.
52. Albrektsson, T., & Wennerberg, A. (2004). Oral implant surfaces: Part 1—Review focusing on topographic and chemical properties of different surfaces and in vivo responses to them. *International Journal of Prosthodontics, 17*(5).
53. Wall, I., et al. (2009). Modified titanium surfaces promote accelerated osteogenic differentiation of mesenchymal stromal cells in vitro. *Bone, 45*(1), 17–26.
54. Scharnweber, D., et al. (2010). How is wettability of titanium surfaces influenced by their preparation and storage conditions? *Journal of Materials Science: Materials in Medicine, 21*(2), 525–532.
55. Hunziker, E. B., et al. (2012). Osseointegration: The slow delivery of BMP-2 enhances osteoinductivity. *Bone, 51*(1), 98–106.
56. Alvis, M., et al. (2003). Osteoinduction by a collagen mineral composite combined with isologous bone marrow in a subcutaneous rat model. *Orthopedics, 26*(1), 77–80.
57. Ohgushi, H., Goldberg, V. M., & Caplan, A. I. (1989). Heterotopic osteogenesis in porous ceramics induced by marrow cells. *Journal of Orthopaedic Research, 7*(4), 568–578.
58. Kain, M. S., & Einhorn, T. A. (2005). Recombinant human bone morphogenetic proteins in the treatment of fractures. *Foot and Ankle Clinics, 10*(4), 639–650.
59. Schmidmaier, G., et al. (2009). Bone morphogenetic proteins in critical-size bone defects: What are the options? *Injury, 40,* S39–S43.
60. Burkus, J. K., et al. (2009). Six-year outcomes of anterior lumbar interbody arthrodesis with use of interbody fusion cages and recombinant human bone morphogenetic protein-2. *JBJS, 91*(5), 1181–1189.
61. Shimer, A. L., Öner, F. C., & Vaccaro, A. R. (2009). Spinal reconstruction and bone morphogenetic proteins: Open questions. *Injury, 40,* S32–S38.
62. Dickerman, R. D., et al. (2007). Rh-BMP-2 can be used safely in the cervical spine: Dose and containment are the keys! *The Spine Journal, 4*(7), 508–509.
63. Barnes, B., et al. (2005). Lower dose of rhBMP-2 achieves spine fusion when combined with an osteoconductive bulking agent in non-human primates. *Spine, 30*(10), 1127–1133.
64. Glassman, S. D., et al. (2008). The perioperative cost of Infuse bone graft in posterolateral lumbar spine fusion. *The Spine Journal, 8*(3), 443–448.
65. Garrison, K. R., et al. (2007). Clinical effectiveness and cost-effectiveness of bone morphogenetic proteins in the non-healing of fractures and spinal fusion: a systematic review. In *NIHR Health Technology Assessment programme: Executive Summaries.* NIHR Journals Library
66. Liu, Y., et al. (2007). Delivery mode and efficacy of BMP-2 in association with implants. *Journal of Dental Research, 86*(1), 84–89.
67. Liu, Y., et al. (2004). Bone morphogenetic protein 2 incorporated into biomimetic coatings retains its biological activity. *Tissue Engineering, 10*(1–2), 101–108.
68. Bourgine, P. E., et al. (2014). Osteoinductivity of engineered cartilaginous templates devitalized by inducible apoptosis. *Proceedings of the National Academy of Sciences, 111*(49), 17426–17431.
69. Sicari, B. M., et al. (2014). An acellular biologic scaffold promotes skeletal muscle formation in mice and humans with volumetric muscle loss. *Science Translational Medicine, 6*(234), 234ra58–234ra58.

70. Grayson, W. L., et al. (2010). Engineering anatomically shaped human bone grafts. *Proceedings of the National Academy of Sciences, 107*(8), 3299–3304.
71. Garcia, P., et al. (2012). Temporal and spatial vascularization patterns of unions and nonunions: Role of vascular endothelial growth factor and bone morphogenetic proteins. *JBJS, 94*(1), 49–58.
72. Yuan, H., et al. (2010). Osteoinductive ceramics as a synthetic alternative to autologous bone grafting. *Proceedings of the National Academy of Sciences, 107*(31), 13614–13619.
73. Hollister, S. J. (2005). Porous scaffold design for tissue engineering. *Nature Materials, 4*(7), 518–524.
74. Griffith, L. G., & Naughton, G. (2002). Tissue engineering—Current challenges and expanding opportunities. *Science, 295*(5557), 1009–1014.
75. Langer, R., & Vacanti, J. (1993). Tissue engineering. *Science, 260*(5110), 920–926.
76. Zhao, S., Li, Y., & Li, D. (2011). Synthesis of $CaO–SiO_2–P_2O_5$ mesoporous bioactive glasses with high P_2O_5 content by evaporation induced self assembly process. *Journal of Materials Science: Materials in Medicine, 22*(2), 201–208.
77. Yun, H.-s., Kim, S.-e., & Hyeon, Y.-t. (2007). Highly ordered mesoporous bioactive glasses with Im3m symmetry. *Materials Letters, 61*(23–24), 4569–4572.
78. López-Noriega, A., et al. (2006). Ordered mesoporous bioactive glasses for bone tissue regeneration. *Chemistry of Materials, 18*(13), 3137–3144.
79. Yan, X., et al. (2006). The in-vitro bioactivity of mesoporous bioactive glasses. *Biomaterials, 27*(18), 3396–3403.
80. Yan, X., et al. (2005). Mesoporous bioactive glasses. I. Synthesis and structural characterization. *Journal of Non-Crystalline Solids, 351*(40–42), 3209–3217.
81. Izquierdo-Barba, I., & Vallet-Regí, M. (2011). Fascinating properties of bioactive templated glasses: A new generation of nanostructured bioceramics. *Solid State Sciences, 13*(4), 773–783.
82. Vallet-Regí, M. (2006). Ordered mesoporous materials in the context of drug delivery systems and bone tissue engineering. *Chemistry–A European Journal, 12*(23), 5934–5943.
83. Vivero-Escoto, J. L., et al. (2010). Mesoporous silica nanoparticles for intracellular controlled drug delivery. *Small (Weinheim an der Bergstrasse, Germany), 6*(18), 1952–1967.
84. Vallet-Regí, M., Balas, F., & Arcos, D. (2007). Mesoporous materials for drug delivery. *Angewandte Chemie International Edition, 46*(40), 7548–7558.
85. Balas, F., et al. (2006). Confinement and controlled release of bisphosphonates on ordered mesoporous silica-based materials. *Journal of the American Chemical Society, 128*(25), 8116–8117.
86. Chen, F.-M., et al. (2011). New insights into and novel applications of release technology for periodontal reconstructive therapies. *Journal of Controlled Release, 149*(2), 92–110.
87. Vallet-Regi, M., Izquierdo-Barba, I., & Colilla, M. (1963). Structure and functionalization of mesoporous bioceramics for bone tissue regeneration and local drug delivery. *Philosophical Transactions of the Royal Society A: Mathematical, Physical and Engineering Sciences, 2012*(370), 1400–1421.
88. Yokoyama, N., et al. (1989). Application of beta-tricalcium phosphate to periodontal therapy. Part 1. Fundamental studies and biological tests of beta-TCP prepared by mechanochemical method. *Nihon Shishubyo Gakkai Kaishi, 31*(1), 213–223.
89. Sandhu, H. S. (2003). Bone morphogenetic proteins and spinal surgery. *Spine, 28*(15S), S64–S73.
90. Sridhar, T. (2010). Nanobioceramic coatings for biomedical applications. *Materials Technology, 25*(3–4), 184–195.
91. Liu, Q., et al. (1996). Surface modification of hydroxyapatite to introduce interfacial bonding with polyactive TM 70/30 in a biodegradable composite. *Journal of Materials Science: Materials in Medicine, 7*(9), 551–557.
92. Knowles, J., et al. (1992). Development of a degradable composite for orthopaedic use: In vivo biomechanical and histological evaluation of two bioactive degradable composites based on the polyhydroxybutyrate polymer. *Biomaterials, 13*(8), 491–496.

93. Doyle, C., Tanner, E., & Bonfield, W. (1991). In vitro and in vivo evaluation of polyhydroxybutyrate and of polyhydroxybutyrate reinforced with hydroxyapatite. *Biomaterials, 12*(9), 841–847.
94. Gristina, A. G. (1987). Biomaterial-centered infection: Microbial adhesion versus tissue integration. *Science, 237*(4822), 1588–1595.
95. Kumar, T. S., & Madhumathi, K. (2016). Antibiotic delivery by nanobioceramics. *Therapeutic Delivery, 7*(8), 573–588.
96. Diefenbeck, M., et al. (2016). Gentamicin coating of plasma chemical oxidized titanium alloy prevents implant-related osteomyelitis in rats. *Biomaterials, 101*, 156–164.
97. Del Curto, B., et al. (2005). Decreased bacterial adhesion to surface-treated titanium. *The International Journal of Artificial Organs, 28*(7), 718–730.
98. Zhang, F., et al. (2008). Silk-functionalized titanium surfaces for enhancing osteoblast functions and reducing bacterial adhesion. *Biomaterials, 29*(36), 4751–4759.
99. Kiran, A., et al. (2018). Antibacterial and bioactive surface modifications of titanium implants by PCL/TiO_2 nanocomposite coatings. *Nanomaterials, 8*(10), 860.
100. Zhao, L., et al. (2011). Antibacterial nano-structured titania coating incorporated with silver nanoparticles. *Biomaterials, 32*(24), 5706–5716.
101. Kiran, A. S. K., et al. (2019). Drug loaded electrospun polymer/ceramic composite nanofibrous coatings on titanium for implant related infections. *Ceramics International, 45*(15), 18710–18720.
102. Pierchala, M. K., et al. (2018). Nanotubes in nanofibers: Antibacterial multilayered polylactic acid/halloysite/gentamicin membranes for bone regeneration application. *Applied Clay Science, 160*, 95–105.
103. Radisavljevic, A., et al. (2018). Cefazolin-loaded polycaprolactone fibers produced via different electrospinning methods: Characterization, drug release and antibacterial effect. *European Journal of Pharmaceutical Sciences, 124*, 26–36.
104. Gilchrist, S. E., et al. (2013). Fusidic acid and rifampicin co-loaded PLGA nanofibers for the prevention of orthopedic implant associated infections. *Journal of Controlled Release, 170*(1), 64–73.
105. Hasnain, M. S., & Nayak, A. K. (2019). Nanocomposites for improved orthopedic and bone tissue engineering applications. *Applications of Nanocomposite Materials in Orthopedics*, 145–177.
106. Sofi, H. S., et al. (2019). Reconstructing nanofibers from natural polymers using surface functionalization approaches for applications in tissue engineering, drug delivery and biosensing devices. *Materials Science and Engineering: C, 94*, 1102–1124.
107. Tafin, U. F., et al. (2015). Staphylococcal biofilm formation on the surface of three different calcium phosphate bone grafts: A qualitative and quantitative in vivo analysis. *Journal of Materials Science: Materials in Medicine, 26*(3), 130.
108. Uskoković, V., & Desai, T. A. (2013). Phase composition control of calcium phosphate nanoparticles for tunable drug delivery kinetics and treatment of osteomyelitis. I. Preparation and drug release. *Journal of Biomedical Materials Research Part A, 101*(5), 1416–1426.
109. Theuretzbacher, U. (2012). Accelerating resistance, inadequate antibacterial drug pipelines and international responses. *International Journal of Antimicrobial Agents, 39*(4), 295–299.
110. Palangkaraya, A., & Yong, J. (2009). Population ageing and its implications on aggregate health care demand: Empirical evidence from 22 OECD countries. *International Journal of Health Care Finance and Economics, 9*(4), 391.
111. Matsumoto, N., et al. (2009). Preparation and characterization of β-tricalcium phosphate co-doped with monovalent and divalent antibacterial metal ions. *Acta Biomaterialia, 5*(8), 3157–3164.
112. Rau, J. V., et al. (2017). The bone building blues: Self-hardening copper-doped calcium phosphate cement and its in vitro assessment against mammalian cells and bacteria. *Materials Science and Engineering: C, 79*, 270–279.
113. Li, Y., Ho, J., & Ooi, C. P. (2010). Antibacterial efficacy and cytotoxicity studies of copper (II) and titanium (IV) substituted hydroxyapatite nanoparticles. *Materials Science and Engineering: C, 30*(8), 1137–1144.

114. Kumar, A., et al. (2019). Additive manufacturing methods for producing hydroxyapatite and hydroxyapatite-based composite scaffolds: A review. *Frontiers in Materials*.
115. Wu, V. M., Tang, S., & Uskoković, V. (2018). Calcium phosphate nanoparticles as intrinsic inorganic antimicrobials: The antibacterial effect. *ACS Applied Materials & Interfaces, 10*(40), 34013–34028.
116. Coelho, C. C., et al. (2019). Antibacterial bone substitute of hydroxyapatite and magnesium oxide to prevent dental and orthopaedic infections. *Materials Science and Engineering: C, 97*, 529–538.
117. Samani, S., et al. (2013). In vitro antibacterial evaluation of sol–gel-derived Zn-, Ag-, and (Zn+ Ag)-doped hydroxyapatite coatings against methicillin-resistant Staphylococcus aureus. *Journal of Biomedical Materials Research Part A, 101*(1), 222–230.
118. Sahithi, K., et al. (2010). Synthesis and characterization of nanoscalehydroxyapatite-copper for antimicrobial activity towards bone tissue engineering applications. *Journal of Biomedical Nanotechnology, 6*(4), 333–339.
119. Gopi, D., et al. (2014). Strontium, cerium co-substituted hydroxyapatite nanoparticles: Synthesis, characterization, antibacterial activity towards prokaryotic strains and in vitro studies. *Colloids and Surfaces A: Physicochemical and Engineering Aspects, 451*, 172–180.
120. Kargozar, S., et al. (2018). Mesoporous bioactive glasses: Promising platforms for antibacterial strategies. *Acta biomaterialia, 81*, 1–19.
121. Law, N., et al. (2018). Characterisation of hyaluronic acid methylcellulose hydrogels for 3D bioprinting. *Journal of the Mechanical Behavior of Biomedical Materials, 77*, 389–399.
122. Pandey, A., et al. (2018). Antioxidant and antibacterial hydroxyapatite-based biocomposite for orthopedic applications. *Materials Science and Engineering: C, 88*, 13–24.
123. Wang, Q., et al. (2019). Experimental and simulation studies of strontium/fluoride-codoped hydroxyapatite nanoparticles with osteogenic and antibacterial activities. *Colloids and Surfaces B: Biointerfaces, 182*, 110359.
124. Bhattacharjee, A., et al. (2019). Site-specific antibacterial efficacy and cyto/hemo-compatibility of zinc substituted hydroxyapatite. *Ceramics International, 45*(9), 12225–12233.
125. Sundarabharathi, L., et al. (2019). Investigation of antimicrobial properties and in-vitro bioactivity of Ce3+-Sr2+ dual-substituted nano hydroxyapatites. *Journal of the American Ceramic Society, 102*(1), 144–157.
126. Safi, S., Karimzadeh, F., & Labbaf, S. (2018). Mesoporous and hollow hydroxyapatite nanostructured particles as a drug delivery vehicle for the local release of ibuprofen. *Materials Science and Engineering: C, 92*, 712–719.
127. Leprêtre, S., et al. (2009). Prolonged local antibiotics delivery from hydroxyapatite functionalised with cyclodextrin polymers. *Biomaterials, 30*(30), 6086–6093.
128. Lazić, V., et al. (2018). Antibacterial ability of supported silver nanoparticles by functionalized hydroxyapatite with 5-aminosalicylic acid. *Vacuum, 148*, 62–68.
129. Rau, J. V., et al. (2015). Bioactive, nanostructured S i-substituted hydroxyapatite coatings on titanium prepared by pulsed laser deposition. *Journal of Biomedical Materials Research Part B: Applied Biomaterials, 103*(8), 1621–1631.
130. Litvinova, L., et al. (2017). Imbalance of morphofunctional responses of Jurkat T lymphoblasts at short-term culturing with relief zinc-or copper-containing calcium phosphate coating on titanium. In *Doklady Biochemistry and Biophysics*. Springer.
131. Huang, Y., et al. (2015). Antibacterial efficacy, corrosion resistance, and cytotoxicity studies of copper-substituted carbonated hydroxyapatite coating on titanium substrate. *Journal of Materials Science, 50*(4), 1688–1700.
132. Koo, H., et al. (2017). Targeting microbial biofilms: Current and prospective therapeutic strategies. *Nature Reviews Microbiology, 15*(12), 740.
133. Graziani, G., Boi, M., & Bianchi, M. (2018). A review on ionic substitutions in hydroxyapatite thin films: Towards complete biomimetism. *Coatings, 8*(8), 269.
134. Galindo, T. G. P., et al. (2016). Preparation of nanocrystalline zinc-substituted hydroxyapatite films and their biological properties. *Colloid and Interface Science Communications, 10*, 15–19.

135. Geng, Z., et al. (2016). Strontium incorporation to optimize the antibacterial and biological characteristics of silver-substituted hydroxyapatite coating. *Materials Science and Engineering: C, 58*, 467–477.
136. Robinson, L., et al. (2017). The deposition of strontium and zinc Co-substituted hydroxyapatite coatings. *Journal of Materials Science: Materials in Medicine, 28*(3), 1–14.
137. Modaresifar, K., et al. (2019). Bactericidal effects of nanopatterns: A systematic review. *Acta biomaterialia, 83*, 29–36.
138. Barranco, A., et al. (2016). Perspectives on oblique angle deposition of thin films: From fundamentals to devices. *Progress in Materials Science, 76*, 59–153.
139. Surmenev, R., et al. (2017). Radio frequency magnetron sputter deposition as a tool for surface modification of medical implants. In *Modern technologies for creating the thin-film systems and coatings* (pp. 213–248). InTech.
140. Surmenev, R. A., et al. (2017). RF magnetron sputtering of a hydroxyapatite target: A comparison study on polytetrafluorethylene and titanium substrates. *Applied Surface Science, 414*, 335–344.
141. Hung, K.-Y., Lai, H.-C., & Feng, H.-P. (2017). Characteristics of RF-sputtered thin films of calcium phosphate on titanium dental implants. *Coatings, 7*(8), 126.
142. Abu-Amer, Y., Darwech, I., & Clohisy, J. C. (2007). Aseptic loosening of total joint replacements: Mechanisms underlying osteolysis and potential therapies. *Arthritis Research & Therapy, 9*(1), 1–7.
143. Zhang, B. G., et al. (2014). Bioactive coatings for orthopaedic implants—Recent trends in development of implant coatings. *International Journal of Molecular Sciences, 15*(7), 11878–11921.
144. Heimann, R. B. (2016). The challenge and promise of low-temperature bioceramic coatings: An editorial. *Surface and Coatings Technology, 301*, 1–5.
145. McLaren, A. C. (2004). Alternative materials to acrylic bone cement for delivery of depot antibiotics in orthopaedic infections. *Clinical Orthopaedics and Related Research®, 427*, 101–106.
146. Verron, E., Bouler, J., & Guicheux, J. (2012). Controlling the biological function of calcium phosphate bone substitutes with drugs. *Acta biomaterialia, 8*(10), 3541–3551.
147. Arcos, D., & Vallet-Regi, M. (2013). Bioceramics for drug delivery. *Acta Materialia, 61*(3), 890–911.
148. Yang, L., Sheldon, B. W., & Webster, T. J. (2010). Nanophase ceramics for improved drug delivery. *American Ceramic Society Bulletin, 89*(2), 24–32.
149. Webster, T. J. (2003). Improved bone tissue engineering materials. *American Ceramic Society Bulletin, 82*(6), 23–28.
150. Verron, E., et al. (2010). Calcium phosphate biomaterials as bone drug delivery systems: A review. *Drug Discovery Today, 15*(13–14), 547–552.
151. Hassan, M. S., et al. (2012). Fabrication, characterization and antibacterial effect of novel electrospun TiO_2 nanorods on a panel of pathogenic bacteria. *Journal of Biomedical Nanotechnology, 8*(3), 394–404.
152. Mohammed Sadiq, I., Chandrasekaran, N., & Mukherjee, A. (2010). Studies on effect of TiO_2 nanoparticles on growth and membrane permeability of *Escherichia coli, Pseudomonas aeruginosa*, and *Bacillus subtilis. Current Nanoscience, 6*(4), 381–387.
153. Swetha, S., et al. (2012). Elucidation of cell killing mechanism by comparative analysis of photoreactions on different types of bacteria. *Photochemistry and Photobiology, 88*(2), 414–422.
154. Rouwkema, J., Rivron, N. C., & van Blitterswijk, C. A. (2008). Vascularization in tissue engineering. *Trends in Biotechnology, 26*(8), 434–441.
155. Malda, J., et al. (2004). Oxygen gradients in tissue-engineered PEGT/PBT cartilaginous constructs: Measurement and modeling. *Biotechnology and Bioengineering, 86*(1), 9–18.
156. Li, H., & Chang, J. (2013). Bioactive silicate materials stimulate angiogenesis in fibroblast and endothelial cell co-culture system through paracrine effect. *Acta Biomaterialia, 9*(6), 6981–6991.

157. Yuan, H., et al. (2001). Material-dependent bone induction by calcium phosphate ceramics: A 2.5-year study in dog. *Biomaterials, 22*(19), 2617–2623.
158. Zhai, W., et al. (2012). Silicate bioceramics induce angiogenesis during bone regeneration. *Acta Biomaterialia, 8*(1), 341–349.
159. Leu, A., et al. (2009). Angiogenic response to bioactive glass promotes bone healing in an irradiated calvarial defect. *Tissue Engineering Part A, 15*(4), 877–885.
160. Gorustovich, A. A., Roether, J. A., & Boccaccini, A. R. (2010). Effect of bioactive glasses on angiogenesis: A review of in vitro and in vivo evidences. *Tissue Engineering Part B: Reviews, 16*(2), 199–207.
161. Day, R. M. (2005). Bioactive glass stimulates the secretion of angiogenic growth factors and angiogenesis in vitro. *Tissue Engineering, 11*(5–6), 768–777.
162. Bernabeu, C., Lopez-Novoa, J. M., & Quintanilla, M. (2009). The emerging role of TGF-β superfamily coreceptors in cancer. *Biochimica et Biophysica Acta -Molecular Basis of Disease, 1792*(10), 954–973.
163. Nakamura, T., et al. (2001). Signals via FGF receptor 2 regulate migration of endothelial cells. *Biochemical and Biophysical Research Communications, 289*(4), 801–806.
164. Coultas, L., Chawengsaksophak, K., & Rossant, J. (2005). Endothelial cells and VEGF in vascular development. *Nature, 438*(7070), 937–945.
165. Miller, T. W., Isenberg, J. S., & Roberts, D. D. (2009). Molecular regulation of tumor angiogenesis and perfusion via redox signaling. *Chemical Reviews, 109*(7), 3099–3124.
166. Xia, L., et al. (2016). Akermanite bioceramics promote osteogenesis, angiogenesis and suppress osteoclastogenesis for osteoporotic bone regeneration. *Scientific Reports, 6*(1), 1–17.
167. Perry, C. R. (1999). Bone repair techniques, bone graft, and bone graft substitutes. *Clinical Orthopaedics and Related Research®, 360*, 71–86.
168. Burg, K. J., Porter, S., & Kellam, J. F. (2000). Biomaterial developments for bone tissue engineering. *Biomaterials, 21*(23), 2347–2359.
169. Vacanti, C. A., Bonassar, L. J., & Vacanti, J. P. (2000). Structural tissue engineering. *Principles of Tissue Engineering*, 671–682.
170. Das, A., & Botchwey, E. (2011). Evaluation of angiogenesis and osteogenesis. *Tissue Engineering Part B: Reviews, 17*(6), 403–414.
171. Mercado-Pagán, Á. E., et al. (2015). Vascularization in bone tissue engineering constructs. *Annals of Biomedical Engineering, 43*(3), 718–729.
172. Kwon, K.-A., et al. (2017). Nanocomposites for bone repair. *Nanobioceramics for Healthcare Applications* (pp. 239–298). World Scientific.
173. Wu, C., et al. (2006). Porous akermanite scaffolds for bone tissue engineering: Preparation, characterization, and in vitro studies. *Journal of Biomedical Materials Research Part B: Applied Biomaterials: An Official Journal of The Society for Biomaterials, The Japanese Society for Biomaterials, and The Australian Society for Biomaterials and the Korean Society for Biomaterials, 78*(1), 47–55.
174. Wu, C., et al. (2006). In vitro bioactivity of akermanite ceramics. *Journal of Biomedical Materials Research Part A: An Official Journal of The Society for Biomaterials, The Japanese Society for Biomaterials, and The Australian Society for Biomaterials and the Korean Society for Biomaterials, 76*(1), 73–80.
175. Wu, C., & Chang, J. (2006). A novel akermanite bioceramic: Preparation and characteristics. *Journal of Biomaterials Applications, 21*(2), 119–129.
176. Wu, C., & Chang, J. (2007). Degradation, bioactivity, and cytocompatibility of diopside, akermanite, and bredigite ceramics. *Journal of Biomedical Materials Research Part B: Applied Biomaterials: An Official Journal of The Society for Biomaterials, The Japanese Society for Biomaterials, and The Australian Society for Biomaterials and the Korean Society for Biomaterials, 83*(1), 153–160.
177. Zhai, W., et al. (2013). Stimulatory effects of the ionic products from Ca–Mg–Si bioceramics on both osteogenesis and angiogenesis in vitro. *Acta Biomaterialia, 9*(8), 8004–8014.
178. Keshaw, H., Forbes, A., & Day, R. M. (2005). Release of angiogenic growth factors from cells encapsulated in alginate beads with bioactive glass. *Biomaterials, 26*(19), 4171–4179.

179. Leu, A., & Leach, J. K. (2008). Proangiogenic potential of a collagen/bioactive glass substrate. *Pharmaceutical Research, 25*(5), 1222–1229.
180. Sorrell, J. M., Baber, M. A., & Caplan, A. I. (2007). A self-assembled fibroblast-endothelial cell co-culture system that supports in vitro vasculogenesis by both human umbilical vein endothelial cells and human dermal microvascular endothelial cells. *Cells, Tissues, Organs, 186*(3), 157–168.
181. Dorozhkin, S. V. (2010). Bioceramics of calcium orthophosphates. *Biomaterials, 31*(7), 1465–1485.
182. Guda, T., et al. (2008). A cellular perspective to bioceramic scaffolds for bone tissue engineering: The state of the art. *Current Topics in Medicinal Chemistry, 8*(4), 290–299.
183. Raines, A. L., et al. (2010). Regulation of angiogenesis during osseointegration by titanium surface microstructure and energy. *Biomaterials, 31*(18), 4909–4917.
184. Schubert, T., et al. (2011). The enhanced performance of bone allografts using osteogenic-differentiated adipose-derived mesenchymal stem cells. *Biomaterials, 32*(34), 8880–8891.
185. Liu, J., et al. (2012). The stimulation of adipose-derived stem cell differentiation and mineralization by ordered rod-like fluorapatite coatings. *Biomaterials, 33*(20), 5036–5046.
186. Xia, L., et al. (2014). Effect of nano-structured bioceramic surface on osteogenic differentiation of adipose derived stem cells. *Biomaterials, 35*(30), 8514–8527.
187. Salahinejad, E., & Vahedifard, R. (2017). Deposition of nanodiopside coatings on metallic biomaterials to stimulate apatite-forming ability. *Materials & Design, 123*, 120–127.
188. Kim, H., et al. (1997). Effect of heat treatment on apatite-forming ability of Ti metal induced by alkali treatment. *Journal of Materials Science: Materials in Medicine, 8*(6), 341–347.
189. Kawashita, M., et al. (2003). Apatite-forming ability of carboxyl group-containing polymer gels in a simulated body fluid. *Biomaterials, 24*(14), 2477–2484.
190. Miyazaki, T., et al. (2001). Induction and acceleration of bonelike apatite formation on tantalum oxide gel in simulated body fluid. *Journal of Sol-Gel Science and Technology, 21*(1–2), 83–88.
191. Miyazaki, T., et al. (2001). Apatite-forming ability of niobium oxide gels in a simulated body fluid. *Journal of the Ceramic Society of Japan, 109*(1275), 929–933.
192. Uchida, M., et al. (2001). Bonelike apatite formation induced on zirconia gel in a simulated body fluid and its modified solutions. *Journal of the American Ceramic Society, 84*(9), 2041–2044.
193. Li, P., et al. (1994). The role of hydrated silica, titania, and alumina in inducing apatite on implants. *Journal of Biomedical Materials Research, 28*(1), 7–15.
194. Li, P., et al. (1992). Apatite formation induced by silica gel in a simulated body fluid. *Journal of the American Ceramic Society, 75*(8), 2094–2097.
195. Tanahashi, M., & Matsuda, T. (1997). Surface functional group dependence on apatite formation on self-assembled monolayers in a simulated body fluid. *Journal of Biomedical Materials Research: An Official Journal of The Society for Biomaterials and The Japanese Society for Biomaterials, 34*(3), 305–315.
196. Uchida, M., et al. (2002). Apatite-forming ability of a zirconia/alumina nano-composite induced by chemical treatment. *Journal of Biomedical Materials Research: An Official Journal of The Society for Biomaterials, The Japanese Society for Biomaterials, and The Australian Society for Biomaterials and the Korean Society for Biomaterials, 60*(2), 277–282.
197. Kokubo, T., et al. (1996). Spontaneous formation of bonelike apatite layer on chemically treated titanium metals. *Journal of the American Ceramic Society, 79*(4), 1127–1129.
198. Kato, H., et al. (2000). Bonding of alkali-and heat-treated tantalum implants to bone. *Journal of Biomedical Materials Research: An Official Journal of The Society for Biomaterials, The Japanese Society for Biomaterials, and The Australian Society for Biomaterials and the Korean Society for Biomaterials, 53*(1), 28–35.
199. Yan, W. Q., et al. (1997). Bonding of chemically treated titanium implants to bone. *Journal of Biomedical Materials Research: An Official Journal of The Society for Biomaterials and The Japanese Society for Biomaterials, 37*(2), 267–275.

200. Nishiguchi, S., et al. (1999). Enhancement of bone-bonding strengths of titanium alloy implants by alkali and heat treatments. *Journal of Biomedical Materials Research: An Official Journal of The Society for Biomaterials, The Japanese Society for Biomaterials, and The Australian Society for Biomaterials and the Korean Society for Biomaterials, 48*(5), 689–696.
201. Sun, L., Berndt, C. C., & Grey, C. P. (2003). Phase, structural and microstructural investigations of plasma sprayed hydroxyapatite coatings. *Materials Science and Engineering: A, 360*(1–2), 70–84.
202. Hu, H., et al. (2002). A study on hybrid bioceramic coatings of HA/poly (vinyl acetate) co-deposited electrochemically on Ti–6Al–4V alloy surface. *Materials Science and Engineering: C, 20*(1–2), 209–214.
203. Nelea, V., et al. (2004). Hydroxyapatite thin films grown by pulsed laser deposition and radio-frequency magnetron sputtering: Comparative study. *Applied Surface Science, 228*(1–4), 346–356.
204. Hata, K., et al. (1995). Growth of a bonelike apatite layer on a substrate by a biomimetic process. *Journal of the American Ceramic Society, 78*(4), 1049–1053.
205. Cortes, D., et al. (2005). Biomimetic apatite coating on Mg-PSZ/Al_2O_3 composites. Effect of the immersion method. *Materials Letters, 59*(11), 1352–1355.
206. Bigi, A., et al. (2005). Nanocrystalline hydroxyapatite coatings on titanium: A new fast biomimetic method. *Biomaterials, 26*(19), 4085–4089.
207. Kawai, T., et al. (2004). Coating of an apatite layer on polyamide films containing sulfonic groups by a biomimetic process. *Biomaterials, 25*(19), 4529–4534.
208. Klopčič, S. B., Kovač, J., & Kosmač, T. (2007). Apatite-forming ability of alumina and zirconia ceramics in a supersaturated Ca/P solution. *Biomolecular Engineering, 24*(5), 467–471.
209. Oyane, A., et al. (1999). Sol–gel modification of silicone to induce apatite-forming ability. *Biomaterials, 20*(1), 79–84.
210. Kokubo, T. (1993). A/W glass-ceramic: Processing and properties. *An introduction to bioceramics* (pp. 75–88). World Scientific.
211. LeGeros, R. Z., & LeGeros, J. P. (1993). Dense hydroxyapatite. *An introduction to bioceramics* (pp. 139–180). World Scientific.
212. Hench, L. L. (1993). *An introduction to bioceramics* (Vol. 1). World scientific.
213. Wilson, J., Yli-Urpo, A., & Happonen, R.-P. (1993). Bioactive glasses: Clinical applications. *An introduction to bioceramics* (pp. 63–73). World Scientific.
214. Yamamuro, T. (1993). A/W glass-ceramic: Clinical applications. *An introduction to bioceramics* (pp. 89–103). World Scientific.
215. Shors, E. C., & Holmes, R. E. (1993). Porous hydroxyapatite. *An introduction to bioceramics* (pp. 181–198). World Scientific.
216. Diba, M., et al. (2012). Magnesium-containing bioactive glasses for biomedical applications. *International Journal of Applied Glass Science, 3*(3), 221–253.
217. Diba, M., et al. (2014). Magnesium-containing bioactive polycrystalline silicate-based ceramics and glass-ceramics for biomedical applications. *Current Opinion in Solid State and Materials Science, 18*(3), 147–167.
218. Salahinejad, E., et al. (2012). Aqueous sol–gel synthesis of zirconium titanate ($ZrTiO_4$) nanoparticles using chloride precursors. *Ceramics International, 38*(8), 6145–6149.
219. Rouhani, P., et al. (2013). Nanostructured zirconium titanate fibers prepared by particulate sol–gel and cellulose templating techniques. *Journal of Alloys and Compounds, 568*, 102–105.
220. Salahinejad, E., et al. (2014). Effect of precursor solution pH on the structural and crystallization characteristics of sol–gel derived nanoparticles. *Journal of Alloys and Compounds, 589*, 182–184.
221. Rastegari, S., et al. (2016). Non-hydrolytic sol-gel processing of chloride precursors loaded at forsterite stoichiometry. *Journal of Alloys and Compounds, 688*, 235–241.
222. Iwata, N. Y., et al. (2004). Preparation of diopside with apatite-forming ability by sol–gel process using metal alkoxide and metal salts. *Colloids and Surfaces B: Biointerfaces, 33*(1), 1–6.

223. Iwata, N. Y., et al. (2004). Sintering behavior and apatite formation of diopside prepared by coprecipitation process. *Colloids and Surfaces B: Biointerfaces, 34*(4), 239–245.
224. Oyane, A., et al. (2002). Bonelike apatite formation on ethylene-vinyl alcohol copolymer modified with a silane coupling agent and titania solution. *Journal of the Ceramic Society of Japan, 110*(1280), 248–254.
225. Oyane, A., et al. (2003). Bonelike apatite formation on ethylene-vinyl alcohol copolymer modified with silane coupling agent and calcium silicate solutions. *Biomaterials, 24*(10), 1729–1735.
226. Salinas, A. J., Esbrit, P., & Vallet-Regí, M. (2013). A tissue engineering approach based on the use of bioceramics for bone repair. *Biomaterials Science, 1*(1), 40–51.
227. Clark Jr, A., Pantano Jr, C., & Hench, L. (1976). Auger spectroscopic analysis of bioglass corrosion films. *Journal of the American Ceramic Society, 59*(1–2), 37–39.
228. Hench, L. L. (1991). Bioceramics: From concept to clinic. *Journal of the American Ceramic Society, 74*(7), 1487–1510.
229. Gunawidjaja, P. N., et al. (1963). Local structures of mesoporous bioactive glasses and their surface alterations in vitro: Inferences from solid-state nuclear magnetic resonance. *Philosophical Transactions of the Royal Society A: Mathematical, Physical and Engineering Sciences, 2012*(370), 1376–1399.
230. Salvioli, S., et al. (2006). Inflamm-aging, cytokines and aging: State of the art, new hypotheses on the role of mitochondria and new perspectives from systems biology. *Current Pharmaceutical Design, 12*(24), 3161–3171.
231. Almeida, M., et al. (2007). Oxidative stress antagonizes Wnt signaling in osteoblast precursors by diverting β-catenin from T cell factor-to forkhead box O-mediated transcription. *Journal of Biological Chemistry, 282*(37), 27298–27305.
232. Vaes, B. L., et al. (2006). Microarray analysis on Runx2-deficient mouse embryos reveals novel Runx2 functions and target genes during intramembranous and endochondral bone formation. *Bone, 39*(4), 724–738.
233. Deschaseaux, F., Sensébé, L., & Heymann, D. (2009). Mechanisms of bone repair and regeneration. *Trends in Molecular Medicine, 15*(9), 417–429.
234. Moorthi, A., et al. (2012). Synthesis, characterization and biological action of nano-bioglass ceramic particles for bone formation. *Journal of Biomaterials and Tissue Engineering, 2*(3), 197–205.
235. Dhivya, S., et al. (2015). Nanohydroxyapatite-reinforced chitosan composite hydrogel for bone tissue repair in vitro and in vivo. *Journal of Nanobiotechnology, 13*(1), 1–13.
236. Kumar, S. D., et al. (2018). Formulation and biological actions of nano-bioglass ceramic particles doped with *Calcarea phosphorica* for bone tissue engineering. *Materials Science and Engineering: C, 83*, 202–209.
237. Sanand, S., Hussain, A., & Kaul, G. (2015). Biosafety of nanomaterials used in nanoceramics and nanocomposites. *Handbook of nanoceramic and nanocomposite coatings and materials* (pp. 471–487). Elsevier.
238. Wu, C., Ramaswamy, Y., & Zreiqat, H. (2010). Porous diopside ($CaMgSi_2O_6$) scaffold: A promising bioactive material for bone tissue engineering. *Acta Biomaterialia, 6*(6), 2237–2245.
239. Wu, C., et al. (2007). The effect of strontium incorporation into $CaSiO_3$ ceramics on their physical and biological properties. *Biomaterials, 28*(21), 3171–3181.
240. Saha, M. C., Kabir, M. E., & Jeelani, S. (2009). Effect of nanoparticles on mode-I fracture toughness of polyurethane foams. *Polymer Composites, 30*(8), 1058–1064.
241. Mahfuz, H., et al. (2008). Reinforcement of nylon 6 with functionalized silica nanoparticles for enhanced tensile strength and modulus. *Nanotechnology Reviews, 19*(44), 445702.
242. No, Y. J., Roohani-Esfahani, S.-I., & Zreiqat, H. (2014). Nanomaterials: The next step in injectable bone cements. *Nanomedicine, 9*(11), 1745–1764.
243. Shalumon, K., et al. (2013). Effect of incorporation of nanoscale bioactive glass and hydroxyapatite in PCL/chitosan nanofibers for bone and periodontal tissue engineering. *Journal of Biomedical Nanotechnology, 9*(3), 430–440.

244. Roohani Esfahani, S., et al. (2011). Effects of bioactive glass nanoparticles on the mechanical and biological behavior of composite coated scaffolds. *Acta Biomaterialia, 7*(3), 1307–1318.
245. Khaled, S., Charpentier, P. A., & Rizkalla, A. S. (2010). Synthesis and characterization of poly (methyl methacrylate)-based experimental bone cements reinforced with TiO_2–SrO nanotubes. *Acta Biomaterialia, 6*(8), 3178–3186.
246. Chen, Y., et al. (2012). In vitro biocompatibility and osteoblast differentiation of an injectable chitosan/nano-hydroxyapatite/collagen scaffold. *Journal of Nanomaterials, 2012.*
247. Hu, N.-M., et al. (2012). Mechanical properties and in vitro bioactivity of injectable and self-setting calcium sulfate/nano-HA/collagen bone graft substitute. *Journal of the Mechanical Behavior of Biomedical Materials, 12*, 119–128.
248. Wang, X., et al. (2007). Reinforcement of calcium phosphate cement by bio-mineralized carbon nanotube. *Journal of the American Ceramic Society, 90*(3), 962–964.
249. Chen, J., et al. (2011). Synthesis, characterization and osteoconductivity properties of bone fillers based on alendronate-loaded poly (ε-caprolactone)/hydroxyapatite microspheres. *Journal of Materials Science: Materials in Medicine, 22*(3), 547–555.
250. SM, Z., et al. (2011). A study on mechanical properties of PMMA/hydroxyapatite nanocomposite. *Engineering, 2011.*
251. Fu, S., et al. (2012). Injectable and thermo-sensitive PEG-PCL-PEG copolymer/collagen/n-HA hydrogel composite for guided bone regeneration. *Biomaterials, 33*(19), 4801–4809.
252. Gutwein, L. G., & Webster, T. J. (2002). Osteoblast and chrondrocyte proliferation in the presence of alumina and titania nanoparticles. *Journal of Nanoparticle Research, 4*(3), 231–238.
253. Gillani, R., et al. (2010). Nanofunctionalized zirconia and barium sulfate particles as bone cement additives. *International Journal of Nanomedicine, 5*, 1.
254. Oyefusi, A., et al. (2014). Hydroxyapatite grafted carbon nanotubes and graphene nanosheets: Promising bone implant materials. *Spectrochimica Acta Part A: Molecular and Biomolecular Spectroscopy, 132*, 410–416.
255. Zanello, L. P., et al. (2006). Bone cell proliferation on carbon nanotubes. *Nano Letters, 6*(3), 562–567.
256. Balani, K., et al. (2007). Plasma-sprayed carbon nanotube reinforced hydroxyapatite coatings and their interaction with human osteoblasts in vitro. *Biomaterials, 28*(4), 618–624.
257. An, S.-H., et al. (2012). Porous zirconia/hydroxyapatite scaffolds for bone reconstruction. *Dental Materials, 28*(12), 1221–1231.
258. Malmström, J., et al. (2008). Bone ingrowth in zirconia and hydroxyapatite scaffolds with identical macroporosity. *Journal of Materials Science: Materials in Medicine, 19*(9), 2983–2992.
259. Róg, G., et al. (2002). Calcium zirconate: Preparation, properties and application to the solid oxide galvanic cells. *Electrochimica Acta, 47*(28), 4523–4529.
260. Jung, G.-Y., Park, Y.-J., & Han, J.-S. (2010). Effects of HA released calcium ion on osteoblast differentiation. *Journal of Materials Science: Materials in Medicine, 21*(5), 1649–1654.
261. Ros-Tárraga, P., et al. (2016). Assessment of effects of si-ca-p biphasic ceramic on the osteogenic differentiation of a population of multipotent adult human stem cells. *Materials, 9*(12), 969.
262. Gaihre, B., & Jayasuriya, A. C. (2018). Comparative investigation of porous nano-hydroxyapaptite/chitosan, nano-zirconia/chitosan and novel nano-calcium zirconate/chitosan composite scaffolds for their potential applications in bone regeneration. *Materials Science and Engineering: C, 91*, 330–339.

Chapter 8
Nano-bioceramics

Abstract Currently, a variety of substances such as composites, synthetic ceramics, corals, polymers, and metals can be utilized for the replacement or repair of bones and teeth. Nano-bioceramics are frequently used as coatings on metal implants to improve their bioactivity, functionality, and resistance to wear and corrosion. The development and progress of nano-bioceramics using novel techniques has garnered considerable attention due to their unique characteristics, which are contingent upon their dimensions and configuration. By altering the morphology and dimensions of nano-bioceramics, as well as including additional elements as impurities or inducing defects by ion implantation, it is possible to induce modifications in their biological, chemical, and physical characteristics. Extensive research has been conducted on the ability of hydroxyapatite to accelerate the regeneration of rigid tissues. However, tricalcium phosphate is a preferable biomaterial compared to HA because it has superior osteoconductivity, improved biodegradability, increased biocompatibility, and enhanced solubility. Recent studies indicate that bioactive ceramics with magnesium can greatly enhance the attachment of bone cells and promote rapid proliferation and multiplication of osteoblasts. Silicate biomaterials, such as bioglasses and silicate bioceramics, have the ability to release ions that stimulate the growth of blood vessels (angiogenesis), the proliferation of fibroblastic cells, and exhibit antimicrobial properties. These actions are essential for the process of tissue repair and regeneration. The utilization of 3D printing is a cutting-edge approach for fabricating implants and bone ceramic materials. Various techniques exist for the 3D printing of ceramics, each with its own set of benefits and drawbacks. By utilizing these techniques, it is possible to fabricate porous scaffolds with precise dimensions and forms, making them suitable for treating fractures and abnormalities of bone. Nanotechnology has significantly enhanced the development of ceramic materials with distinct biological characteristics. This chapter will provide a comprehensive overview of nano-bioceramics, with a specific focus on their 3D printing capabilities and the significant role played by nanotechnology.

R. Gholami and S. M. Naghib, *Nanobioceramics for Bone Tissue Engineering and Regenerative Biomedicine*, Biological and Medical Physics, Biomedical Engineering,
https://doi.org/10.1007/978-981-96-0041-0_8

Abbreviations

3D	Three-dimensional
ACP	Amorphous calcium phosphate
ALP	Alkaline phosphatase
BCP	Biphasic calcium phosphate
BGs	Bioactive glass
BTE	Bone tissue engineering
CAD	Computer-aided design
CaP	Calcium phosphate
CC	Calcium carbonate
CDHA	Calcium-deficient HA
CTE	Coefficient of thermal expansion
DCP	Dicalcium phosphate
DCPA	Monetite
DCPD	Dicalcium phosphate dehydrate
Dio	Diopside
HA and HAP	Hydroxyapatite
HOB	Human osteoblast-like cells
HUVECs	Human umbilical vein endothelial cells
LDH	Layered double hydroxide
MAO	Micro-arc oxidation
MCPM	Monocalcium phosphate monohydrate
MSCs	Mesenchymal stem cells
OCP	Octacalcium phosphate
PHA	Precipitated HA
SBF	Simulated body fluid
TCP	Tricalcium phosphate
TEOS	Tetraethyl orthosilicate
TTCP	Tetracalcium phosphate
α-TCP	α-Tricalcium phosphate
β-TCP	β-Tricalcium phosphate

8.1 Introduction

Nanomaterials including nano-bioceramics are employed in the field of nanomedicine for the purpose of disease diagnosis and treatment [1–6]. Nanoceramics are ceramic materials that have a size ranging from 1 to 100 nm in at least one dimension. Advanced nanoceramics exhibit significant surface characteristics due to their high surface-to-volume ratio, which exceeds 10% when their particle size is reduced to less than 100 nm [7]. Nanoscale bioceramics have garnered increased attention in bone tissue engineering due to their suitable surface characteristics and improved biomechanical capabilities compared to biopolymers. These nanomaterials offer numerous benefits for bone regeneration, such as osteoinductivity, osteoconductivity, appropriate biomechanical strength, biodegradability, biocompatibility, and high bioactivity. Nevertheless, nanomaterials are linked to certain issues including cytotoxicity potentials [1, 8–11]. The design and advancement of nano-bioceramics using innovative methods has attracted significant interest due to their distinct features, which vary based on their size and shape. Thus far, nano-bioceramics of various dimensions and forms, including nanofluids, nanosheets, nanotubes, nanoribbons, nanorods, and nanoparticles have been synthesized. Modifying the shape and size of nano-bioceramics, as well as introducing other elements as impurities or producing defects through ion implantation, can lead to changes in their biological, chemical, and physical properties. There are three main categories of advanced nanoceramics: (i) oxides of beryllium, cerium, titanium, zirconia, alumina, etc., (ii) non-oxides including silicide, nitrides, carbides, borides, etc., and (iii) composite ceramics which are combinations of oxides and non-oxides, including particulate reinforced, fiber reinforced ceramics [7].

Nano-bioceramics are commonly employed as coatings for metal implants to enhance their bioactivity, functionality, and wear and corrosion resistance. Nano-bioceramic coatings serve a dual purpose as an in situ drug delivery system, delivering drugs directly to the implantation site and assisting in the acceleration of healing and the formation of a stronger bone-implant interface. Nano-bioceramics are consistently utilized in numerous applications to exploit their beneficial features for the development of functional biomaterials. Nano-bioceramics are employed in diagnostics and medical imaging due to their radiopacity and magnetic characteristics. They also are effectively utilized in dental applications due to their superior compressive wear resistance and strength. The biodegradation property of nano-bioceramics can be altered by transforming them into nanocomposites with biopolymeric materials, which are effectively employed as scaffold materials. The utilization of scaffolds as implant devices is on the rise because of their capacity to biodegrade and facilitate the formation of new bone in areas with defects. Nano-bioceramics are being reimagined as fillers in tissue engineering applications, and the field is rapidly growing because of their ability to interact well with living organisms [7]. This chapter will comprehensively review the characteristics and behavior of available nano-bioceramics.

8.2 Calcium Phosphate-Based Biomaterials

Autografts are often regarded as the most reliable method for bone augmentation, replacement, and rehabilitation. This is because they include the transplantation of bone from one part of an individual's body to another. The next most viable alternative is allografts, which are organs harvested from a different person, such as cadaver bones [12]. There is a pressing need for synthetic biomaterials that can have an impact on biological processes, expedite tissue regeneration, and accurately mimic the structure of replacement tissues. These advanced materials are expected to provide pharmacological or morphological signals that control important cellular interactions during tissue regeneration. Bioactive ceramics and glasses show great potential since they can produce hydroxyapatite (HA) layers and strongly bond with host tissues. The effectiveness of these materials depends on several aspects, including the media type, test conditions, surface charge, and glass composition [13–18]. Currently, a range of substances including as composites, synthetic ceramics (such as calcium carbonate, calcium sulfates, calcium phosphates, and calcium carbonate), corals, polymers, and metals may be used to substitute or repair bones and teeth [12, 19–22].

Calcium phosphate (CaP) is characterized by its structural similarity to bone mineral and its resemblances to osteoconductivity, bioactivity, and biodegradability. As a result, biomaterials including CaP have been developed. To improve these characteristics, connected porosity may be included throughout the production process. However, CaP biomaterials are unsuitable for load-bearing zones because they have a restricted fracture strength [14]. Porosity, size of large pores, content, shape, and surface features are all characteristics that are linked to the ability to induce bone formation. These qualities provide osteoinductive capabilities by aiding in the attraction and buildup of osteoprogenitor cells and bone growth factors [12]. The progress in creating osteoinductive CaP materials is hindered by the difficulties in generating controlled release of osteogenic factors (such as RGDs, Ops, and BMPs) and choosing suitable carriers or scaffolds. When dealing with materials that have natural bone-forming characteristics, it is crucial to determine the best structure (both in terms of large and small pores) for effectively gathering and concentrating bone-forming components. Additionally, ensuring that the manufacturing process can be repeated accurately is also of utmost importance [12]. Another significant application of CaP cements is as a drug delivery system. They enhance the desirable characteristics of a drug carrier, including a large surface area, biodegradability, and the capacity to be injected. These characteristics are necessary for the drug carrier to effectively integrate, retain, and slowly release specific medications or compounds (Fig. 8.1) [23].

Calcium phosphate-based biomaterials are used in several medical and dental operations, including the treatment of periodontal problems, spinal fusion, tooth replacement, alveolar bone augmentation, and sinus lifts. Moreover, they function as supportive structures for the regrowth of bone and dentin, the administration of medicine, the creation of substances that fight against microbes, and the development of bone tissue engineering (BTE) techniques. Biomaterials containing calcium phosphate promote osteoblast cell adhesion, proliferation, and specialization. These cells cling to and break down CaPs both in laboratory conditions (in vitro) and in living organisms (in vivo) [24]. The cellular response is affected by many parameters, including particle size, chemistry, and surface topography. Calcium phosphates include fluoride [25] and zinc [26, 27] ions, which might hinder osteoclast activity and promote the production of collagen and matrix proteins that contribute to bone mineralization [12]. To fully comprehend the osteoinductive potential and mechanical properties of biomaterials made from calcium phosphate, more inquiry

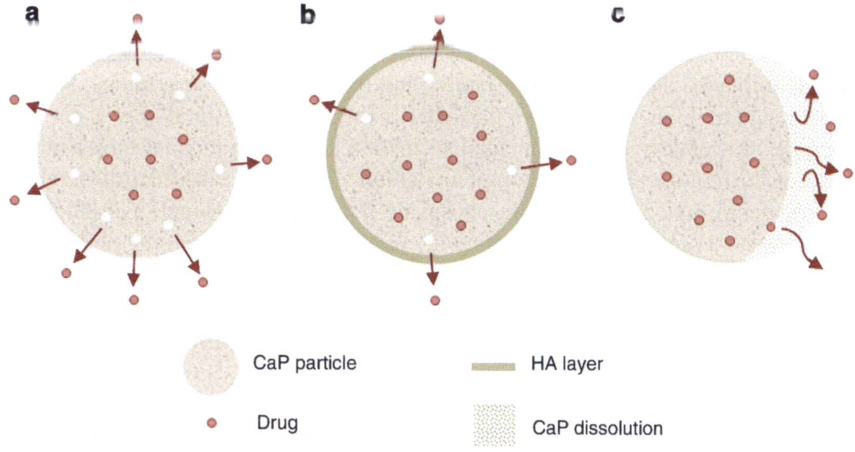

Fig. 8.1 Various mechanisms of drug release by utilizing calcium phosphate cements as carriers, **a** diffusion mechanism, **b** diffusion mechanism characterized by a slower diffusion rate caused by the formation of a HA layer as a result of bioactivity, **c** release caused by the biodegradation process of the particles. Reprinted with permission from [23]

and improvement are required. In contrast, these materials provide a versatile and supportive method for bone restoration.

8.2.1 Hydroxyapatite

Artificial materials used in orthopedics must possess enough degrees of osteogenic activity, osteointegrativity, osteoinductivity, and osteoconductivity [28]. Considerable study has been carried out on the capacity of hydroxyapatite, a calcium phosphate biomaterial with the chemical formula $Ca_{10}(PO_4)_6(OH)_2$, to expedite the regeneration of inflexible tissues. HA is a very suitable substitute for bone transplants because it has osteogenic bioactivity, biodegradability, and cytocompatibility. Additionally, HA closely mimics the mineral composition of bone [29]. The osteoconductivity of HA facilitates bone development and osseointegration. The composition of bone structure, which makes up 60–70% of it, helps to reduce inflammation and immune reactions at the location of the implant. The mechanical properties, morphology, and porous architecture of pure HA are crucial after it is inserted into bone defect regions [30].

HA and its derivatives provide many benefits for delivering medicines [31], including proteins, nucleic acids, antibiotics, and anticancer therapies. These advantages stem from their large surface-to-volume ratio, strong ability to absorb biomacromolecules, and surface bioactivity [8, 9]. Nanoscale HA greatly promotes the proliferation of healthy bone cells while effectively suppressing the proliferation of different cancer cells, including gliomas, melanoma, lung, breast, gastric, colon, osteosarcoma, and hepatoma [29, 32–34]. HA is used in the application of chemicals for metallic implants, strengthening of bones, and repairing bone deformities. Within the pH range of 4.2 to 8.0, calcium phosphate exhibits greater stability in water and undergoes slower degradation. Fluctuations in the Ca/P molar ratio may have a major impact on the biological response and composition of HA. This can lead to the production of calcium oxide or tricalcium phosphate phases [35]. Sol–gel, hydrothermal processing, and precipitation are among the several ceramic processing procedures used to manufacture HA. Designing porous HA scaffolds that efficiently promote bone formation and regeneration continues to be a difficult task. Additive manufacturing methods facilitate the fabrication of bioceramic scaffolds with precision, efficiency, and scalability [29, 35].

HA-based ceramic composites have been thoroughly studied as potential alternatives in the area of bone tissue engineering [36–38]. HA has more efficacy in promoting the differentiation and proliferation of bone cells compared to other bioactive materials, despite its higher bioactivity and biocompatibility [39, 40]. The combination of osteoconductive scaffolds with osteoinductive biomolecules, such as growth hormones, improves bone healing. The incorporation of HA with additional biomaterials, including as carbon nanostructures, biopolymers, and bioceramics, is augmenting its physical and biological characteristics to a greater extent [41–43]. By mixing HA with other bioactive ceramics or materials capable of ion replacement, the material's mechanical characteristics, cell compatibility, and biological activity may be improved. The combination of HA and bredigite exhibits enhanced bioactivity and biocompatibility compared to their individual usage [44]. Magnesium-containing biocomposites have the potential to be efficient bioactive materials for bone tissue replacement [4]. Recent advancements in HA-polymer composites have coupled the bioactivity of HA with the flexibility, robustness, and biodegradability of the polymer network. These composites are being improved for use in bone replacement and tissue engineering applications [45]. Establishing a constant dispersion and solid adherence between the polymer matrix and HA continues to be a tough task [28, 35]. Stoichiometric HA lacks crucial ions like as Mg^{2+}, Sr^{2+}, and Si^{4+}, which are often found in real bone tissue. The goal is to create bone replacements that are more efficient by integrating ion-doped HA into three-dimensional scaffolds [35, 46, 47]. The main reason for the significant interest in using HA-based composites in bone replacement and tissue engineering is their ability to improve biological responses and mechanical properties. Further study and innovation are required to fully harness their medicinal potential [3].

8.2.2 Tricalcium Phosphates

Tricalcium phosphate (TCP) is a more suitable biomaterial than HA [48] due to its superior osteoconductivity, better biodegradability, higher biocompatibility, and enhanced solubility, as shown by many researches [9, 49]. Tricalcium phosphates exist in three different forms, known as polymorphs, which may be categorized into two groups: the high-temperature α- and α'-TCP, and the low-temperature β-TCP. Although α- and β-TCP have the same chemical composition, their solubility, density, and structure differ dramatically, indicating that they possess distinct biological properties and potential use in medicine [50].

8.2.2.1 α-TCP

Currently, α-tricalcium phosphate (α-TCP, $\alpha - Ca_3(PO_4)_2$) is considered the primary substance used in injectable hydraulic bone cements for bone regeneration, biodegradable bioceramics, and diverse composites. α-TCP has a higher solubility and undergoes quick hydrolysis to generate calcium-deficient HA, which is identical to β-TCP. This characteristic confers advantages in the production of biodegradable composites and bioceramics, as well as self-regulating osteotransductive bone cements for the regeneration of shattered bones. α-TCP exhibits osteoconductive and bioactive characteristics in both laboratory and living organism settings, along with its non-toxic features, within the realm of biology. The main driver behind the increasing attention in α-TCP as a material for bone implants is its biodegradation potential [50]. The technique entails the thermal crystallization of amorphous precursors with the suitable composition above the transition temperature or the heating of the low-temperature polymorph β-TCP. The α-TCP phase may exist in a metastable form at room temperature, and the stability of this phase is greatly affected by ionic substitutions [50]. Table 8.1 lists some commercial CaP bone cements having α-TCP in their formulas.

The main approach to produce α-TCP involves the reaction of a combination of solid precursors in a solid state at high temperatures. Alternatively, β-TCP, also

Table 8.1 Commercial α-TCP-based bone cements [50]

Powder composition	Manufacturer	Brand name
α-TCP + CC + MCPM	Skeletal Kinetics, USA	Callos™
α-TCP + ACP + BCP	Biomatlante, France	MCPC
α-TCP + TTCP + DCPD + HA	Mitsubishi Materials Co., Japan	Biopex®
α-TCP + CC + MCPM	Norian, USA	Norian SRS®
α-TCP + DCP + CC + PHA	Biomet, Europe	Calcibon®
α-TCP + TTCP	Biomet, USA	Mimix®
α-TCP + TTCP	Teknimed SA, France	Cementek®

BCP biphasic calcium phosphate, HA + β-TCP; *ACP* amorphous calcium phosphate; *PHA* precipitated HA, $Ca_{10-x}(HPO_4)_x(PO_4)_{6-x}(OH)_{2-x}$; *MCPM* monocalcium phosphate monohydrate, $Ca(H_2PO_4)_2.H_2O$; *HA* hydroxyapatite, $Ca_{10}(PO_4)_6(OH)_2$; *CC* calcium carbonate, $CaCO_3$; *DCPD* dicalcium phosphate dehydrate; *DCP* dicalcium phosphate $CaHPO_4$; *TTCP* tetracalcium phosphate, $Ca_4(PO_4)_2O$

known as calcium-deficient HA (CDHA), may be created by heating a precursor with a Ca/P molar ratio of about 1.5. In addition, they have been used in self-propagating high-temperature synthesis [51] and combustion synthesis [50].

8.2.2.2 β-TCP

Due to its synthetic origin, osteoinductive and osteoconductive properties, and ability to be absorbed by cells, β-TCP is a very successful substitute for bone grafts [52]. In therapeutic settings, β-TCP is often used. In contrast to sintered HA, β-TCP may be quickly replaced by new bone and absorbed by the body. Because of its solubility, which is the same as that of bone mineral, β-TCP is insoluble under physiological conditions [53]. Rather, it causes a localized acidity that cells, mostly osteoclasts, absorb, which aids in the disintegration of β-TCP. In addition to its potential to promote bone formation, several investigations have shown the heightened ability of β-TCP to stimulate bone growth [54]. Therefore, β-TCP is considered to be one of the most attractive alternatives for bone transplantation [52]. β-TCP is a white solid when it is pure, but its color may be altered by the quantity of impurities it contains. It is delicate. The color of β-TCP doped with manganese is vibrant pink, copper-doped β-TCP is blue and may occasionally exhibit purple or violet tones [55], and chromium-doped β-TCP is green [56].

8.2.3 Crystal Structure and Thermal Stability

The two most crucial physical features for bone repair are the crystal structure and the temperature stability range [52]. Tricalcium phosphates are often formed by thermally converting amorphous CaP or calcium-deficient HA ($Ca_9(PO_4)_5(HPO_4)OH$; CDHA) at temperatures between 650 and 750 °C. As mentioned earlier, there are three distinct forms of calcium phosphates that may be classified as follows: α-TCP, α′-TCP, and β-TCP. According to the findings of research, the formation of α-TCP occurs at temperatures ranging from 1115 to 1150 °C, while the appearance of α′-TCP occurs between 1430 and 1470 °C. Additionally, β-TCP is shown to be present at temperatures ranging from 650 to 750 °C [57–60]. The crystal structures of the two variants of TCP are shown in Fig. 8.2. The unit cell of α-TCP comprises a total of 312 atoms, where the numbers of crystallographically inequivalent sites are forty-eight for oxygen, twelve for phosphorus, and eighteen for calcium. Its lattice parameters are $a = 12.89$ Å, $b = 27.28$ Å, and $c = 15.22$ Å, and it has the space group of P21/a [61].

The crystal structure in which β-TCP is located is the rhombohedral space group R3c [62]. A triangle is frequently employed to illustrate the concept. The anticipated measurements of chemically pure β-TCP unit cells are $a = 1.0435$–1.0438 nm and $c = 3.739$–3.743 nm [62–65]. The "*c*" axis spans from 3.689 to 3.763 nm, whereas the "*a*" axis ranges from 1.035 to 1.053 nm, depending on the particular research.

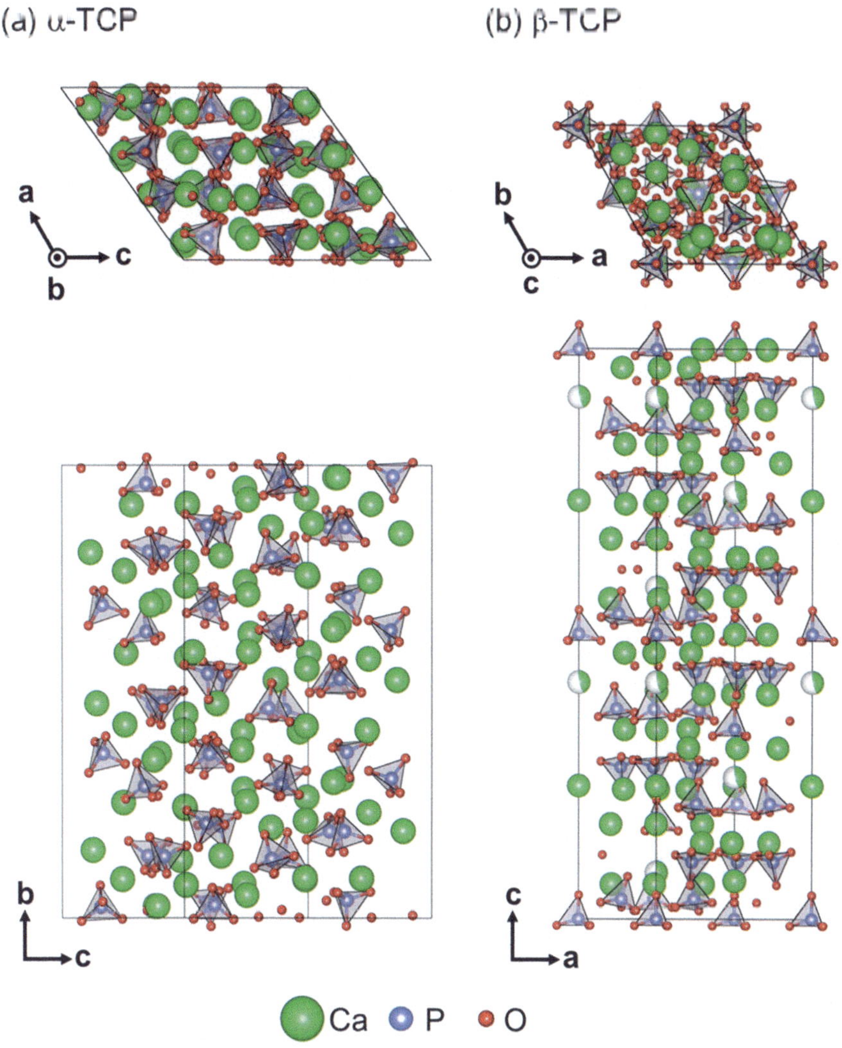

Fig. 8.2 Crystal structures of **a** α-TCP and **b** β-TCP. Tetrahedra are used to represent PO_4^{3-} groups for the purpose of enhancing clarity. Reprinted with permission from [61]

The notable alterations seen in many buildings being examined are likely attributed to the existence of elemental pollutants. The unit cell of β-TCP contains 63 calcium atoms and 42 PO_4 groups. β-TCP consists of two planar domains that are arranged in a specific manner. The molar ratio of Ca/P in a specific domain is 1.571 and in the second domain is 1.429. The observed occupancy pattern may vary as a result of the limited number of computational choices in their simulations [66]. Because XRD measurements have a large range, the structure may include many tiny, organized domains that seem to be disordered [52]. The thin columns in the crystal structure are

Table 8.2 Structural data of α-TCP, α′-TCP, and β-TCP [50]

Property	$Ca_3(PO_4)_2$ polymorph		
	α-TCP	α′-TCP	β-TCP
Symmetry	Monoclinic	Hexagonal	Rhombohedral
Space group	$P2_1/a$	$P6_3/mmc$	R3C
a (nm)	1.2859(2)	0.53507(8)	1.04352(2)
b (nm)	2.7354(2)	0.53507(8)	1.04352(2)
c (nm)	1.5222(3)	0.7684(1)	3.74029(5)
α (°)	90	90	90
β (°)	126.35(1)	90	90
γ (°)	90	120	120
D_{th} (g cm^{-3})	2.866	2.702	3.066
V (nm^3)	4.31(6)	0.19052(8)	3.5272(2)
V_0 (nm^3)	0.180(6)	0.19052(8)	0.1680(2)
Z	24	1	21

aligned with the *c*-axis due to the osteoclastic resorption of the β-TCP. Characterizing a morphology that suggests a uniform network composition is a difficult undertaking [52]. Cell parameters (a, b, c, α, β and γ), theoretical density (D_{th}), volume per formula unit (V_0), number of formula units per cell (Z), and cell volume (V) are presented in Table 8.2 for $Ca_3(PO_4)_2$ polymorphs [50].

8.2.4 Solubility of CaPs

The solubility of calcium phosphates may vary greatly in response to changes in pH, sometimes by many orders of magnitude. Solubility may seem to be an intricate function that depends on the qualities of the solution. However, our grasp of the solubility properties of CaPs may be improved by gaining a deeper understanding of the underlying principles that govern the balance between solids and liquids. The solubility isotherm is a unidimensional curve that represents the saturated solutions of a particular salt at a consistent temperature and pressure, in contrast to a three-dimensional surface in a phase diagram. This is because these solutions possess just a single degree of freedom [67]. Table 8.3 shows the solubility of CaP products and other calcium orthophosphates.

Table 8.3 Calcium orthophosphates and their solubility [50, 68]

Chemical formula	Compound	Ca/P	Solubility (mg L^{-1})		$-\log K_{ps}$	
			25 °C	37 °C	25 °C	37 °C
$\alpha-Ca_3(PO_4)_2$	α-tricalcium phosphate (α-TCP)	1.5	0.97	0.24	25.5	28.5
$\beta-Ca_3(PO_4)_2$	β-tricalcium phosphate (β-TCP)	1.5	0.20	0.15	28.9	29.6
$Ca_4(PO_4)_2O$	Tetracalcium phosphate	2.0	0.28–0.038	0.39–0.075	38–44	37–42
$Ca_{10}(PO_4)_6(OH)_2$	Hydroxyapatite (HAP)	1.67	0.00010	0.000096	116.8	117.2
$Ca_8(HPO_4)_2(PO_4)_4.5H_2O$	Octacalcium phosphate (OCP)	1.33	0.025	0.018	96.6	98.6
$CaHPO_4$	Monetite (DCPA)	1.0	48	41	6.90	7.04
$CaHPO_4.2H_2O$	Brushite	1.0	87	74	6.59	6.73

8.3 Calcium-Magnesium-Based Bioceramics

Magnesium (Mg) and its alloys are favorable materials for biodegradable applications because they are compatible with biological systems and have the ability to degrade over time. These materials have a little impact on stress shielding since they have the same density and elastic modulus as real bone [69, 70]. Magnesium is the second most common cation in cells and the fourth most abundant cation in the organism [71]. Recent study suggests that bioactive ceramics containing magnesium may significantly improve the adherence of bone cells and stimulate the fast growth and multiplication of osteoblasts [6]. Magnesium is essential for a multitude of enzymatic reactions in the body [72]. Alloying magnesium with elements like Mn, Zn, Li, Ca, and RE (such as La and Y) may generally improve its resistance to corrosion [73, 74]. The mechanical strength of magnesium alloys is specifically affected by the presence of calcium. Calcium helps to disperse the secondary Mg_2Ca phase, enhance the solid solution, and decrease the size of particles, hence increasing the overall strength. Additionally, it improves the quality of the granules, thereby boosting the microscopic composition of the alloys. Calcium is an essential component of the skeletal structure of humans. Adding lithium to HCP-magnesium alloys may improve their malleability and impact resistance, as well as decrease their weight, by reducing the c/a ratio. To a greater extent than magnesium-lithium-calcium (Mg-Li-Ca) alloys, magnesium-calcium (Mg-Ca) and magnesium-lithium-calcium (Mg-Li-Ca) alloys exhibit lesser levels of mechanical capabilities and corrosion resistance. Moreover, prior studies have shown that the magnesium alloy LAE442, including 4% lithium by weight, 4% aluminum by weight, and 2% rare earth elements by weight, displays exceptional resistance to corrosion when used in biological applications within live beings [75].

To enhance the corrosion resistance and biocompatibility of magnesium alloys, several surface modification methods may be used. These options include calcium phosphate and its chemical conversion coatings, micro-arc oxidation (MAO) coatings, sol–gel films, layer-by-layer built coatings, and layered double hydroxide (LDH) coatings [75]. Prior studies have shown that TiO_2 may stimulate the development of a calcium phosphate layer on magnesium alloys. The similar crystal structures of SnO_2 and TiO_2 are expected to have equivalent impacts on the development of the CaP layer. Furthermore, studies have shown that nanocrystalline-doped SnO_2 has antimicrobial characteristics. Titanium dioxide also has antibacterial properties. However, it is important to acknowledge that it is often activated by UV radiation [75]. Coating the magnesium-lithium-cathode alloy with calcium-doped zinc phosphate (Ca-P-Zn) improves its blood biocompatibility and corrosion resistance. This coating is created via a chemical conversion procedure. However, the Ca-P-Zn coating does not meet the necessary requirements for density and adhesive strength when applied to magnesium substrates. Ca-P coatings may be effectively strengthened by hydrothermal synthesis. Furthermore, the structure and level of crystallization of the coatings may be tailored by modifying the hydrothermal settings [75].

8.4 Calcium Silicate-Based Bioceramics

There has been a significant rise in the fascination in using bioceramics containing magnesium (Mg) and silicon (Si) for tissue engineering applications in recent years. Their acceptable microstructural and physicochemical features, together with their capacity to promote cell proliferation and differentiation, are the reasons for this [6]. Silicon is widely acknowledged as a crucial trace element for the process of bone regeneration and repair. Because it has the ability to accelerate the differentiation of bone marrow mesenchymal stem cells into bone-forming cells, activate signaling pathways related with the production of bone tissue, and raise the expression of genes involved in bone formation, it is able to do all of these things [76, 77]. Significant advancements have been achieved in the last ten years in the field of silicate-based bioactive ceramics. Si ions may be liberated from these ceramics, leading to a considerable enhancement in bone cell proliferation and differentiation, as shown by many experiments. Studies have shown that silicate-based bioceramics have higher osteogenic activity and in vivo degradability compared to calcium phosphate ceramics. This indicates that they have considerable promise for improving bone regeneration [78, 79].

Silicate biomaterials, including bioglasses and silicate bioceramics, may release ions that promote angiogenesis, fibroblastic cell proliferation, and antibacterial effectiveness. These activities are crucial for tissue repair and regeneration [80, 81]. CaPs doped with silicon (Si) have shown the most notable improvements in bioactivity, surface chemistry, morphology, and solubility. The presence of Si^{4+} ions considerably increased the proliferation and propagation of osteoblast cells in CaPs [82, 83]. Many techniques that may be used to cause ions to form a solid structure known as HA. These approaches include solid-state, hydrothermal, precipitation, wet chemical, sol–gel, and hydrothermal processes. All of these approaches need high temperatures and are often linked with lengthy manufacturing processes [84]. The in vitro biological efficacy of the Si-doped granules was evaluated, and it was determined that they were cytocompatible. Further, the utilization of Si-doped granules significantly enhances the likelihood of calcium nodule development and bone growth. Improved structural and physicochemical characteristics, as well as exceptional biological performance, were observed in the glycine sample that was doped with 0.1 mol of Si^{4+} ions. The substance exhibited a zeta potential of – 27 mV, a particle size of 26 nm, and a favorable surface area of 146 m^2/g [84].

8.4.1 *Akermanite*

Akermanite $(Ca_2MgSi_2O_7)$, a bioactive glass, is formed from silica and has enhanced hardness and density [85]. Akermanite, has the ability to release cations due to its calcium, magnesium, and silicon components under physiological settings [81]. Furthermore, studies have shown that the presence of Ca, Mg, and Si in akermanite aids in the process of calcium deposition and the development of adult stem cells, leading to the creation of fully developed osteoblasts. The release of silicon ions may cause regression [85]. Numerous studies have shown that Aker is capable of enhancing the process of division and proliferation of human umbilical vein endothelial cells (HUVECs), as well as stimulating the production of new blood vessels in live animals [81, 86]. Akermanite enhances the attachment, growth, and specialization of MC3T3-E1 cells by the release of silicon, magnesium, and calcium ions [85]. Akermanite, a commonly found kind of silicate-based bioceramics, has the ability to release silicate, calcium, and magnesium ions in a natural environment. Furthermore, the introduction of porous akermanite into the fractured femurs of rabbits has been shown to enhance the formation of new blood vessels and the growth of new bone tissue, when compared with rabbits treated with β-TCP [87]. The findings from these experiments suggest that akermanite bioceramics, a material with notable bioactivity, has great potential as a viable option for bone restoration. Nevertheless, it is crucial to recognize that these researches have mostly concentrated on investigating the influence of the material on the process of osteogenic differentiation and the regeneration of bone tissue. The subchronic systemic toxicity of the material's degradation compounds, which is crucial for its possible therapeutic use, is not well-recognized [78].

While the akermanite bioceramic has shown promising osteogenic activity, there is still a lack of knowledge on its potential subchronic systemic toxicity when tested in living organisms [302]. This information is crucial for assessing the biocompatibility and clinical uses of the material in bone implants [78]. Subchronic systemic toxicity testing aims to assess the possible impacts on body parts or organs that are not directly exposed to a substance. This is done by examining the absorption, distribution, and metabolism of substances derived from the material. It differs from testing the ability of an implant to promote bone formation. The possible influence of akermanite bioceramics on blood and organ function may raise concerns because of their biodegradability and the likelihood of their ionic breakdown products, such as Ca, Mg, and Si ions, being present in the body. Overdosing on silicate or magnesium has been shown to impair several organs such as the kidney, liver, lung, and cardiovascular system. It has been found to inhibit cell proliferation and even trigger apoptosis [88–92]. Therefore, more study is required to determine the biological safety of akermanite, despite its good osteogenicity, since the distribution and long-term effects of its breakdown products are not yet fully characterized [78]. A study was done to explore the subchronic systemic toxicity of the ionic breakdown products of akermanite bioceramic, in compliance with ISO standard [78]. The administration of the ionic extract derived from akermanite bioceramics did not have a significant impact on the count and function of primary blood cells. What was discovered was that. In addition, the liver and kidney, both of which are essential components in the metabolic process, continued to operate normally even after the extracts were administered to the body. Additionally, there were no pathological changes identified in any of the main organs of the animals. The findings of a study demonstrate that akermanite bioceramics have great promise as bioactive bone graft materials in orthopedic applications. These bioceramics showed no subchronic systemic toxicity and are well-known for their ability to promote bone growth [78].

8.4.2 Bredigite

Bredigite, with chemical formula of $Ca_7MgSi_4O_{16}$, is a kind of silicate bioceramic. It was developed with the purpose of controlling the pace at which it breaks down and promoting the development of apatite in physiological saline. Bredigite is an orthorhombic calcium-magnesium silicate. Bredigite has enhanced bioresorbability and bioactivity as a result of its elevated concentration of calcium ions compared to diopside and akermanite [93]. Over the last several years, bredigite has been used for BTE because of the exceptional mechanical, biodegradable, and bioactive properties that it has [94, 95]. An example of the methodology used to produce a pure bredigite scaffold with specific properties is the use of a polymer sponge method. This method resulted in a scaffold with 90% porosity, a pore size ranging from 300 to 500 μm, a high degradation rate, and satisfactory mechanical properties (compressive strength = 233 ± 14 kPa) [95, 96]. Bredigite has low biocompatibility since it is rapidly absorbed by the body, leading to an undesirable elevation in physiological pH upon implantation [94, 97]. Moreover, there is a need to enhance the mechanical characteristics of ceramic scaffolds in order to facilitate the development of hard tissue in various applications, as stated in a general reference [98].

Unlike bioglass and AW-GC, bredigite ceramics demonstrated an appropriate mechanical strength and a fair deterioration rate. This indicates that they have the potential to be effective as innovative implant materials with biological activity. Bredigite is a mineral with a density of 3.40 g/cm^3, however it has not proved to be economically valuable so far. Naturally occurring bredigite often contains impurities from other minerals, making it less pure. The researchers Wu et al. conducted a study to assess the in vitro bioactivity of bredigite. They measured its ability to generate HA in simulated body fluid (SBF) and also examined the effects of ionic products released by bredigite dissolution on the proliferation of osteoblasts [99]. The pure bredigite powders were produced chemically using the sol–gel process. Bredigite ceramics may stimulate the formation of bone-like apatite in SBF. The breakdown of bredigite ceramics produces ionic products including Si and Ca, which can enhance cell proliferation [94, 95]. Bredigite scaffolds, coated with a biomimetic apatite layer, have potential uses in bone tissue creation. This is because they possess the qualities of being biodegradable, promoting cell adhesion, and stimulating cell proliferation and differentiation [95].

8.4.3 Monticellite

The bioceramics with the highest magnesium content are monticellite $(CaMgSiO_4)$, which contains approximately 33%. It has been proposed that the ions generated by the monticellite constructions may directly stimulate osteoblast growth and significantly enhance osteoblast adhesion [79, 100, 101]. The sol–gel method, which is based on the polycondensation of tetraethyl orthosilicate (TEOS) with nitric acid, is frequently cited as the method used to produce this category of bioceramics. Nevertheless, this approach has certain disadvantages, such as the necessity of increasing the dehydrating gel's temperature and time, raising the heat treatment temperature to conserve the final phase, and performing additional milling procedures to produce particles smaller than 100 nm [94, 101–103]. The nanostructure of monticellite produced by the sol–gel process using ethanol as a green solvent is relatively unknown, despite the numerous reports of this type of bioactive bioceramics. The release of Ca and Si ions from specific monticellite varieties has been demonstrated to provide substantial advantages over other bioceramics, as per Chen et al. [104]. Kalantari et al. [5] utilized a novel sol–gel green method to sinter powder compacts, resulting in the production of nanostructured monticellite.

Initially, the sol–gel method was employed to generate monticellite particles in a single study. Ultimately, the single-phase ceramics were produced by effectively sintering the monticellite powder compacts. The monticellite ceramics' bioactivity, biocompatibility, mechanical properties, and coefficient of thermal expansion (CTE) were assessed in vitro. The findings indicated that the materials exhibited a CTE more akin to the Ti-6Al-4V alloy, a lower Young's modulus, and a superior fracture durability compared to HAp [104]. Although the bioactivity and biocompatibility of certain Mg-containing silicate bioceramics, including diopside [105, 106], wollastonite [71], akermanite [87], and bredigite [94], have been thoroughly investigated and optimized in previous studies using the ALP activity and MTT assay test, other Mg-containing bioactive ceramics, such as monticellite, which contain more Mg^{2+} ions, have not been systematically examined, optimized, or acknowledged. In their investigation, Kalantari et al. efficiently investigated the physicochemical properties and morphology of monticellite nanoparticles [5, 101]. Consequently, monticellite was chosen for additional investigation in the $CaMgSiO_4$ system. The investigation and optimization of monticellite's cytocompatibility, bioactivity, osteogenic activity, and antibacterial and antibiofilm activity at various concentrations were conducted in the study [6].

8.4.4 Hardystonite

Zinc is a trace element that is also essential for the metabolism of bones. Yamaguchi et al. [107, 108] have investigated the stimulatory effects of zinc on alkaline phosphatase activity, bone formation, and bone protein in both in vitro and in vivo contexts. In this regard, zinc-containing ceramic composites offer novel research opportunities in the field of BTE [109]. Zinc has the capacity to enhance bone formation and increase alkaline phosphatase activity, calcium content, and bone proteins in both animals and humans. Wu et al. [110] introduced Zn to calcium silicate to produce hardystonite ($Ca_2ZnSi_2O_7$), which was mechanically more robust than HA. $Ca_2ZnSi_2O_7$ ceramics were found to enhance the adhesion, proliferation, and differentiation of human osteoblast-like cells (HOB) and raised alkaline phosphatase activity, as per Ramaswamy et al. [111]. The mineral hardystonite is composed of Si, Zn, and Ca. It has a melting point of 1425 °C and a density of 3.40 g/cm^3. It has not yet been implemented in a substantial commercial capacity. Hardystonite is biocompatible and can be employed to produce biomaterials due to its chemical composition [110].

Hardystonite is pertinent for biomedical applications due to its biological properties. Research has demonstrated that it stimulates osteogenic differentiation and mesenchymal stem cell proliferation [112, 113]. Furthermore, it stimulates the production of alkaline phosphatase, osteocalcin, and type I collagen when it interacts with human osteoblast-like cells [111, 112]. Additionally, the mechanical properties of hardystonite are intriguing. Dense HT exhibits deformation capabilities ranging from 50 to 140 MPa, which are comparable to the 50–150 MPa of human cortical bone. Its Young's modulus is between 20 and 40 GPa, which is comparable to the range of 7 to 30 GPa in cortical bone [110, 114]. Nevertheless, the material's strength is diminished by the necessity of a relatively high porosity to facilitate cell colonization and cytocompatibility [115]. There is a significant difference in the biocompatibility, fracture toughness, and bending strength of hardystonite and those of HA ($Ca_{10}(PO_4)_6(OH)_2$).

Calcium and silicate are quite plentiful in the bioceramic hardystonite. Through the incorporation of zinc into the Ca–Si oxide system, the chemical stability of the system was improved [110]. Ceramics made from hardystonite that are generated from sol–gel have a significant amount of promise for bone tissue engineering [111, 113]. It has been shown that these materials may improve osteogenic differentiation and speed up the proliferation of mesenchymal stem cells (MSCs) that are produced from human bone marrow [113]. Furthermore, in comparison with $CaSiO_3$, it has been shown that ceramics composed of $Ca_2ZnSi_2O_7$ exhibit enhanced cellular adhesion, as well as increased cellular proliferation and differentiation of cells. Additionally, when $Ca_2ZnSi_2O_7$ comes into contact with HOB cells, there is an increase in the production of alkaline phosphatase (ALP), osteocalcin, and collagen type I, respectively [111]. In comparison with $CaSiO_3$ and hydroxyapatite, hardystonite ceramics also exhibit superior biocompatibility, fracture durability, and bending strength [109]. In order to overcome the difficulties associated with sintering, silicate-based ceramics are often manufactured via the use of complex techniques. These techniques include the hydrothermal technique, glass devitrification, solution combustion methods, spark plasma sintering, and sol–gel processing. In order to get pure quaternary-oxide and ternary-oxide silicate granules, it is of the utmost importance to choose the appropriate reaction circumstances, synthesis processes, and ingredients [112].

8.4.5 Diopside

The $CaO - MgO - SiO_2$ system contains diopside, a member of the pyroxene family. Recently, there has been considerable interest in diopside as a bioceramic substitute for biomedical applications. Diopside has been demonstrated to be bioactive, biocompatible, and non-toxic, which facilitates cell division and proliferation through numerous mechanisms. Over the course of three days, apatite is generated on the surface of diopside, which is an indication that it has a greater level of bioactivity than forsterite [106, 116, 117]. Furthermore, diopside (Dio) is recognized as the original pyroxene mineral group, possessing superior biological and mechanical properties in comparison with HT [106]. A comparison is made between the bending strength of dense Dio (3.2 g/cm^3) and the fracture toughness of HT (3.4 g/cm^3) [118]. Dense Dio has a bending strength of 300 MPa, a young's modulus of 170 GPa, and a fracture toughness of 3.5 MPa $m^{1/2}$. These values were approximately two to three times higher than those of HAp. Diopside lost less weight than HAp in an aqueous solution of physiological salt solution and lactic acid. Diopside is expected to be beneficial for a variety of applications, including dental roots and prosthetic bones [118]. Dio has been primarily employed as a bone infill, coating, and scaffold for orthopedic implants, as of this date [79, 117]. HA formation, a bioactivity phenomenon, has also been demonstrated in prior research to occur on the surface of bulk Dio three days after immersion in SBF [79, 119].

It has been established that the diopside bioceramic is both bioactive and biocompatible, and it has the capacity to promote osteogenesis in vivo, as well as stimulate osteoblast and bone marrow mesenchymal stem cell proliferation and osteogenic differentiation [120, 121]. In order to generate single-phase diopside and forsterite scaffolds, a variety of methods can be implemented, including frothing [105, 122], gel-casting [123, 124], and space holder [125]. The space holder approach is of great interest to a large number of scientists because it has the capacity to produce linked porosities with regulated dimensions and appropriate mechanical characteristics [126]. In addition, the space holder method was used in order to produce bioglass and bioceramics. This was accomplished by using a broad variety of porogen agents, including sodium chloride, gelatin, sucrose, ammonium nitrate, rice husk, and polyethylene [116, 127].

8.4.6 Other Ceramics

Olivine, $(Mg.Fe.Mn.Ca)_2SiO_4$ is a family of orthosilicate minerals with a relatively simple structure. The magnesium octahedra are connected to the silicate tetrahedra by oxygen bonding. It is a mineral that is found in a substantial quantity of the Earth's mantle and oceanic crust, as well as in a variety of terrestrial environments [128]. Pyroxenes are also essential in a variety of extraterrestrial and geological environments, despite their lower prevalence. For instance, the granulite facies are identified by the presence of orthopyroxenes and clinopyroxenes in metamorphosed mafic rocks. Pigeonites were identified as significant components of lunar samples, with these minerals being present in a diverse assemblage of igneous materials. The production of glass–ceramics and phosphors [129] has been accomplished by the use of pyroline compositions. The structure of pyroxene is more polymerized than that of olivines. It is composed of a single-chain silicate that has two octahedral sites that are located in between the chains. The enstatite ($MgSiO_3$)—ferrosilite ($FeSiO_3$) series is compositionally comparable to the well-known olivine series forsterite (Mg_2SiO_4)—fayalite (Fe_2SiO_4) [128]. This is despite the fact that pyroxenes span a wider variety of solid solutions that contain a multitude of prospective and known end-member phases. Bioactive mineral additives are frequently incorporated into degradable polymers during the fabrication of composite scaffolds. Polyester/silica scaffolds are particularly well-suited for the production of bone tissues. In addition to improving the mechanical properties of the scaffold, appropriate mineral compounds can also substantially contribute to the enhancement of cell adhesion, proliferation, and differentiation in these scaffolds [130]. Additionally, bioactive additives have the ability to neutralize deleterious acidic products, which prevents the inflammatory response and speeds up the formation of the bone-calcium phosphate interface layer [131].

However, the compounds' non-uniform distribution is a consequence of the significant disparity in surface tension [132]. Significant attention has been paid to the bioactive and mechanical properties of silica-based glass–ceramic [85].

8.5 Bioglass

In addition to crystalline bioceramics, bioactive glass (BGs) is a synthetic, biocompatible, vitreous-structured material that has been employed as a bone substitute. Bioactive glasses have shown very favorable characteristics as scaffold materials, such as their capacity to establish a strong connection with bone tissue and their ability to bind with bone at a faster pace compared to other bioactive ceramics [133, 134]. Various biomedical applications for bioactive glasses include scaffolds and bone fillers that facilitate the formation of new bones, and small implants in non-load-bearing purposes [135]. Table 8.4 presents some examples of these BGs and their respective compositions. The bioactivity of BGs is frequently associated with their capacity to form a HA layer on their surface, which closely resembles the composition and structure of natural bone. This occurs when BGs are implanted in the human body or soaked in SBF [133–135].

Ceramic and bioglass scaffolds can be manufactured using several methods, each possessing distinct benefits and drawbacks. Some of the characteristics of a scaffold may be influenced by the methodology used to acquire it. Three-dimensional (3D) printing, foam replica, and freeze-drying are the most frequently employed techniques [134, 136–140].

Table 8.4 Composition of different types of BGs in %mol [134]

BG name	SrO	B_2O_3	MgO	K_2O	P_2O_5	Na_2O	CaO	SiO_2
PSr40	20.0	–	–	–	50.0	10.0	20.0	–
P40B10	20.0	10.0	–	–	40.0	10.0	20.0	–
S53B50	–	26.9	–	–	1.7	22.7	21.8	26.9
13-93B3	–	54.0	8.0	8.0	2.0	6.0	22.0	–
13-93B1	–	18.0	8.0	8.0	2.0	6.0	22.0	36.0
13-93	–	–	8.0	8.0	2.0	6.0	22.0	54.0
45S5	–	–	–	–	2.6	24.3	26.9	46.1

8.6 3D Printing

A diverse range of methods have been employed in the construction of 3D scaffolds, occasionally in combination. Conventional approaches typically struggle to produce intricate scaffold microarchitectures with accurate control. Nevertheless, the incorporation of 3D printing with computer-aided design (CAD) modeling into BTE has significantly enhanced repeatability and accuracy of scaffold manufacturing, allowing for precise control over both micro and macroporosity of the scaffolds. Table 8.5 summarizes the benefits and drawbacks of regular scaffold production methods and newer 3D printing processes [138]. Figure 8.3 displays a schematic depiction of the development of printing technology [35].

Figure 8.4 illustrates the mechanisms of some newly developed 3D printing techniques employed in the printing of biomaterials [35]. Laser-assisted printers utilize a concentrated pulse laser beam to target an absorbing substrate, causing the formation of a pressure bubble. This bubble then propels biological material onto the collecting substrate. Thermal inkjet printers utilize electrical heating of the print head to generate pressure pulses, which in turn propel droplets of biological material via a nozzle. Acoustic inkjet printers employ piezoelectric pressure to generate pulses that divide liquids into droplets [141].

For the advancement of numerous applications, such as the production of injectable and highly efficient calcium phosphate cements [143, 144] and the production of durable and long-lasting composites [145], precise control over particle morphology is essential. Currently, the sole methods available for modifying the small-scale structure of β-TCP particles are hydrothermal and solution-based methods [52]. β-TCP scaffolds without macropores are readily degraded and replaced by bone due to their resorbable nature. However, this activity is much enhanced when there are connected large holes (larger than 100 μm) [146]. To avoid nozzle blockage during robocasting or 3D printing (paste) extrusion, it is essential to use β-TCP granules that are highly de-agglomerated, spherical, and fine [143, 144]. The existence of agglomerates in the β-TCP granules, which are usually several dozen micrometers in size, results in printed rods with a diameter greater than 100 μm [147]. On the other hand, pastes and slurries that have the right rheological properties need a large amount of liquid to be created by very small particles [148, 149]. As a result, the strength of hydraulic cements may decrease [149] and it may become more challenging to employ slip-casting processes to make ceramics containing β-TCP [148]. It is recommended that the materials used in 3D printing have a particle size that is neither too fine nor too coarse [150]. This image underscores the significance of the ability to regulate the size and morphology of β-TCP particles, particularly in the sub-micrometer region. The biological response at the macroscale is determined by the unique structure and form of β-TCP, which is why geometry control is necessary [151, 152].

The size of the β-TCP particle can be regulated using either a bottom-up or top-down approach. The β-TCP units that are produced during phase conversion or solid-state sintering are fractured using the top-down method [52]. Rapid purification of

Table 8.5 Comparison of scaffold manufacturing techniques [130]

Manufacturing	Benefits	Potential limitations
3D printing • Microextrusion • Microvalve • Laser-assisted • Inkjet • FDM • SLS • SLA	Cells can be incorporated into scaffold materials at high concentrations, depending on technique used Complex 3D shapes with controlled internal structures, morphology and pore size, and high resolution are possible to create Enhanced ability to integrate vascular structures into constructions	The initial investment required for machines might be costly Certain procedures are constrained by the available range of printable materials
Electrospinning	Low-cost and simple technique Produce scaffold that provides large surface area for cells to adhere to	It is challenging to integrate specific microarchitecture into constructs Restricted mechanical characteristics Cells may be adversely affected by the use of organic solvents, which can be toxic
Phase separation	Efficiently can be integrated with other methods Removes the leaching step of porogen	Utilization of organic solvents hinders the application of bioactive compounds or cells in process of scaffold construction The utilization is restricted due to the presence of narrow pore diameters
Emulsification freeze-drying	Does not necessitate the utilization of a solid porogen	Extended duration of processing Porosity is frequently irregular Limited pore dimensions Necessitates the utilization of organic solvents

(continued)

Table 8.5 (continued)

Manufacturing	Benefits	Potential limitations
Gas foaming	Reduces reliance on chemical solvents	Challenging to regulate the dimensions of pores and guarantee their connection Temperature-sensitive materials may undergo denaturation during the compression molding process The incorporation of cells and bioactive compounds directly into scaffolds is prohibited due to the high pressures involved
Solvent casting/particulate leaching	A relatively straightforward method that enables the fabrication of scaffolds with precise control over pore size and composition, as well as uniform porosity	The constructions have a limited thickness and it is difficult to achieve desired mechanical qualities Controlling the interconnectivity and form of pores might pose challenges in biomolecules and cells cannot be directly integrated in scaffolds due to the use of organic solvents

large amounts of material may be achieved; however, the fine texture of the powder is lost and it becomes amorphous when the particle size exceeds a certain threshold. In order to get powders that are much finer, it is important to use a bottom-up method that includes the precipitation or thermal conversion of nanosized ACP or CDHA particles, even though attrition may result in somewhat finer powders compared to ball milling [52]. Micropores, with diameters ranging from 0.1 to 100 μm, improve the process of bone formation and promote the breakdown of material, hence enhancing the biological reaction of β-TCP. As the temperature and sintering time increase, the microporosity of a ceramic decreases. Micropores expand concurrently. Consequently, it is difficult to modify the microporosity without also affecting the dimensions of the micropores [52]. The theoretical tensile strength of β-TCP is 1–10 GPa, as indicated by Refs. [123, 145]. This figure is equivalent to a range of 1–10% of the anticipated Young's modulus, which is 110 GPa. Nevertheless, the experimental tensile properties are significantly diminished as a result of the intrinsic fragility of ceramics. A value of 154 MPa was reported by Jarcho et al. [153].

Despite substantial study, the processing, manufacturing, and synthesis of bio-inks using HA still face major obstacles in order to attain the desired characteristics in natural hard tissue environments. This is especially applicable to load-bearing implants composed of biomaterials produced from HA, which differ according to the specific hard tissue or organ being addressed. Therefore, it is crucial to improve the design and development of bio-inks to enable the successful regeneration of tissues that closely mimic their natural counterparts. This involves the use of bioprinted

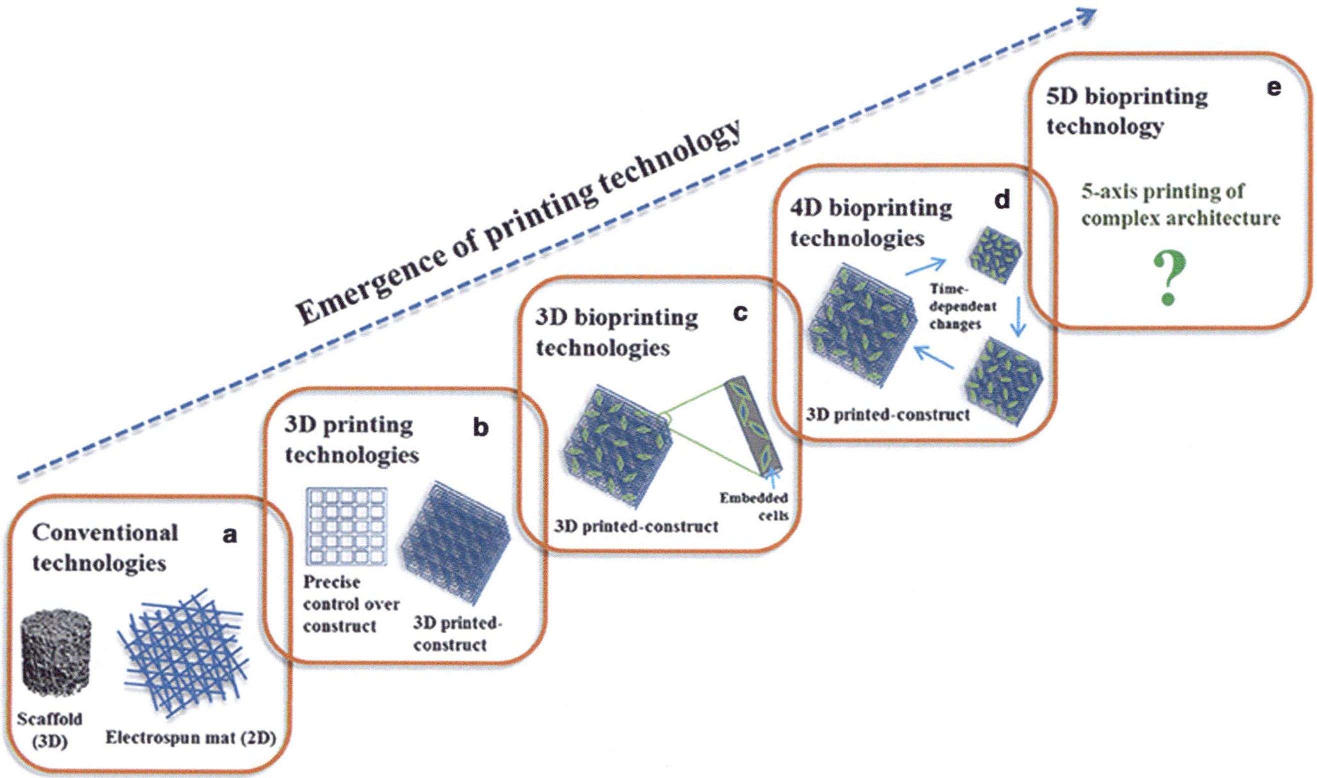

Fig. 8.3 Emergence of printing technology: **a** 2D/3D conventional, **b** 3D printing, **c** 3D bioprinting, **d** 4D bioprinting, and **e** 5D bioprinting. Reprinted with permission from [35]

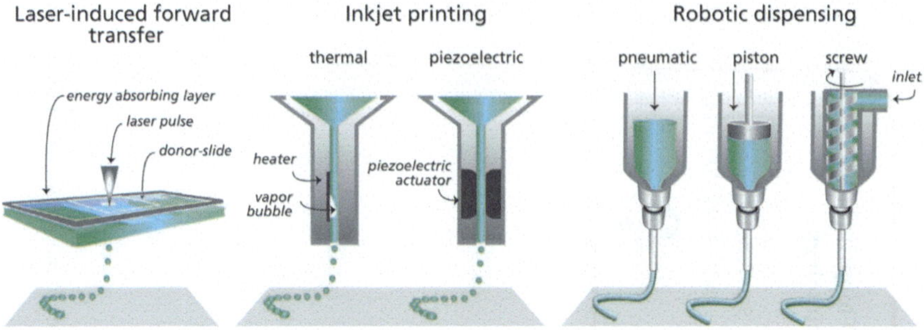

Fig. 8.4 Popular additive manufacturing techniques used for the fabrication of intricate structures. Reprinted with permission from [142]

implants to promote the integration, gradual restructuring, development of blood vessels, and maturation of the implant. This optimization entails evaluating printing-related concerns relating to cells, materials, and processing parameters, both before and after printing [154]. This requires expertise in the living biological environment, biomaterials, cells, and printing techniques. Nozzle obstruction and the achievement of structurally sound designs are important factors in the printing process. To achieve cellular activity and diversity in a live organism, it is necessary to use a range of printing processes, cell types, and materials to build increasingly intricate 3D structures [155]. Obtaining well-structured hydroxyapatite scaffolds through printing is a challenging task. To do this, sacrificial materials such as polymers or other complex substances like ceramics and bioactive glasses are often used. These materials help in creating 3D-printed constructions with precise architecture and porosity. In this development, different additive manufacturing techniques have been employed to create either pure HA or HA-based three-dimensional printed structures for the purpose of regenerating hard tissues. The resulting inks may be classified into two categories: cell-free biomaterial inks and cell-encapsulated biomaterial inks. Figure 8.5 will outline the current advancements achieved by utilizing either HA or HA as one of the constituents [35].

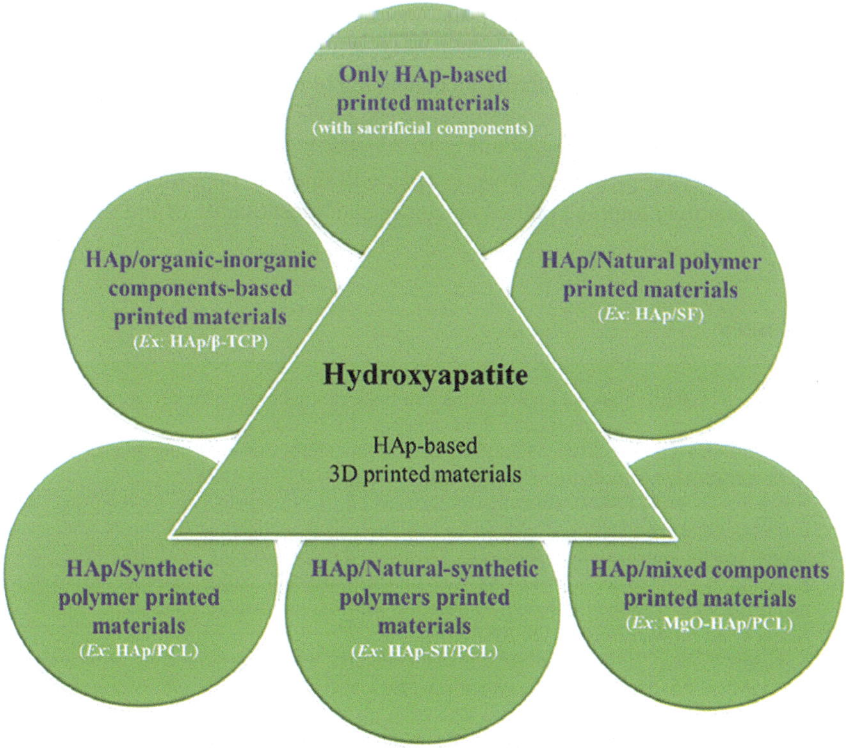

Fig. 8.5 Diagram illustrating an overview of hydroxyapatite-based printed materials used in hard tissue regeneration. Reprinted with permission from [35]

8.7 Conclusion

Calcium phosphate-based biomaterials, namely HA and tricalcium phosphates, are considered extremely promising for bone regeneration. This is because they closely resemble natural bone in terms of chemical composition and exhibit remarkable biocompatibility. HA, which imitates bone mineral, promotes the integration of bone and acts as a framework for bone formation. Tricalcium phosphate, on the other hand, has degradation rates that can be adjusted, making it more suitable for a variety of medical uses. Doping calcium phosphates with other chemicals, along with other variants, offers further options in the area of bone tissue engineering. This broadens the scope of potential options for tailored therapies. In addition to conventional calcium phosphates, bioceramics based on calcium-magnesium and calcium silicate, such as enstatite, forsterite, akermanite, bredigite, monticellite, hardystonite, and diopside, possess distinct characteristics that are favorable for bone regeneration. These materials improve the process of protein adsorption, aid in the spreading of cells, and stimulate the expression of genes related to bone formation. Consequently, they are highly suitable for a range of uses, such as nanocomposite scaffolds and bone cavity transplants. Bioglass, a vital biomaterial, enhances bone bonding, stimulates new bone tissue creation, and promotes blood vessel growth via compositional modifications. This presents further prospects for enhancing biological responses in the realm of regenerative medicine. The continuous investigation of nano-bioceramics has the potential to improve bone regeneration treatments by clarifying cellular connections and underlying processes. This, in turn, will open up opportunities for novel techniques that may improve patient outcomes and boost quality of life.

References

1. Alidadi, S. (2020). Nanoscale bioceramics in bone tissue engineering-an overview. *Indian Journal of Veterinary Sciences and Biotechnology, 16*(2, 3, 4), 7–11.
2. Kalantari, E., et al. (2018). Nanostructured monticellite: An emerging player in tissue engineering. *Materials Today: Proceedings, 5*(7, Part 3), 15744–15753.
3. Kalantari, E., et al. (2019). Biocomposites based on hydroxyapatite matrix reinforced with nanostructured monticellite ($CaMgSiO_4$) for biomedical application: Synthesis, characterization, and biological studies. *Materials Science and Engineering: C, 105*, 109912.
4. Kalantari, E., & Naghib, S. M. (2019). A comparative study on biological properties of novel nanostructured monticellite-based composites with hydroxyapatite bioceramic. *Materials Science and Engineering: C, 98*, 1087–1096.
5. Kalantari, E., et al. (2018). Nanostructured monticellite for tissue engineering applications—Part I: Microstructural and physicochemical characteristics. *Ceramics International, 44*(11), 12731–12738.
6. Kalantari, E., et al. (2018). Nanostructured monticellite for tissue engineering applications—Part II: Molecular and biological characteristics. *Ceramics International, 44*(12), 14704–14711.
7. Pauline, S. A. (2021). Nanoceramics: Synthesis, characterizations and applications. *Nanomaterials and Their Biomedical Applications*, 131–156.
8. Seyfoori, A., et al. (2019). Multifunctional magnetic $ZnFe_2O_4$-hydroxyapatite nanocomposite particles for local anti-cancer drug delivery and bacterial infection inhibition: An in vitro study. *Journal of the Taiwan Institute of Chemical Engineers, 96*, 503–508.
9. Rahmanian, M., et al. (2019). Multifunctional gelatin–tricalcium phosphate porous nanocomposite scaffolds for tissue engineering and local drug delivery: In vitro and in vivo studies. *Journal of the Taiwan Institute of Chemical Engineers, 101*, 214–220.
10. Askari, E., et al. (2021). Local delivery of chemotherapeutic agent in tissue engineering based on gelatin/graphene hydrogel. *Journal of Materials Research and Technology, 12*, 412–422.
11. Askari, E., et al. (2022). Highly bioactive akermanite-monticellite nanocomposites for bone tissue engineering: A tunable three-dimensional biological study. *Journal of Materials Research and Technology, 20*, 4066–4076.
12. LeGeros, R. Z. (2008). Calcium phosphate-based osteoinductive materials. *Chemical Reviews, 108*(11), 4742–4753.
13. Moioli, E. K., et al. (2007). Matrices and scaffolds for drug delivery in dental, oral and craniofacial tissue engineering. *Advanced Drug Delivery Reviews, 59*(4–5), 308–324.
14. Islam, M. T., et al. (2017). Bioactive calcium phosphate–based glasses and ceramics and their biomedical applications: A review. *Journal of Tissue Engineering, 8*, 2041731417719170.
15. Shah, R., et al. (2005). Craniofacial muscle engineering using a 3-dimensional phosphate glass fibre construct. *Biomaterials, 26*(13), 1497–1505.
16. Kangarshahi, B. M., et al. (2024). Bioprinting of self-healing materials and nanostructures for biomedical applications: Recent advances and progresses on fabrication and characterization techniques. *Bioprinting, 38*, e00335.

17. Soleymani, S., & Naghib, S. M. (2023). 3D and 4D printing hydroxyapatite-based scaffolds for bone tissue engineering and regeneration. *Heliyon, 9*(9), e19363.
18. Bagherpour, I., et al. (2023). Synthesis and investigation on microstructural, mechanical features of mesoporous hardystonite/reduced graphene oxide nanocomposite for medical applications. *Frontiers in Bioengineering and Biotechnology, 11*, 1073435.
19. Safarzadeh, S., Mozafari, M. R., & Naghib, S. M. (2024). Chitosan-incorporated bioceramic-based nanomaterials for localized release of therapeutics and bone regeneration: An overview of recent advances and progresses. *Current Organic Chemistry, 28*(15), 1190–1214.
20. Soleymani, S., Naghib, S. M., & Mozafari, M. R. (2024). An overview of cultured meat and stem cell bioprinting: How to make it, challenges and prospects, environmental effects, society's culture and the influence of religions. *Journal of Agriculture and Food Research*, 101307.
21. Naghib, S. M., Zarrineh, M., & Mozafari, M. R. (2024). 3D printing chitosan-based nanobiomaterials for biomedicine and drug delivery: Recent advances on the promising bioactive agents and technologies. *Current Organic Chemistry, 28*(7), 510–525.
22. Mikaeeli Kangarshahi, B., & Naghib, S. M. (2024). Multicomponent 3D-printed Collagen-based Scaffolds for cartilage regeneration: Recent progress, developments, and emerging technologies. *Current Organic Chemistry, 28*(8), 576–594.
23. Canillas, M., et al. (2017). Calcium phosphates for biomedical applications. *Boletín de la Sociedad Española de Cerámica Y vidrio, 56*(3), 91–112.
24. Heymann, D., Guicheux, J., & Rousselle, A. (2001). Ultrastructural evidence in vitro of osteoclastinduced degradation of calcium phosphate ceramic by simultaneous resorption and phagocytosis mechanisms. *Histology and Histopathology, 16*(1), 37–44.
25. Sakae, T., et al. (2003). Bone formation induced by several carbonate-and fluoride-containing apatite implanted in dog mandible. In *Key Engineering Materials*. Trans Tech Publ.
26. Ito, A., et al. (2006). Dissolution rate of zinc-containing β-tricalcium phosphate ceramics. *Biomedical Materials, 1*(3), 134.
27. Yamada, Y., et al. (2008). Inhibitory effect of Zn2+ in zinc-containing β-tricalcium phosphate on resorbing activity of mature osteoclasts. *Journal of Biomedical Materials Research Part A, 84*(2), 344–352.
28. Szcześ, A., Hołysz, L., & Chibowski, E. (2017). Synthesis of hydroxyapatite for biomedical applications. *Advances in Colloid and Interface Science, 249*, 321–330.
29. Ansari, M., et al. (2011). Synthesis and characterisation of hydroxyapatite-calcium hydroxide for dental composites. *Ceramics-Silikáty, 55*(2), 123–126.
30. Khoshsima, S., et al. (2016). Structural, mechanical and biological properties of hydroxyapatite-zirconia-lanthanum oxide composites. *Ceramics International, 42*(14), 15773–15779.
31. Seyfoori, A., Imani Fooladi, A. A., & Mahmoodzadeh Hosseini, H. (2017). Calcium phosphate-based nanocomposite carriers for local antibiotic delivery against an osteomyelitis agent. *Advances in Applied Ceramics, 116*(6), 316–324.
32. Saikia, K., et al. (2008). Calcium phosphate ceramics as bone graft substitutes in filling bone tumor defects. *Indian Journal of Orthopaedics, 42*(2), 169.
33. Thamaraiselvi, T., Prabakaran, K., & Rajeswari, S. (2006). Synthesis of hydroxyapatite that mimic bone minerology. *Trends Biomater Artif Organs, 19*(2), 81–83.
34. Sopyan, I., et al. (2007). Porous hydroxyapatite for artificial bone applications. *Science and Technology of Advanced Materials, 8*(1–2), 116.
35. Kumar, A., et al. (2019). Additive manufacturing methods for producing hydroxyapatite and hydroxyapatite-based composite scaffolds: A review. *Frontiers in Materials*.
36. Radha, G., et al. (2015). Evaluation of hemocompatibility and in vitro immersion on microwave-assisted hydroxyapatite–alumina nanocomposites. *Materials Science and Engineering: C, 50*, 143–150.
37. Pandey, A., et al. (2018). Antioxidant and antibacterial hydroxyapatite-based biocomposite for orthopedic applications. *Materials Science and Engineering: C, 88*, 13–24.

38. Bellucci, D., et al. (2013). A new hydroxyapatite-based biocomposite for bone replacement. *Materials Science and Engineering: C, 33*(3), 1091–1101.
39. Amerio, P., et al. (2010). The effect of deproteinized bovine bone on osteoblast growth factors and proinflammatory cytokine production. *Clinical Oral Implants Research, 21*(6), 650–655.
40. Remya, N., et al. (2017). Investigation of chronic toxicity of hydroxyapatite nanoparticles administered orally for one year in wistar rats. *Materials Science and Engineering: C, 76*, 518–527.
41. Gonçalves, G., et al. (2012). Graphene oxide versus functionalized carbon nanotubes as a reinforcing agent in a PMMA/HA bone cement. *Nanoscale, 4*(9), 2937–2945.
42. Liu, Y., et al. (2014). Hydroxyapatite/graphene-nanosheet composite coatings deposited by vacuum cold spraying for biomedical applications: Inherited nanostructures and enhanced properties. *Carbon, 67*, 250–259.
43. Li, M., et al. (2014). Graphene oxide/hydroxyapatite composite coatings fabricated by electrophoretic nanotechnology for biological applications. *Carbon, 67*, 185–197.
44. Kouhi, M., et al. (2016). Synthesis, characterization, in vitro bioactivity and biocompatibility evaluation of hydroxyapatite/bredigite (Ca7 MgSi4O16) composite nanoparticles. *JOM Journal of the Minerals Metals and Materials Society, 68*(4), 1061–1070.
45. Ramesh, N., Moratti, S. C., & Dias, G. J. (2018). Hydroxyapatite–polymer biocomposites for bone regeneration: A review of current trends. *Journal of Biomedical Materials Research Part B: Applied Biomaterials, 106*(5), 2046–2057.
46. Luo, Y., et al. (2018). 3D printing of strontium-doped hydroxyapatite based composite scaffolds for repairing critical-sized rabbit calvarial defects. *Biomedical Materials, 13*(6), 065004.
47. Chen, S., et al. (2019). Biomimetic synthesis of Mg-substituted hydroxyapatite nanocomposites and three-dimensional printing of composite scaffolds for bone regeneration. *Journal of Biomedical Materials Research Part A, 107*(11), 2512–2521.
48. Hu, M., et al. (2018). Enhancement mechanism of hydroxyapatite for photocatalytic degradation of gaseous formaldehyde over TiO_2/hydroxyapatite. *Journal of the Taiwan Institute of Chemical Engineers, 85*, 91–97.
49. Rahmanian, M., et al. (2017). Tricalcium phosphate nanostructures loaded with bisphosphonate as potential anticancer agents. *Ceramic Science and Technology, 8*, 505–512.
50. Carrodeguas, R. G., & De Aza, S. (2011). α-Tricalcium phosphate: Synthesis, properties and biomedical applications. *Acta Biomaterialia, 7*(10), 3536–3546.
51. Burkes, D. E., Moore, J. J., & Ayers, R. A. (2008). Method for producing calcium phosphate powders using an auto-ignition combustion synthesis reaction. *Google Patents.*
52. Bohner, M., Santoni, B. L. G., & Döbelin, N. (2020). β-tricalcium phosphate for bone substitution: Synthesis and properties. *Acta Biomaterialia.*
53. Bohner, M., & Lemaitre, J. (2009). Can bioactivity be tested in vitro with SBF solution? *Biomaterials, 30*(12), 2175–2179.
54. Yuan, H., et al. (2010). Osteoinductive ceramics as a synthetic alternative to autologous bone grafting. *Proceedings of the National Academy of Sciences, 107*(31), 13614–13619.
55. Gomes, S., et al. (2018). Cu-doping of calcium phosphate bioceramics: From mechanism to the control of cytotoxicity. *Acta Biomaterialia, 65*, 462–474.
56. Alshemary, A. Z., et al. (2015). Synthesis, characterization and optical properties of chromium doped β-Tricalcium phosphate. *Ceramics International, 41*(1), 1663–1669.
57. Destainville, A., et al. (2003). Synthesis, characterization and thermal behavior of apatitic tricalcium phosphate. *Materials Chemistry and Physics, 80*(1), 269–277.
58. Hudon, P., & Jung, I.-H. (2015). Critical evaluation and thermodynamic optimization of the CaO-P 2 O 5 system. *Metallurgical and Materials Transactions B, 46*(1), 494–522.
59. Carrodeguas, R. G., et al. (2008). New approach to the β→ α polymorphic transformation in magnesium-substituted tricalcium phosphate and its practical implications. *Journal of the American Ceramic Society, 91*(4), 1281–1286.
60. Carbajal, L., et al. (2017). Phase and melting relationships of β, α and α′-Ca_3 (PO_4) 2 polymorphs in the Ca_3 (PO_4) 2-Zn3 (PO_4) 2 system. *Journal of the European Ceramic Society, 37*(5), 2277–2283.

61. Matsunaga, K., et al. (2015). First-principles calculations of divalent substitution of Ca^{2+} in tricalcium phosphates. *Acta Biomaterialia, 23*, 329–337.
62. Yashima, M., et al. (2003). Crystal structure analysis of β-tricalcium phosphate $Ca_3 (PO_4) 2$ by neutron powder diffraction. *Journal of Solid State Chemistry, 175*(2), 272–277.
63. Frasnelli, M., & Sglavo, V. M. (2016). Effect of Mg2+ doping on beta–alpha phase transition in tricalcium phosphate (TCP) bioceramics. *Acta Biomaterialia, 33*, 283–289.
64. Kannan, S., et al. (2010). Synthesis and structural characterization of strontium-and magnesium-co-substituted β-tricalcium phosphate. *Acta Biomaterialia, 6*(2), 571–576.
65. Hesaraki, S., et al. (2012). Phase evaluation and lattice parameters of strontium/magnesium incorporated beta-tricalcium phosphate bioceramics. *Journal of the Australian Ceramic Society, 48*(2), 166–172.
66. Jay, E., et al. (2010). Predicted energies and structures of β-Ca_3 (PO_4) 2. *Journal of Solid State Chemistry, 183*(10), 2261–2267.
67. Chow, L. C. (2009). Next generation calcium phosphate-based biomaterials. *Dental Materials Journal, 28*(1), 1–10.
68. Wang, L., & Nancollas, G. H. (2008). Calcium orthophosphates: Crystallization and dissolution. *Chemical Reviews, 108*(11), 4628–4669.
69. Zeng, R.-C., et al. (2015). In vitro degradation of pure Mg in response to glucose. *Scientific Reports, 5*(1), 1–14.
70. Zeng, R., et al. (2008). Progress and challenge for magnesium alloys as biomaterials. *Advanced Engineering Materials, 10*(8), B3–B14.
71. Noronha, J. L., & Matuschak, G. M. (2006). Magnesium in critical illness: Metabolism, assessment, and treatment. *Applied Physiology in Intensive Care Medicine*, 157–169.
72. Wolf, F. I., & Cittadini, A. (2003). Chemistry and biochemistry of magnesium. *Molecular Aspects of Medicine, 24*(1–3), 3–9.
73. Zhang, X., et al. (2016). Microstructures and corrosion behavior of biodegradable Mg–6Gd–xZn–0.4 Zr alloys with and without long period stacking ordered structure. *Corrosion Science, 105*, 68–77.
74. Cheng, J., et al. (2013). Comparative in vitro study on pure metals (Fe, Mn, Mg, Zn and W) as biodegradable metals. *Journal of Materials Science & Technology, 29*(7), 619–627.
75. Cui, L.-Y., et al. (2019). In vitro corrosion resistance and antibacterial performance of novel tin dioxide-doped calcium phosphate coating on degradable Mg-1Li-1Ca alloy. *Journal of Materials Science & Technology, 35*(3), 254–265.
76. Lu, J., et al. (2011). Preparation and preliminary cytocompatibility of magnesium doped apatite cement with degradability for bone regeneration. *Journal of Materials Science: Materials in Medicine, 22*(3), 607–615.
77. Yang, X., et al. (2014). Rational design and fabrication of a β-dicalcium silicate-based multifunctional cement with potential for root canal filling treatment. *Journal of Materials Chemistry B, 2*(24), 3830–3838.
78. Ma, N., et al. (2019). In vivo evaluation of the subchronic systemic toxicity of akermanite bioceramic for bone regeneration following ISO standard methods. *RSC Advances, 9*(31), 17530–17536.
79. Diba, M., et al. (2014). Magnesium-containing bioactive polycrystalline silicate-based ceramics and glass-ceramics for biomedical applications. *Current Opinion in Solid State and Materials Science, 18*(3), 147–167.
80. Wu, C., et al. (2014). Stimulation of osteogenic and angiogenic ability of cells on polymers by pulsed laser deposition of uniform akermanite-glass nanolayer. *Acta Biomaterialia, 10*(7), 3295–3306.
81. Wang, F., et al. (2020). Akermanite bioceramic enhances wound healing with accelerated reepithelialization by promoting proliferation, migration, and stemness of epidermal cells. *Wound Repair and Regeneration, 28*(1), 16–25.
82. Šupová, M. (2015). Substituted hydroxyapatites for biomedical applications: A review. *Ceramics International, 41*(8), 9203–9231.

83. Camaioni, A., et al. (2015). Silicon-substituted hydroxyapatite for biomedical applications. *Hydroxyapatite (HAp) for Biomedical Applications* (pp. 343–373). Elsevier.
84. Kermani, F., et al. (2020). Silicon-doped calcium phosphates; the critical effect of synthesis routes on the biological performance. *Materials Science and Engineering: C, 111*, 110828.
85. Arastouei, M., et al. (2020). Improving the properties of the porous polylactic acid scaffold by akermanite nanoparticles for bone tissue engineering. *Journal of Advanced Materials and Processing, 8*(2), 11–19.
86. Dong, X., Chang, J., & Li, H. (2017). Bioglass promotes wound healing through modulating the paracrine effects between macrophages and repairing cells. *Journal of Materials Chemistry B, 5*(26), 5240–5250.
87. Huang, Y., et al. (2009). In vitro and in vivo evaluation of akermanite bioceramics for bone regeneration. *Biomaterials, 30*(28), 5041–5048.
88. Ewais, E. M. M., et al. (2015). Bioactivity investigations with calcia magnesia based composites. *Journal of Materials Science and Engineering A, 5*(1–2), 21–36.
89. Amin, A., et al. (2016). Zirconia effect on the bioactivity and the mechanical properties of calcium magnesium silicate ceramics at (Cao + Mgo)/SiO_2 molar ratio close to unity. *Bioceramics Development and Applications, 6*(088), 2.
90. Kumar, R., et al. (2010). In vivo biodistribution and clearance studies using multimodal organically modified silica nanoparticles. *ACS Nano, 4*(2), 699–708.
91. Yu, T., et al. (2012). In vivo biodistribution and pharmacokinetics of silica nanoparticles as a function of geometry, porosity and surface characteristics. *Journal of Controlled Release, 163*(1), 46–54.
92. Shie, M.-Y., Ding, S.-J., & Chang, H.-C. (2011). The role of silicon in osteoblast-like cell proliferation and apoptosis. *Acta Biomaterialia, 7*(6), 2604–2614.
93. Jadidi, A., & Salahinejad, E. (2020). Mechanical strength and biocompatibility of bredigite ($Ca_7MgSi_4O_{16}$) tissue-engineering scaffolds modified by aliphatic polyester coatings. *Ceramics International, 46*(10), 16439–16446.
94. Wu, C., et al. (2005). Preparation and characteristics of a calcium magnesium silicate (bredigite) bioactive ceramic. *Biomaterials, 26*(16), 2925–2931.
95. Wu, C., et al. (2007). A novel bioactive porous bredigite ($Ca_7MgSi_4O_{16}$) scaffold with biomimetic apatite layer for bone tissue engineering. *Journal of Materials Science: Materials in Medicine, 18*(5), 857–864.
96. Eilbagi, M., et al. (2016). Mechanical and cytotoxicity evaluation of nanostructured hydroxyapatite-bredigite scaffolds for bone regeneration. *Materials Science and Engineering: C, 68*, 603–612.
97. Wu, C., & Chang, J. (2007). Degradation, bioactivity, and cytocompatibility of diopside, akermanite, and bredigite ceramics. *Journal of Biomedical Materials Research Part B: Applied Biomaterials: An Official Journal of The Society for Biomaterials, The Japanese Society for Biomaterials, and The Australian Society for Biomaterials and the Korean Society for Biomaterials, 83*(1), 153–160.
98. Keihan, R., et al. (2020). Biomineralization, strength and cytocompatibility improvement of bredigite scaffolds through doping/coating. *Ceramics International, 46*(13), 21056–21063.
99. Wu, C., & Chang, J. (2007). Synthesis and in vitro bioactivity of bredigite powders. *Journal of Biomaterials Applications, 21*(3), 251–263.
100. Webster, T. J., et al. (2002). Hydroxylapatite with substituted magnesium, zinc, cadmium, and yttrium. II. Mechanisms of osteoblast adhesion. *Journal of Biomedical Materials Research: An Official Journal of The Society for Biomaterials, The Japanese Society for Biomaterials, and The Australian Society for Biomaterials and the Korean Society for Biomaterials, 59*(2), 312–317.
101. Kalantari, E., et al. (2017). Green solvent-based sol–gel synthesis of monticellite nanoparticles: A rapid and efficient approach. *Journal of Sol-Gel Science and Technology, 84*(1), 87–95.
102. Hafezi-Ardakani, M., et al. (2011). Synthesis and characterization of nanocrystalline merwinite ($Ca_3Mg(SiO_4)_2$) via sol–gel method. *Ceramics International, 37*(1), 175–180.

103. Wu, C., & Chang, J. (2004). Synthesis and apatite-formation ability of akermanite. *Materials Letters, 58*(19), 2415–2417.
104. Chen, X., et al. (2008). Synthesis and characteristics of monticellite bioactive ceramic. *Journal of Materials Science: Materials in Medicine, 19*(3), 1257–1263.
105. Wu, C., Ramaswamy, Y., & Zreiqat, H. (2010). Porous diopside ($CaMgSi_2O_6$) scaffold: A promising bioactive material for bone tissue engineering. *Acta Biomaterialia, 6*(6), 2237–2245.
106. Iwata, N. Y., et al. (2004). Sintering behavior and apatite formation of diopside prepared by coprecipitation process. *Colloids and Surfaces B: Biointerfaces, 34*(4), 239–245.
107. Yamaguchi, M., & Yamaguchi, R. (1986). Action of zinc on bone metabolism in rats: Increases in alkaline phosphatase activity and DNA content. *Biochemical Pharmacology, 35*(5), 773–777.
108. Yamaguchi, M., Oishi, H., & Suketa, Y. (1987). Stimulatory effect of zinc on bone formation in tissue culture. *Biochemical Pharmacology, 36*(22), 4007–4012.
109. Jaiswal, A. K., et al. (2013). Hardystonite improves biocompatibility and strength of electrospun polycaprolactone nanofibers over hydroxyapatite: A comparative study. *Materials Science and Engineering: C, 33*(5), 2926–2936.
110. Wu, C., Chang, J., & Zhai, W. (2005). A novel hardystonite bioceramic: Preparation and characteristics. *Ceramics International, 31*(1), 27–31.
111. Ramaswamy, Y., et al. (2008). Biological response of human bone cells to zinc-modified Ca–Si-based ceramics. *Acta Biomaterialia, 4*(5), 1487–1497.
112. Elsayed, H., et al. (2016). Hardystonite bioceramics from preceramic polymers. *Journal of the European Ceramic Society, 36*(3), 829–835.
113. Lu, H., et al. (2010). In vitro proliferation and osteogenic differentiation of human bone marrow-derived mesenchymal stem cells cultured with hardystonite ($Ca_2ZnSi_2O_7$) and β-TCP ceramics. *Journal of Biomaterials Applications, 25*(1), 39–56.
114. Zhang, M., Lin, K., & Chang, J. (2012). Preparation and characterization of Sr–hardystonite ($Sr_2ZnSi_2O_7$) for bone repair applications. *Materials Science and Engineering: C, 32*(2), 184–188.
115. Studart, A. R., et al. (2006). Processing routes to macroporous ceramics: A review. *Journal of the American Ceramic Society, 89*(6), 1771–1789.
116. Sadeghzade, S., et al. (2017). Fabrication and evaluation of silica-based ceramic scaffolds for hard tissue engineering applications. *Materials Science and Engineering: C, 71*, 431–438.
117. Razavi, M., et al. (2014). In vitro study of nanostructured diopside coating on Mg alloy orthopedic implants. *Materials Science and Engineering: C, 41*, 168–177.
118. Nonami, T., & Tsutsumi, S. (1999). Study of diopside ceramics for biomaterials. *Journal of Materials Science: Materials in Medicine, 10*(8), 475–479.
119. Sadeghzade, S., Emadi, R., & Labbaf, S. (2017). Hardystonite-diopside nanocomposite scaffolds for bone tissue engineering applications. *Materials Chemistry and Physics, 202*, 95–103.
120. Tsigkou, O., et al. (2009). Differentiation of fetal osteoblasts and formation of mineralized bone nodules by 45S5 Bioglass® conditioned medium in the absence of osteogenic supplements. *Biomaterials, 30*(21), 3542–3550.
121. Ba, Z., et al. (2018). Nanoporous diopside modulates biocompatibility, degradability and osteogenesis of bioactive scaffolds of gliadin-based composites for new bone formation. *International Journal of Nanomedicine, 13*, 3883.
122. Mirhadi, S. (2014). Synthesis and characterization of nanostructured forsterite scaffolds using two step sintering method. *Journal of Alloys and Compounds, 610*, 399–401.
123. Ghomi, H., Emadi, R., & Javanmard, S. H. (2016). Preparation of nanostructure bioactive diopside scaffolds for bone tissue engineering by two near net shape manufacturing techniques. *Materials Letters, 167*, 157–160.
124. Ghomi, H., Jaberzadeh, M., & Fathi, M. (2011). Novel fabrication of forsterite scaffold with improved mechanical properties. *Journal of Alloys and Compounds, 509*(5), L63–L68.

125. Ghomi, H., Emadi, R., & Javanmard, S. H. (2016). Fabrication and characterization of nanostructure diopside scaffolds using the space holder method: Effect of different space holders and compaction pressures. *Materials & Design, 91*, 193–200.
126. Arifvianto, B., & Zhou, J. (2014). Fabrication of metallic biomedical scaffolds with the space holder method: A review. *Materials, 7*(5), 3588–3622.
127. Baino, F., Verné, E., & Vitale-Brovarone, C. (2009). 3-D high-strength glass–ceramic scaffolds containing fluoroapatite for load-bearing bone portions replacement. *Materials Science and Engineering: C, 29*(6), 2055–2062.
128. Anovitz, L. M., et al. (2017). Nano-scale synthesis of the complex silicate minerals forsterite and enstatite. *Journal of Colloid and Interface Science, 495*, 94–101.
129. Xia, Z., et al. (2013). Linear structural evolution induced tunable photoluminescence in clinopyroxene solid-solution phosphors. *Scientific Reports, 3*(1), 1–7.
130. Nejati, E., Mirzadeh, H., & Zandi, M. (2008). Synthesis and characterization of nano-hydroxyapatite rods/poly (l-lactide acid) composite scaffolds for bone tissue engineering. *Composites Part A: Applied Science and Manufacturing, 39*(10), 1589–1596.
131. Correia, D. M., et al. (2016). Strategies for the development of three dimensional scaffolds from piezoelectric poly (vinylidene fluoride). *Materials & Design, 92*, 674–681.
132. Deng, Y., et al. (2017). A novel akermanite/poly (lactic-co-glycolic acid) porous composite scaffold fabricated via a solvent casting-particulate leaching method improved by solvent self-proliferating process. *Regenerative Biomaterials, 4*(4), 233–242.
133. Erasmus, E., et al. (2018). In vitro evaluation of porous borosilicate, borophosphate and phosphate bioactive glasses scaffolds fabricated using foaming agent for bone regeneration. *Scientific Reports, 8*(1), 1–13.
134. Ribas, R. G., et al. (2019). Current advances in bone tissue engineering concerning ceramic and bioglass scaffolds: A review. *Ceramics International, 45*(17), 21051–21061.
135. Baino, F. (2018). Bioactive glasses–when glass science and technology meet regenerative medicine. *Ceramics International, 44*(13), 14953–14966.
136. Luo, G., et al. (2017). 13–93 bioactive glass/alginate composite scaffolds 3D printed under mild conditions for bone regeneration. *RSC Advances, 7*(20), 11880–11889.
137. Jin, Z., et al. (2018). Nonstoichiometric wollastonite bioceramic scaffolds with core-shell pore struts and adjustable mechanical and biodegradable properties. *Journal of the Mechanical Behavior of Biomedical Materials, 88*, 140–149.
138. Turnbull, G., et al. (2018). 3D bioactive composite scaffolds for bone tissue engineering. *Bioactive Materials, 3*(3), 278–314.
139. Fu, S., et al. (2018). 3D printed porous β-Ca_2SiO_4 scaffolds derived from preceramic resin and their physicochemical and biological properties. *Science and Technology of Advanced Materials, 19*(1), 495–506.
140. Meißner, R., et al. (2019). Bioprinted 3D calcium phosphate scaffolds with gentamicin releasing capability. *Ceramics International, 45*(6), 7090–7094.
141. Dzobo, K., et al. (2018). Advances in regenerative medicine and tissue engineering: Innovation and transformation of medicine. *Stem Cells International, 2018*(1), 2495848.
142. Malda, J., et al. (2013). 25th anniversary article: Engineering hydrogels for biofabrication. *Advanced Materials, 25*(36), 5011–5028.
143. O'Neill, R., et al. (2017). Critical review: Injectability of calcium phosphate pastes and cements. *Acta Biomaterialia, 50*, 1–19.
144. Bohner, M., & Baroud, G. (2005). Injectability of calcium phosphate pastes. *Biomaterials, 26*(13), 1553–1563.
145. Galea, L., et al. (2013). Control of the size, shape and composition of highly uniform, non-agglomerated, sub-micrometer β-tricalcium phosphate and dicalcium phosphate platelets. *Biomaterials, 34*(27), 6388–6401.
146. Bohner, M., et al. (2017). Characterization and distribution of mechanically competent mineralized tissue in micropores of β-tricalcium phosphate bone substitutes. *Materials Today, 20*(3), 106–115.

147. Miranda, P., et al. (2006). Sintering and robocasting of β-tricalcium phosphate scaffolds for orthopaedic applications. *Acta Biomaterialia, 2*(4), 457–466.
148. Descamps, M., Hornez, J., & Leriche, A. (2007). Effects of powder stoichiometry on the sintering of β-tricalcium phosphate. *Journal of the European Ceramic Society, 27*(6), 2401–2406.
149. Brunner, T. J., et al. (2007). Comparison of amorphous TCP nanoparticles to micron-sized α-TCP as starting materials for calcium phosphate cements. *Journal of Biomedical Materials Research Part B: Applied Biomaterials: An Official Journal of The Society for Biomaterials, The Japanese Society for Biomaterials, and The Australian Society for Biomaterials and the Korean Society for Biomaterials, 83*(2), 400–407.
150. Butscher, A., et al. (2012). Printability of calcium phosphate powders for three-dimensional printing of tissue engineering scaffolds. *Acta Biomaterialia, 8*(1), 373–385.
151. Feng, Y.-F., et al. (2012). Influence of architecture of β-tricalcium phosphate scaffolds on biological performance in repairing segmental bone defects. *PLoS ONE, 7*(11), e49955.
152. Karageorgiou, V., & Kaplan, D. (2005). Porosity of 3D biomaterial scaffolds and osteogenesis. *Biomaterials, 26*(27), 5474–5491.
153. Jarcho, M., et al. (1979). Synthesis and fabrication of β-tricalcium phosphate (whitlockite) ceramics for potential prosthetic applications. *Journal of Materials Science, 14*(1), 142–150.
154. Singh, S., & Ramakrishna, S. (2017). Biomedical applications of additive manufacturing: Present and future. *Current Opinion in Biomedical Engineering, 2*, 105–115.
155. Derakhshanfar, S., et al. (2018). 3D bioprinting for biomedical devices and tissue engineering: A review of recent trends and advances. *Bioactive Materials, 3*(2), 144–156.

Chapter 9
Composites for BTE

Abstract Despite possessing impressive qualities and the ability to imitate bone properties, the exclusive use of ceramic materials still encounters significant limitations. By combining ceramic materials with other biological materials, it is possible to mitigate their drawbacks and introduce novel and distinctive characteristics. The qualities of biological materials utilized for bone can be altered in a flexible and unrestricted manner. Furthermore, the enhancement of the biological characteristics of ceramics can be achieved by including various biological components. These materials can be fabricated as a composite consisting of ceramic-polymer, ceramic-hydrogel, and even ceramic-ceramic nanocomposites. Polymers have great promise for being used with bioceramics. Biodegradable polymers are utilized in several fields such as tissue engineering, medication delivery, cardiovascular applications, dental applications, orthopedic devices, and wound treatment. They can be classified into two distinct categories including natural polymers and synthetic polymers. Significant drawbacks of natural polymers are their quick breakdown rates and subpar mechanical properties. Various synthetic polymers have undergone substantial research for a wide range of biomedical applications due to their advantageous properties of being able to degrade naturally and being compatible with living organisms. Each polymer possesses unique features and characteristics, which will be thoroughly discussed in this chapter. Then, we will review the various combinations of diverse materials with commonly used nano-bioceramics.

Abbreviations

ALP	Alkaline phosphatase
BCP	Biphasic calcium phosphates
BGs	Bioactive glasses
COL	Collagen
CPCs	Calcium phosphate cements
CSCs	Calcium silicate cements
CSCs	Calcium sulfate cements

R. Gholami and S. M. Naghib, *Nanobioceramics for Bone Tissue Engineering and Regenerative Biomedicine*, Biological and Medical Physics, Biomedical Engineering,
https://doi.org/10.1007/978-981-96-0041-0_9

ePTFE	Expanded polytetrafluoroethylene
GA	Glutaraldehyde
HA	Hydroxyapatite
nHA	Nanohydroxyapatite
NPs	Nanoparticles
PCL	Poly(e-caprolactone)
PDLLA	Poly-DL-lactide
PEEK	Polyether ether ketone
PEG	Polyethylene glycol
PLA	Poly(lactic acid)
PLGA	Poly(lactic-co-glycolic) acid
PLLA	Poly-L-lactic acid
PMMA	Polymethyl methacrylate
PURs	Polyurethanes
PVA	Poly(vinyl alcohol)

9.1 Introduction

The utilization of diverse materials in biomedical treatments has recently undergone significant expansion. A topic currently garnering considerable focus from the scientific community is the utilization of biodegradable polymer-ceramic composites for treating various disorders and diseases related to bones [1–4]. Materials designed for a specific class may not meet all the necessary criteria for a particular implant application. Hence, by merging two or more classes, it is possible to create composites with multiscale structures that possess the desired features for particular applications [1]. Polymers, specifically poly-α-hydroxy acids, are commonly employed as alternative materials for making resorbable bone implants [5–9]. The degradation kinetics of these substances can be altered through chemical alteration of their molecular weight, either by copolymerization or by blending. Polymer implants are highly manageable and may be easily shaped or carved to suit specific needs. Nevertheless, their osteoconductivity is significantly limited and they often fail to facilitate direct contact between the implant and bone due to their inadequate hydrophilic characteristics. Therefore, composite materials consisting of polymers and ceramics are of interest due to their potential to enhance the osteoconductive features [10] of the polymer component in such implants. Additionally, these materials offer improved malleability for shaping the brittle ceramic part, which can be challenging to handle when customizing it to fit an individual defect [11–13].

The often employed polymers in these applications include poly(trimethylene carbonate), poly(ethylene oxide), polydioxanone, poly(lactic acid) (PLA), poly(glycolic acid), and poly(lactide-*co*-glycolide) (PLGA) [14, 15]. Furthermore, polyurethanes, poly(vinyl alcohol) (PVA), polyanhydrides, and poly(e-caprolactone)

(PCL) have been studied for their potential in bone regeneration [14, 16]. Incorporating a soft biodegradable polymer into inorganic scaffolds can offer two benefits. (1) Under specific circumstances, the composite's strength or toughness may surpass that of the inorganic scaffold by itself. (2) The biodegradable polymer has the capability to serve as a transporter for precise and localized release of bioactive substances, such as adhesion peptides and growth factors [17–20]. Presently, the advancement of novel barrier materials relies on the utilization of both degradable and non-degradable polymers, with or without the inclusion of growth factors. The frequently employed polymers include nanocellulose [21, 22], chitosan [23–28], the expanded polytetrafluoroethylene (ePTFE) [29–31], the polyethylene glycol hydrogel (PEG) with/without arginylglycylaspartic acid [32], and collagen derived from animal or human tissues [29]. Several materials including precious metals (platinum, silver, gold) [22], calcium phosphates [32], and bioactive glasses [29, 33–36], can be used to strengthen them even more. These materials have a well-established purpose in the field of bone regeneration. Furthermore, metals such as titanium were specifically selected for synthesis of the entire barrier materials or for use as reinforcement material [31, 37–39].

Biodegradable polymers can be classified into two groups: natural and synthetic polymers. Both synthetic and natural polymers are crucial components in contemporary medical practices. Natural polymers, including collagen (a protein found in living organisms [40]) and gelatin (obtained from collagen [41]), have exhibited issues including limited biodegradability, immunogenicity, incompatible properties, and instability. Nevertheless, because of the alterable characteristics of certain synthetic polymers like polyurethanes (PURs) and PLGA, they possess exceptional versatility [42, 43]. These biodegradable polymers have uses in tissue engineering, drug delivery, cardiovascular applications, dental applications, orthopedic devices, and wound management [1, 44–47].

9.2 Natural Polymers

The careful selection of the optimal material for scaffold production is an essential stage in the development of a tissue-engineered product. The qualities of the scaffold are mostly determined by the material characteristics in most cases. Due to their potential bioactivity, inherent biodegradability, and biocompatibility, natural polymers provide several benefits for tissue-engineering scaffolds. In general, the biodegradation of natural polymers involves the enzymatic cleavage of bonds, resulting in the erosion of the polymer. Thus, natural polymers are categorized as enzymatically degradable materials. This distinguishes them from hydrolytically degradable manufactured polymers, which undergo debonding as a result of hydrolysis. Major limitations of natural polymers include rapid degradation rates and suboptimal mechanical characteristics. Furthermore, a significant number of them possess a restricted accessibility, resulting in a substantial expense. Due to these factors, natural polymers are subjected to chemical modification by crosslinking or frequently employed in composites, in order to decrease the rate of degradation and enhance their mechanical characteristics. Nevertheless, these alterations in structure could potentially diminish the compatibility with living organisms and even trigger harmful effects on cells [48–50].

9.2.1 Collagen

The primary constituents of human bone consist of collagen and a substituted hydroxyapatite (HA). While each of these components individually contributes to enhancing bone formation to some extent, the combination of these two natural elements surpasses this level of accomplishment [51]. Collagen, as a natural polymer, is becoming more commonly utilized as a material for devices in the field of tissue engineering. Collagen undergoes rapid degradation and absorption by the body, facilitating strong cellular adhesion. Nevertheless, the mechanical capabilities of this material are somewhat inferior (Young's Modulus of about 100 MPa) when compared with bone (Young's Modulus of about 2-50GPa [52]). As a result, it is extensively crosslinked or included into composites, including collagen-hydroxyapatite for bone remodeling [53], or collagen-glycosaminoglycans for skin regeneration [54]. Composites of collagen and hydroxyapatite (Col-HA) have the capability to imitate and substitute skeletal bones. The significance of collagen type, crosslinking, production circumstance, porosity, and mineralization conditions is demonstrated by both in vitro and in vivo research. Both hydroxyapatite and collagen devices also effectively suppressed the proliferation of bacterial pathogens, which are the primary culprits of prosthesis-related infections, in comparison with PLGA devices [51, 55]. The data presented regarding the bone growth, cell culturing, and mechanical qualities of these devices pertain to their effectiveness as potential bone implants in the future [51].

Both HA and collagen type I (COL-I) were discovered to promote differentiation of osteoblasts [56]. However, when used in combination, they were seen to expedite osteogenesis. When human-like osteoblasts were incorporated into a composite matrix, the composite exhibited superior osteoconductive characteristics compared to HA material. Additionally, it resulted in the formation of a bone matrix with comparable calcification [57, 58]. By incorporating a calcium/phosphate compound into collagen sheets, their stability was improved, resulting in increased resistance to 3D swelling compared to the collagen reference [59]. Additionally, the mechanical "wet" characteristics of them were enhanced [60]. This occurred even in instances when collagen was extensively crosslinked. Furthermore, these composites exhibited enhanced mechanical behavior compared to the individual components. The ductile characteristics of collagen contribute to enhancing the low fracture toughness of HA [51]. Polymeric scaffolds have a degradation time of up to 2 years, but Col-HA has a more suitable degradation rate for clinical application, ranging from 2 months to a year [61]. In addition, osteogenic cells exhibited superior adhesion to collagen surfaces in vitro when compared with PGA and PLLA implants [51, 62].

A comparison between ceramic composite scaffolds and ceramic scaffolds showed that Col-HA composites outperformed individual TCP or HA scaffolds [63]. Incorporating collagen into a ceramic structure has several benefits for surgical applications, including improved adhesion to defect walls and particles, spatial adaptation, precise shaping, and the potential to promote clot stabilization and formation [51, 64]. Integrating collagen into biphasic calcium phosphates (BCP) ceramics can offer a practical method for enhancing the organic component of these materials and acquiring the organic characteristics found in biological calcified tissues [65].

9.2.2 Gelatin

Gelatin is a polymer that is formed naturally through the process of collagen hydrolysis. These scaffold materials are appealing due to their affordability, accessibility, and ability to be easily distributed in water-based solutions. These substances are non-toxic and can be simply applied as thin coatings [66–68]. Nevertheless, they exhibit a deficiency in bioactivity, their mechanical characteristics are unsatisfactory, and they have demonstrated unmanageable degradation in water-based environments [69–71]. The rapid solubility of gelatin in water is a major disadvantage of gelatin-based scaffolds, which restricts their use in BTE [70]. To mitigate this undesirable characteristic, one can employ efficient crosslinking using an appropriate crosslinking agent. Crosslinking is a widely employed approach in tissue engineering scaffolds to enhance the stability and biomechanical properties of the scaffolds [72]. Crosslinking refers to the creation of a chemical or physical connection between the active functional groups of a polymer chain [73]. Glutaraldehyde (GA) is extensively employed as a chemical crosslinking agent for the purpose of crosslinking various types of scaffolds and hydrogels. GA can improve the mechanical properties of scaffolds [66]. Furthermore, it has been observed that GA provides superior stability to the scaffolds in comparison with other aldehydes such as glyceraldehyde and formaldehyde [74]. In general, GA reacts with the hydroxyl lysine or ε-amine groups of lysine in gelatin by a Schiff's Base reaction. The stability of these Schiff's Bases is enhanced through their reactivity with additional GA molecules during the process of crosslink production [69].

The interaction between the organic gelatin and the inorganic ceramics have an essential function in creating porous scaffolds that promote cell growth, nutrition transmission, and waste excretion [75]. HA-embedded gelatin nanostructure has been synthesized as an artificial bone substitute, using a biomimetic co-precipitation process of HA nanocrystals within a soluble gelatin matrix. To enhance toughness of the materials, HA-gelatin nanocomposite was crosslinked using glutaraldehyde [75, 76]. According to studies, the mechanical strength of HA-polymer composites can only be attained when the HA particles are tiny and evenly distributed throughout the composite. Consequently, nanoapatite crystals have been employed, as reinforcements in the fabrication of these types of composites. The predictable biodegradation kinetics, low toxicity, and mechanical characteristics of gelatin make it suitable for the organic phase of a native bone [77]. Potential engineering biomaterials can be created by combining the cost-effectiveness and flexibility of gelatin with the osteoconductivity and bioactivity of nano-HA (nHA). The incorporation of bioinert ceramics into the nHA-gelatin composite is done to achieve superior biocompatibility, high porosity, and enhanced fracture toughness [75, 78, 79].

9.2.3 Chitin and Chitosan

The biocompatible natural polymers chitin and chitosan are currently being studied for their potential applications in wound healing, drug delivery [80], and tissue engineering (such as skin, cartilage, and bone) [81, 82]. Chitin is obtained from fungi, insects, shells of marine crustaceans, and endoskeleton of cephalopods [83, 84]. Therefore, the process of extracting chitin can be regarded as the utilization of waste materials derived from fish industries. It is composed of 2-acetamido-2-deoxy-β-D-glucose through α β (1 → 4) linkage. Chitin attained from porifera (sponges) possesses distinctive biological characteristics that make it an appropriate option for biomimetic approaches [85, 86]. Chitosan is a biodegradable and non-toxic natural polymer [81, 87–90]. It is a linear polysaccharide that is obtained through the process of partially deacetylated chitin (Fig. 9.1). The substance is a copolymer consisting of (1 → 4)-2-amino-2-deoxy-β-D-glucan (D-glucosamine) and (1 → 4)-2-acetamido-2-deoxy-β-D-glucan (*N*-acetyl D-glucosamine) units. These units are dispersed randomly or in blocks throughout the chain of the biopolymer, depending on the particular method of deacetylation, throughout the biopolymer chain. Chitosan and chitin have a similar structure to glycosaminoglycans, which are the main component of the bone ECM. Because of this reason, they are considered promising materials for creating scaffolds in tissue engineering [91].

Chitin is nitrogenous, inelastic, rigid, and white polysaccharide. It is hydrophobic and does not dissolve in water and most organic solvents, unlike chloro-alcohols, hexafluoroacetone and hexafluoroisopropanol [81]. However, chitosan is soluble in weak organic acids including acetic acid and formic acid. Chitosan can be chemically modified more easily than chitin because it has an amino group at the C-2 position of the monomer ring [93]. Chitosan is recognized for its ability to stimulate the formation of mesenchymal cells and osteoblasts, as well as in vivo vascularization [94–96]. Particularly, chitosan-based hydrogels exhibit favorable characteristics for bone replacement, due to their low toxicity, antimicrobial activity, biodegradability, and angiogenicity [81, 97–100]. Chitosan is soluble in acidic aqueous solutions, because it contains protonated amine groups that facilitate hydrophilic interactions [87, 101–105]. Gelation of a chitosan solution occurs when the protonated amine groups of the polymer are neutralized, provided that the concentration of polymer exceeds a critical value [18, 106]. Gelation is the process that takes place when a solution of chitosan in water comes into contact with a base, such as sodium hydroxide [101] or ammonia [106]. The physical gelation of a chitosan solution can occur at alkaline or neutral pH levels, resulting from a shift in the equilibrium

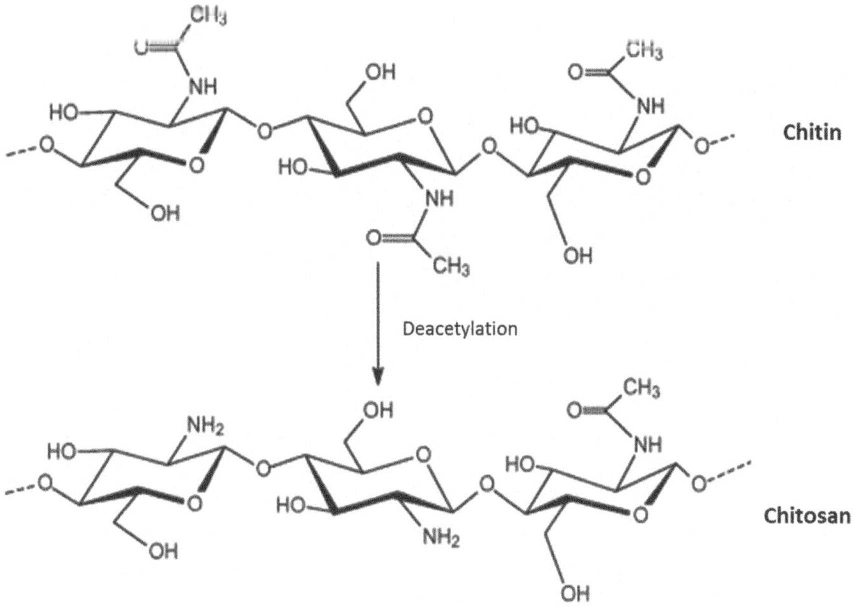

Fig. 9.1 Structure of chitin and chitosan. Reprinted with permission from [92]

between hydrophobic and hydrophilic interactions within the solution [106]. There is no need for an external crosslinking agent in this gelation procedure [18, 106].

Chitosan and chitin offer a significant benefit in their ability to be readily transformed into various forms such as scaffolds, nanofibers, nanoparticles, microparticles, gels, and beads. Capacity to promote osteoconductivity and increase the formation of bone both in vitro and in vivo, as well as the construction of extremely porous scaffolds with interconnected pores are additional benefits of chitosan and chitin scaffolds for tissue engineering [93]. Chitosan and chitin also possess excellent antibacterial properties and are highly biocompatible materials that balances their drawback of providing mechanical integrity turning them into suitable materials for wound healing [80, 81] and tissue engineering [8, 9, 81, 82, 107]. Chitin dressings have the capability to speed up the healing process of wounds, and control the release of substances that cause inflammation, including IL-1 β, prostaglandin E, and IL-8 [81]. Several investigations have been carried out on chitosan and chitin nHA composites for tissue engineering, leveraging the resemblance between chitosan and chitin to native ECM polymer and the benefits of nHA as the mineral constituent of bone [82]. Nanoceramics have the benefit of establishing an osteointegrative surface with the natural bone during the process of bone regeneration. Chitin and chitosan, as extracellular mimicking biopolymers, offer a good environment for cell loading. Studying tissue engineering structures that utilize cell-laden chitosan/chitin nanoceramics composites might be a more advantageous focus. The increased loading capability of nano-bioceramics also presents the possibility of utilizing them as carriers for drug delivery. Another potential application, in addition to combining antimicrobial drugs, is the utilization of pharmaceuticals that promote formation of bone, such as growth factors like FGF-18 and BMP-2, or strontium ranelate [93].

9.3 Synthetic Polymers

Synthetic polymers commonly used in medical applications are polyether ether ketone (PEEK), polymethyl methacrylate (PMMA), poly(lactic-co-glycolic) acid (PLGA), poly-L-lactic acid (PLLA), and the polylactic acid (PLA). These polymers are used because of their advantageous features [43, 108, 109]. In recent times, several synthetic polymers, including PLLA and PLGA have been extensively researched for diverse biomedical purposes because of their favorable characteristics of biodegradability and biocompatibility [1]. Polyesters have garnered greater interest compared to other synthetic polymers. They can be classified into three significant categories, including aliphatic, semi-aromatic, and aromatic. PHBV, PBS, PEA, PHB, PHA, PCL, PLA, and PGA are included in the category of aliphatic polyesters. Semi-aromatic polyesters include PEN, PBT, PTT, and PEN. Vectran is an example of aromatic polyesters. An inventory of the mechanical and physical characteristics of typical polymers that have been investigated in recent years is presented in Table 9.1 [1].

The micro-HA-PLLA composite has the most extensive surface roughness and the highest rate of critical energy release. On the other hand, the nano-HA-PLLA composites display a brittle fracture surface due to the nanoscale interactions between primary HA particles and PLLA fibrils [1, 111].

9.3.1 PLA

Polylactic acid (PLA), also known as polylactide, is a highly adaptable biopolymer. PLA was first produced in 1932 by Wallace Carother. Subsequently, the utilization

Table 9.1 Characteristics of typically used polymers in terms of their mechanical and physical qualities [1, 110]

Polymers	Melting temperature, °C	Glass transition temperature, °C	Tensile modulus, GPa	Tensile strength, MPa	Density, g/cm^3
PCL	58–65	(− 60)–(− 65)	0.21–0.44	20.7–42	1.11–1.14
PGA	220–233	35–45	6.0–7.0	60–99.7	1.50–1.71
PLLA	170–200	55–65	2.7–4.14	15.5–150	1.24–1.30
PLA	150–162	45–60	0.35–3.5	21–60	1.21–1.25

of PLA has significantly increased. PLA has the ability to undergo biodegradation and exhibits thermoplastic characteristics. The advantages of PLA include its minimal carcinogenic effects, biocompatibility, compostability, recyclability, ease of manufacture, and environmental friendliness. PLA has emerged as a significant material in the field of bioengineering, whereas also finding current uses in industries like packaging and textiles. The FDA has granted approval for direct interaction between PLA and biological fluids. Medical applications might range from dental, cardiac, and orthopedic usage to tissue engineering and regenerative medicine. PLA possesses intrinsic features that make it highly suitable for efficient manufacturing and rapid prototyping in 3D-printed structures. This can be utilized to create tissue engineering scaffolds customized for individual patients or efficiently manufacture medical devices, including the personal protection equipment necessary to ensure the safety of healthcare professionals during the Covid-19 pandemic [112, 113].

PLA can be obtained from sustainable sources including rice, corn, wheat, and carbon dioxide. PLA consists of lactic acid monomers that create its polymeric structure. Its monomers can also be referred to as cyclic diester lactide or 2-hydroxypropionic acid. The byproducts of PLA degradation are non-toxic for the environment or humans. Ease of producing PLA is attributed to the affordability and widespread availability of raw materials. It requires 25–55% less energy for production compared to polymers derived from petroleum [112, 113]. Hydroxyapatite (HA) incorporated into PLA is regarded as a highly favorable composite material, particularly for bone regeneration. By the addition of HA into the PLA matrix, the composites become more hydrophilic and develop a microrelief on the surface of the polylactide samples. This alteration also affected the tensile characteristics of the polylactide. The Young's modulus of the PLA scaffolds enhanced when HA particles were added. Consequently, the inclusion of HA particles led to a decrease in both the ultimate strength and the ability of the scaffold to stretch before breaking. Porous PLA-HA scaffolds might be well-suited for applications in soft tissue engineering, particularly where materials with elastomeric properties are highly valued. The slight deviations in the absorption bands of the composites indicate the presence of interference between the HA and the polymer chain. Nevertheless, there are no observable addition of novel functional groups in the PLA-HA scaffolds. The cellular investigations demonstrated that the scaffolds exhibited the capacity to promote growth and adhesion of cells [114].

9.3.2 PLLA

Poly-L-lactic acid (PLLA) is a synthetic polymer that is both biodegradable and biocompatible. It undergoes degradation in a safe manner, following the same metabolic pathway as lactic acid. It has been extensively utilized in several therapeutic applications for more than three decades, such as soft tissue implants, intrabone implants, and dissolvable sutures, without any reported safety concerns. PLLA was initially offered in 2004 as a treatment for "facial filling" of lipotropic HIV patients [115], following its availability in Europe since 1999. Since its first licensure in the United States for this challenging patient population, clinical experience has resulted in the advancement of treatment regiments that reduce the occurrence of side events reported in the early clinical studies. PLLA microparticles have the capability to induce subclinical inflammation in the host, hence facilitating the production of collagen. Recent technological advancements, combined with a better comprehension of how volume loss affects face aging [116], have resulted in the rise of PLLA as a reliable and efficient solution for addressing the volume loss that causes a deflated or sagging look, which is a characteristic sign of facial age [117].

9.3.3 PLGA

Poly (lactic-co-glycolic acid) (PLGA) is a widely studied and highly representative degradable biopolymer that has garnered significant consideration from investigators. It has been approved by the FDA for use in biomedical applications. PLGA-based devices are favored due to their tunable mechanical characteristics and degradation, excellent biocompatibility, and the versatility in various manufacturing processes. Consequently, PLGA-based biomaterials with diverse structures have been advanced and employed in diverse fields of tissue repair. Several commercial PLGA-based products are available, including bone graft, bone screw, drug delivery platform, and absorbable suture. PLGA is an aliphatic amorphous polymer, which is synthesized by polymerizing two monomers, glycolic acid, and lactic acid [118]. The chemical structure of PLGA and its monomers is shown in Fig. 9.2 [1].

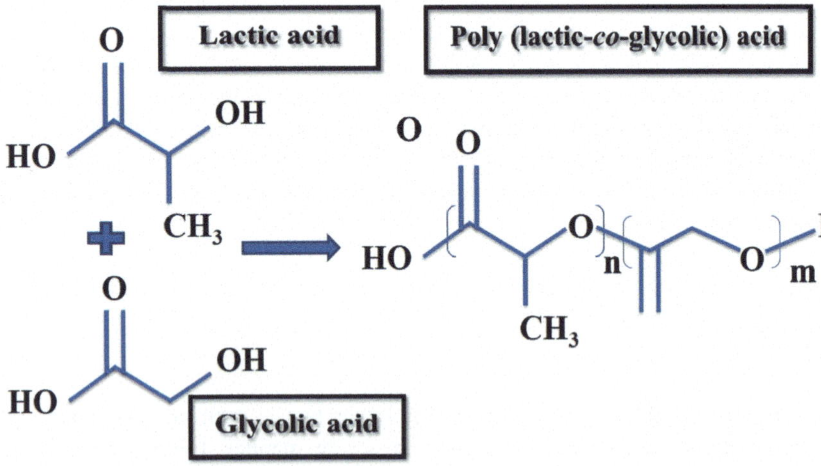

Fig. 9.2 Chemical structure of PLGA and its monomers (the variables n and m indicate the number of repetitions of each unit). Reprinted with permission from [1]

At present, there are two conventional methods used to produce PLGA: (1) direct polycondensation process and (2) polyaddition or ring-opening polymerization. Despite its comparatively low cost, the direct polycondensation reaction is limited in its applicability because of the low molecular weight of produced PLGA. Consequently, there has been increased focus on ring-opening polymerization, which is a favorable method for synthesizing PLGA with higher molecular weight. Nevertheless, chemical synthesis procedures unavoidably need the elimination of residual monomers and the utilization of metal catalysts, both of which are crucial for ensuring the biosafety and biocompatibility of materials used in biomedical applications [118]. Nevertheless, PLGA exhibits inadequate bioactivity and mechanical characteristics, which may hinder its ability to achieve robust osseointegration [119]. To resolve these concerns, it is advisable to utilize ceramics, specifically HA, as a coating on PLGA [120, 121].

9.3.4 PCL

Polycaprolactone (PCL) is a semi-crystalline synthetic polymer that has been approved by the FDA. It is being studied as a tissue engineering scaffold due to its good mechanical characteristics, permeability, biocompatibility, and slow biodegradability (taking 2–4 years). Despite possessing the necessary bulk qualities, the high surface hydrophobicity of PCL (poor wettability) restricts its use in BTE applications [122]. Consequently, a secondary phase of the element is combined to enhance the hydrophilic characteristics of PCL [123]. The utilization of a PCL nanofiber scaffold has the potential to enhance the process of osteogenic differentiation in hMSCs, hence demonstrating a promising capability for repairing bone tissue using MSCs. The slow degrading and bioinert properties of PCL scaffold make it challenging for HA production to occur on its surface. However, it may have limitations in terms of its application. Co-electrospinning and blending PCL with bioactive ceramics and inorganic materials such as HA, silica, and β-tricalcium phosphate, as well as surface modification techniques can enhance the characteristics of PCL-based scaffolds, which is not only beneficial for cartilage regeneration but also for bone repair [124].

9.4 Nanocomposites for Bone Tissue Engineering

Bone tissue is an inherent composite consisting of a mineral phase and a collagenous matrix. Hence, it is imperative to create novel therapeutic materials that possess the same hybrid construction, integrating the resorbability, toughness, and flexibility of an organic phase, together with the osteoconductivity, stiffness, and strength of an inorganic component [16, 125, 126]. The initial bone regeneration materials utilized collagen scaffolds that also administered recombinant proteins to enhance targeted cell signaling processes for enhanced curing. The next generation of scaffolds was designed using type I collagen, TCP and demineralized bone matrix particles to create composites with inherent bioactivity that can stimulate a positive osteogenic response and considerably enhance and expedite regeneration. The upcoming scaffolds will include many advancements in imitating the bone connective tissue, such as synthetics utilizing peptide and protein sequences, nanocomposites and nanoarchitecture [15]. Nanophase ceramics, metals, and polymers all enhance activity of bone cells in comparison with their microstructured equivalents [127]. The optimum materials must possess specified characteristics including excellent regulated biodegradability, biocompatibility, mechanical qualities, and osteointegration. Such materials must promote the formation of ligamentous tissue in order to meet the necessary requirements. Bioresorbable and biodegradable materials with suitable mechanical qualities have the potential to retain their structural stability and integrity [16]. Researchers have studied hybrid scaffolds made of organic–inorganic compounds [128] for the purpose of bone regeneration. Loading and release of drugs are crucial features of optimal BTE scaffolds [129]. Composites are materials composed of two or more components that have distinct characteristics, which are maintained in the final structure. Monophasic materials may not always meet the necessary criteria for a successful bone graft substitution [15]. Researchers have shown interest in composite materials because they have the potential to design novel materials that combine the beneficial qualities of both constituents, resulting in a synergistic impact [130]. Due to the fact that natural bone is a biocomposite, much effort has been dedicated to developing composite materials for the regeneration of bone tissue [131, 132].

Currently, nanocomposites are widely acknowledged in the field of BTE due to their resemblance to natural bone tissue, which displays a complex hierarchical structure consisting of a fibrous collagen organic medium, as well as HA mineral and nanocrystals medium [133]. Nanocomposite biomaterials are a novel type of materials that combine a matrix structure made of biodegradable and biopolymeric materials with fillers that are both easily resorbable and bioactive, and are at the nanoscale [134]. The utilization of nanotechnology in tissue engineering enables the replication of the intricate structure of natural bone. There is significant potential to design sophisticated scaffolds and enhance existing biomaterials using advanced nanocomposite systems, by incorporating osteoinductive and osteoconductive factors, as well as cells [15]. Various materials, including natural and synthetic polymers, carbon nanotubes, hydroxyapatite, and silicates have been utilized for the design and production of nanocomposites to meet various requirements in BTE. Different bioglasses and ceramics offer excellent qualities for scaffold applications, either because they promote bone formation or due to their comparable structure to bone mineral. Calcium phosphates exhibit advantageous properties such as biocompatibility, resorbability, and osteoconductivity. Nevertheless, they have certain disadvantages, including rapid degradation rate and inadequate mechanical qualities. Nanocomposites that involve bioactive nanosized fillers and biopolymeric matrices are particularly interesting in the field of BTE. Biopolymers offer advantages because they closely resemble ECM, have a particular rate of degradation, and exhibit superior biological performance [92, 131]. In summary, it may be inferred that a perfect natural nanocomposite scaffold, which accurately replicates the hierarchical morphology and structure of bone while also serving a temporary purpose, has not yet been advanced [92]. However, the literature extensively documents the impregnation of ceramics with polymers as a common method for creating composite scaffolds [17, 135–137]. Ceramic scaffolds have mostly been infused with artificial polymers including poly (lactic-co-glycolic acid) (PLGA), poly-DL-lactide (PDLLA), or polycaprolactone (PCL). Lately, polymers derived from natural sources, including chitosan, collagen, alginate, silk, and gelatin, have also found applications [17, 18].

9.4.1 Ceramic-Polymer Nanocomposites

Nanocomposites consisting of bioactive inorganic nanoparticles (NPs) and biocompatible polymers have garnered significant interest for their potential applications as materials for bone tissue regeneration and even long-lasting bone replacements because of their outstanding biocompatibility and mechanical characteristics [138, 139]. Biomimetic approaches for developing nanocomposites for tissue engineering and regenerative medicine involve utilizing bioactive structures with precise geometry that can mimic the natural ECM found in living organisms and their capability to guide interactions between cells and the matrix, as well as between cells themselves [92]. The use of nanocomposites, which consist of resorbable and bioactive nanofillers combined with biodegradable and biopolymeric matrices, has been proposed as an approach for BTE (Fig. 9.3). Nanoscale fillers have the ability to significantly alter the physical characteristics of the polymeric matrix, enabling the development of enhanced biomaterials that cannot be achieved with individual components alone [92].

Polymers that are biodegradable and come from natural sources, such as proteins (e.g., elastin, fibrinogen, silk, collagen) and polysaccharides (e.g., glycosaminoglycans, chitin, cellulose), have many several similarities with the ECM. They are chemically versatile and have good biological performance, without causing immunological reactions or toxicity [141, 142]. Bioresorbable fillers like CaPs possess advantageous properties including biocompatibility, resorbability, and osteoconductivity [143]. These nanocomposites are composed of a polymer matrix that is coupled with nanosized CaPs. This combination enables the customization of the desired resorption and degradation rates of the matrix. In addition, nanosized CaPs have the capability to enhance the tissue bonding characteristics of polymeric materials, as well as, promote differentiation and adhesion of cells [92]. The incorporation of nano-HA into natural bone enhances its mechanical strength and promotes the proliferation and adhesion of osteoblasts [144]. Injectable bone defect fillers produced from nanosized HA particles and polymeric microspheres are highly appealing nanocomposites. This instance closely resembles the composition and structure of the bone ECM, specifically when collagen is utilized as microsphere matrix [127].

Scaffolds are constructed to serve as a 3D support structure for the surrounding bone tissue, imitating the ECM behavior. They possess several advantageous,

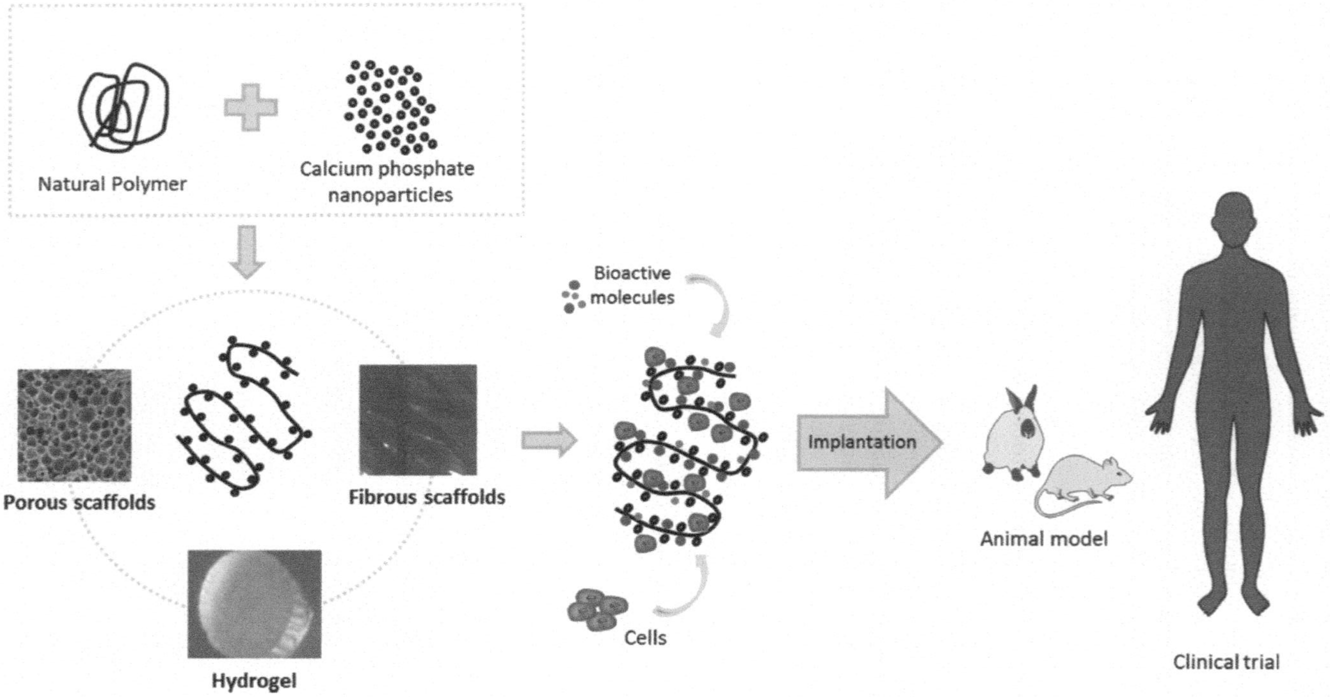

Fig. 9.3 Approach of tissue engineering using nanocomposites composed of calcium phosphates and biopolymers. Reprinted with permission from [92] and [140]

including: (i) minimal toxicity or inflammation in vivo, formation of new bone with a uniform distribution to prevent necrosis, controlled degradation, appropriate mechanical characteristics including elasticity and tensile strength, interconnected pores to simplify transport of regulatory factors, nutrients, and mass, enabling differentiation, proliferation, and survival of cells, and porous structure that stimulates migration, growth, and adhesion of cells as well as interactions between cells and biomaterials [145]. Moreover, scaffolds possess favorable attributes for the transplantation of cells into a damaged area and to minimize cell loss, as opposed to the straightforward injection of cells into the flaws [146]. Nanocomposite scaffolds have been constructed using various techniques such as electrospinning [147], rapid prototyping [148], gas foaming [149], phase separation [150], freeze drying [151], solvent casting and particulate-leaching [152], and foam replica method [153]. These scaffolds possess specific characteristics including crystallinity, surface-area-to-volume ratio, porosity, and nanosized pores.

Freeze drying process is capable of creating high-quality pores in the polymer structure through the formation of ice crystals and their growth along heat gradients. Nevertheless, the alignment of the pores can be restricted by manipulating the thermal gradients' geometry in the mold during the freezing process [154]. Chitosan can be transformed into several structures, including microspheres, nanofibers, membranes, and 3D scaffolds. As mentioned earlier, chitosan has demonstrated its exceptional effectiveness in BTE [154] because of its biodegradation, antibacterial properties, capability to interact with anions, and ability to create pores. Nevertheless, it is crucial to improve the mechanical properties of chitosan-based composites, particularly for their application in BTE [87]. The nHA-chitosan composite scaffolds exhibited superior biocompatibility and favorable pore size compared to pure chitosan scaffolds [155]. Incorporating nHA into chitosan enhances the mechanical characteristics of nanocomposites, hence replicating the natural structure of bone. Various techniques have been utilized to design nHA-chitosan nanocomposites [82].

Furthermore, the combination of nHA-chitosan has a higher propensity for apatite formation during the biomimetic process compared to pure chitosan. This indicates that the composite has superior mineralization activity. The cells cultured on composite scaffolds exhibited elevated alkaline phosphatase activity, indicating a greater degree of differentiation [156]. The i*n situ* chemical approach is a frequently employed procedure for the preparation of the nHA-chitosan nanocomposite [155, 157]. Incorporating HA and BMP-2 into polymer scaffolds is an effective method for creating bone regenerative materials with exceptional performance [158]. Furthermore, the BMP-2 enhanced the regenerative activities of porous microspheres in promoting the growth of bone tissue, specifically in terms of its ability to induce osteogenesis [16]. The issue lies in the fabrication of scaffolds that maintain a high level of compatibility between the phases, while still preserving mechanical characteristics and porous structure. Furthermore, it is crucial to attain a uniform dispersion of nanoscale CaP particles within the polymer. Techniques employed to optimize the dispersion of CaP nanoparticles involve either utilizing dispersants like sodium citrate or precipitating the mineral crystals inside the polymer matrix in situ [159–161].

9.4.2 Ceramic-Hydrogel Nanocomposites

Hydrogels are composed of a 3D structure that is extensively hydrated, but mechanically weak because of the random arrangement of their chemically crosslinked network. Hydrogel materials have compositional and structural resemblances with the ECM, and are appealing scaffolds due to their capacity to encapsulate biomolecules and cells, efficient mass transfer, biocompatibility, and swelling network structure [162]. The qualities of these materials are determined by the characteristics of the polymer chains, the level of water absorption, the molecular structure, and the degree of crosslinking [163]. Hydrogels often have a hydrophilic porous structure with interconnected pores larger than 10 μm, which allows cells to infiltrate and spread. This structure also provides a larger surface area for cells to attach to and interact with. Furthermore, the porosity of hydrogels can be regulated using several methods [164]. Hydrogel networks can be intentionally designed to have several shapes and sizes, including bellows, hollow tubes, rods, spheres, sheets, and thin films, thanks to their distinctive physical characteristics [165]. From the tissue engineering perspective, hydrogels serve as 3D structures that arrange cells and provide stimuli to guide the development of certain tissues. They also function as agents to fill space and as carriers for bioactive molecules [166].

Nanocomposite hydrogels consist of an organic–inorganic network structure, which is crosslinked with NPs through physical or chemical means [167]. They possess superior biological, electrical, physical, and chemical characteristics compared to the typical hydrogels. These improvements can be primarily attributable to the enhanced interactions between the NPs and the polymeric network. However, hydrogel nanocomposites do not possess certain crucial characteristics, including biodegradability and reactivity to stimuli, which can be improved by including various phases into a nanocomposite hydrogel [168]. Hydrogels made from inorganic NPs, such as nanosized CaPs, can be combined with a synthetic or natural polymeric hydrogel matrix. This combination enhances both mechanical strength and the ability to adjust the bioactive properties of the network [92].

9.4.3 Ceramic-Ceramics Composites

CaP ceramic materials are appealing in nanocomposite design due to their predominant presence in bone. When the substrates are porous, they provide a larger surface area for attachment of cells and adsorption of proteins [169]. Hydroxyapatite, β-TCP and bioactive glasses (BGs) are commonly employed as inorganic biomaterials because of their exceptional osteoconductivity [129, 170]. These bioceramics can be printed together by mixing their powder forms in the ink or printing bed. These biphasic bioceramic scaffolds have demonstrated increased cell proliferation and comparable cell viability when compared to scaffolds made solely of β-TCP or HA [171]. Furthermore, the amalgamation of two or more bioceramics can balance the drawbacks of single component materials [129]. Bioactive bone cements are considered as promising options for BTE. They have the potential to fill bone defects of various shapes and serve as a support for cell adhesion [172, 173]. The primary bioactive bone cements consist of calcium silicate cements (CSCs), calcium phosphate cements (CPCs), and calcium sulfate cements (CSCs) [174–176]. The production of cement scaffolds involves a hydration process that enhances their biological, chemical, and physical properties [177–179].

Despite the advantageous properties of bone cements such as biocompatibility, fast resorption, and quick curing time, the use of mono bone cement materials in clinical applications is restricted because of their excessively rapid resorption rate, lack of bioactivity, and the unstable environment they create in the body. Therefore, significant endeavors have been made to design composite cements by blending them with other bioactive substances in order to minimize the rate of resorption, enhance bioactivity, diminish inflammation, and customize the inject ability and setting time [129].

9.4.4 Decorated Metal Particle-Based Composites

Metallic biomaterials experience premature mechanical failure due to the creation of cyclic stress. Several studies have been conducted to comprehend the means between corrosion formation and fatigue in metallic implants within a physiological situation [180]. Metallic NPs, including zinc, silver, titanium, and zirconia, exhibit antifungal and antibacterial properties, making them widely utilized in tissue engineering [181]. Recent studies have shown that these NPs improve mechanical properties and promote cell adhesion [182]. Zinc is an essential element that is also present in the human body, particularly in the bones [183]. This nanomaterial demonstrates excellent antifungal and antibacterial activity, chemical stability, and compatibility, without while remaining without any toxicity in both laboratory and living organism experiments. Zinc also improves the capacity to inhibit osteoclastic reabsorption process [184]. It has been extensively utilized in preclinical research due to its easy accessibility, affordability, and exhibition of fluorescence. Zinc also aids in the catalysis, architectural, and monitoring of alkaline phosphatase (ALP), hence playing an important part in mineralization and osteogenesis [185].

The elastic modulus of magnesium (41–45 GPa) is more similar to the elastic moduli of natural bone (3–20 GPa) than to those of zinc (90 GPa) or iron (211.4 GPa) [186]. When there is a mismatch in the elastic moduli, the implant carries a significantly large portion of the load, leading to bone stress shielding [186, 187]. In addition, Mg effectively interacts with the surface of bone minerals, resulting in reconstruction and proliferation of bone [188, 189]. Moreover, Mg demonstrates excellent biodegradability along with biocompatibility, making it suitable for use in implants. This eliminates the need for a separate procedure to remove the implant [18]. However, Mg-based alloys are not commonly used since they have poor corrosion resistance due to their high reactivity [187, 190]. Moreover, gold and silver

NPs are widely utilized in the field of biomedicine to treat cancer, because of their antitumor properties [191]. Therefore, the application of these nanomaterials in bone tissue regeneration has significant potential due to their ability to simultaneously treat bone cancer, which is responsible for bone deterioration and following fracturing [184]. By incorporating gold NPs into fibrous PCL fibers, there is enhanced detection of a-sarcomeric-actinin, a marker associated with the contraction of the cardiac muscle, which interacts to significantly decrease electrical fields when compared with the controlling group after one week of culture [185, 191].

While metal-based implants and operations have notable benefits, it is crucial to acknowledge the significant drawbacks associated with them before choosing them for the ultimate application. They need additional advancements and improvement. As life-expectancy continues to rise, there is a growing demand for implants that are both durable and long-lasting, without causing any negative reactions. The advantages and disadvantages of metallic biomaterials for implants are presented in Table 9.2 [180].

Zirconium oxide (ZrO_2) has been utilized as a biomaterial due to its multifunctionality [185]. It is abundant on earth, and can be found in bone tissue in the human body at concentrations ranging from 2 to 20 mg/kg [192]. Titanium oxide (TiO_2) is extensively used in tissue engineering due to its bactericidal attributes, photocatalytic properties, low cost, and non-toxic nature. It is an organic ceramic nanomaterial that displays a tricrystalline structure, namely: rutile (tetragonal), anatase (tetragonal), and brookite (orthorhombic). Nevertheless, the brookite structure lacks stability and is challenging to manufacture, whereas both anatase and rutile have been extensively studied and can be made in the laboratory, demonstrating both catalytic activity and semi-conductivity [185, 192].

Magnetic therapy has emerged as a viable alternative for treating several disorders, particularly those affecting the bones. According to previous studies, magnetic fields have been found to speed up the formation of new bone, improve osteointegration, enhance growth factors expression, and stimulate osteoblast proliferation [193, 194]. Utilizing magnetic fields has additional benefits in promoting the integration of implants and bone, increasing calcium content and bone density, and accelerating the healing process of bone fractures [195–198]. Magnetic NPs have garnered significant attention due to their distinctive magnetic features, including non-toxicity and biocompatibility. In addition, they exhibit superparamagnetic properties when their size is smaller than 50 nm [199]. Superparamagnetic iron oxide nanoparticles have found extensive use in several iomedical applications, including magnetic hyperthermia, enzyme immobilization, bioseparation, magnetic resonance imaging reagents, and drug delivery [200–203], making them appropriate for in vivo applications [204]. Nevertheless, the utilization of these materials in the process of bone tissue regeneration has received little attention. Each magnetic domain possesses an inherent magnetic therapeutic effect that is further amplified by the presence of an external magnetic field [205–207]. Therefore, the development of a well-designed scaffold incorporating nanomagnetic components for regeneration of bone has emerged as a crucial area of research. In this connection, a few researchers

Table 9.2 Specific properties of metallic biomaterials utilized in implants [180]

Metal-based	Advantages	Disadvantages
Noble metals (gold, silver, platinum, and iridium)	High resistance to both chemical and electrochemical corrosion	Prone to corrosion both in vitro and in vivo
NiTi shape memory alloy	Biocompatible Superelastic Capable of returning to its original shape even after being subjected to a change in temperature Low Young's modulus	Ni induces allergic reaction as a result of the wear debris
Mg-based alloys	Biocompatible Highly robust	Formation of hydrogen bubbles during the corrosion Presence of an unshielded oxide film results in the corrosion of the implant
Ti-Nb-Ta-Zr (TZNT) alloy	Low elastic modulus Excellent corrosion resistance Biocompatibility Improved wear properties Exceptional biological features Excellent mechanical characteristics	Lower strength
Ti-6Al-4V	Applying appropriate surface treatments can enhance fatigue strength	Limited mobility Implant breakage Bone resorption Poisoning Ion release
Cobalt-chromium-based alloy	Load-bearing materials Biocompatibility High wear resistance Exceptional corrosion resistant with a low susceptibility	Cr and Ni allergies High processing costs Challenging to machine Restricted application Early implant loosening rate
Stainless steel (bio-steel)	Strong fatigue resistance Excellent wear resistance High elastic modulus Cost-effective to manufacture	Stress shielding effect Long-term uses can lead to pitting and crevice corrosion, as well as the development of Cr and Ni allergies
Titanium	Reduced Young's modulus Minimal weight Excellent biocompatibility Exceptional mechanical characteristics	Reduced shear strength Costly Inadequate tribological qualities Triggers allergic reactions

have combined nano – Fe_3O_4 and HA in presence of natural and synthetic polymer to enhance biocompatibility and mechanical qualities, respectively [144, 208–215].

The demand for materials that are both high-strength and lightweight has led to the development of metal matrix composites [216–218]. A composite scaffold made of nHA-chitin hydrogel has been developed using a freeze drying method. This scaffold is intended for use in wound and BTE applications [155, 219]. Li et al. successfully designed nanocomposites called nHA-chitosan-graphene oxide. These nanocomposites showed an enhanced hardness and elastic modulus, as well as higher rate of cell proliferation for MG-63 and L929 cell lines [220]. Incorporating nano copper–zinc (nCu–Zn) into nHA-chitosan scaffolds led to elevations in antibacterial activity, augmentations in protein adsorption, reductions in degradation, and notable enhancements in swelling. The nHA-chitosan-nCu–Zn scaffolds exhibit no harmful effects on rat osteoprogenitor cells [221, 222]. A biocomposite consisting of nHA-chitosan-silver particles was created using freeze drying method, followed by introduction of silver ions in controlled proportion was achieved through a reduction process facilitated by the functional groups present in chitosan [223].

9.5 Conclusion

Bone tissue is a natural combination of a mineral phase and a collagenous matrix. Therefore, it is crucial to develop innovative therapeutic materials that exhibit a hybrid structure, combining the ability to be absorbed, durability, and flexibility of an organic phase, together with the ability to promote bone growth, rigidity, and strength of an inorganic component. Composites are materials that consist of two or more components with individual features that are preserved in the final structure. Presently, nanocomposites are well-recognized in the field of bone tissue engineering due to their similarity to natural bone tissue. Natural bone tissue exhibits a complex hierarchical structure comprising a fibrous collagen organic medium, as well as a medium of HA mineral and nanocrystals. A wide range of materials, such as natural and synthetic polymers, carbon nanotubes, hydroxyapatite, and silicates, have been used to create nanocomposites in order to fulfill different needs in the field of BTE. Materials tailored for a given class may not fulfill all the essential requirements for a specific implant application. Therefore, by the process of combining two or more classes, it becomes feasible to generate composites that exhibit multiscale structures and contain the specific characteristics required for certain applications. Resorbable bone implants often utilize polymers, notably poly-α-hydroxy acids, as alternative materials. Biodegradable polymeric materials can be categorized into two distinct groups: natural polymers and synthetic polymers. Natural polymers offer numerous advantages for tissue-engineering scaffolds due to their intrinsic bioactivity, biodegradability, and biocompatibility. Collagen, as a natural polymer, is increasingly being used as a material for devices in the field of tissue engineering. Collagen and hydroxyapatite composites have the ability to mimic and replace skeletal bones. Gelatin is a naturally occurring polymer that is generated through the process of collagen hydrolysis. The attractiveness of these scaffold materials lies in their cost-effectiveness, availability, and capacity to be conveniently dispersed in aqueous solutions. However, they lack bioactivity, their mechanical properties are inadequate, and they have shown uncontrollable deterioration in water-based environments. Chitin and chitosan are currently being researched for their potential uses in wound healing, drug delivery, and tissue engineering. Chitosan and chitin scaffolds have the ability to enhance the growth of bone cells and stimulate the production of bone, both in laboratory settings and in living organisms. Additionally, these scaffolds may be constructed to have a highly porous structure with interconnected pores, which is advantageous for tissue engineering purposes. Recently, there has been significant study on several synthetic polymers for a wide range of biomedical applications due to their advantageous properties of being able to degrade naturally and being compatible with living organisms. PLA is a versatile biopolymer. To summarize, PLA has several benefits such as its low carcinogenicity, biocompatibility, compostability, recyclability, ease of production, and environmental friendliness. PLLA is a man-made polymer that can be broken down by natural processes and is compatible with living organisms. It undergoes safe breakdown through the same metabolic mechanism as lactic acid. PLGA-based devices are preferred because of their adjustable mechanical properties and degradation, outstanding biocompatibility, and versatility in different manufacturing methods. PCL is an FDA-approved semi-crystalline synthetic polymer. Due to its favorable mechanical properties, permeability, biocompatibility, and slow biodegradability, it is currently being investigated as a scaffold for tissue engineering.

References

1. Alizadeh-Osgouei, M., Li, Y., & Wen, C. (2019). A comprehensive review of biodegradable synthetic polymer-ceramic composites and their manufacture for biomedical applications. *Bioactive Materials, 4*, 22–36.
2. Seyfoori, A., et al. (2019). Multifunctional magnetic $ZnFe_2O_4$-hydroxyapatite nanocomposite particles for local anti-cancer drug delivery and bacterial infection inhibition: An in vitro study. *Journal of the Taiwan Institute of Chemical Engineers, 96*, 503–508.
3. Askari, E., et al. (2022). Highly bioactive Akermanite-Monticellite nanocomposites for bone tissue engineering: A tunable three-dimensional biological study. *Journal of Materials Research and Technology, 20*, 4066–4076.
4. Soleymani, S., Naghib, S. M., & Mozafari, M. R. (2024). An overview of cultured meat and stem cell bioprinting: How to make it, challenges and prospects, environmental effects, society's culture and the influence of religions. *Journal of Agriculture and Food Research*, 101307.
5. Rozema, F. R., et al. (1990). Poly (L-lactide) implants in repair of defects of the orbital floor: An animal study. *Journal of Oral and Maxillofacial Surgery, 48*(12), 1305–1309.
6. Soleymani, S., & Naghib, S. M. (2023). 3D and 4D printing hydroxyapatite-based scaffolds for bone tissue engineering and regeneration. *Heliyon, 9*(9), e19363.
7. Bagherpour, I., et al. (2023). Synthesis and investigation on microstructural, mechanical features of mesoporous hardystonite/reduced graphene oxide nanocomposite for medical applications. *Frontiers in Bioengineering and Biotechnology, 11*, 1073435.

8. Safarzadeh, S., Mozafari, M. R., & Naghib, S. M. (2024). Chitosan-incorporated bioceramic-based nanomaterials for localized release of therapeutics and bone regeneration: An overview of recent advances and progresses. *Current Organic Chemistry, 28*(15), 1190–1214.
9. Naghib, S. M., Zarrineh, M., & Mozafari, M. R. (2024). 3D Printing chitosan-based nanobiomaterials for biomedicine and drug delivery: Recent advances on the promising bioactive agents and technologies. *Current Organic Chemistry, 28*(7), 510–525.
10. Nichter, L. S., et al. (1992). Demineralized bone matrix polydioxanone composite as a substitute for bone graft: A comparative study in rats. *The Journal of Craniofacial Surgery, 3*(2), 63–69.
11. Schliephake, H., & Kage, T. (2001). Enhancement of bone regeneration using resorbable ceramics and a polymer–ceramic composite material. *Journal of Biomedical Materials Research: An Official Journal of The Society for Biomaterials, The Japanese Society for Biomaterials, and The Australian Society for Biomaterials and the Korean Society for Biomaterials, 56*(1), 128–136.
12. Mikaeeli Kangarshahi, B., & Naghib, S. M. (2024). Multicomponent 3D-printed collagen-based scaffolds for cartilage regeneration: Recent progress, developments, and emerging technologies. *Current Organic Chemistry, 28*(8), 576–594.
13. Kangarshahi, B. M., et al. (2024). Bioprinting of self-healing materials and nanostructures for biomedical applications: Recent advances and progresses on fabrication and characterization techniques. *Bioprinting, 38*, e00335.
14. Pielichowska, K., & Blazewicz, S. (2010). Bioactive polymer/hydroxyapatite (nano) composites for bone tissue regeneration. *Biopolymers*, 97–207.
15. James, R., et al. (2011). Nanocomposites and bone regeneration. *Frontiers of Materials Science, 5*(4), 342–357.
16. Sahoo, N. G., et al. (2013). Nanocomposites for bone tissue regeneration. *Nanomedicine, 8*(4), 639–653.
17. Philippart, A., et al. (2015). Toughening and functionalization of bioactive ceramic and glass bone scaffolds by biopolymer coatings and infiltration: A review of the last 5 years. *Expert Review of Medical Devices, 12*(1), 93–111.
18. Caballero, S. S. R., et al. (2019). Fabrication and characterization of hardystonite-chitosan biocomposite scaffolds. *Ceramics International, 45*(7), 8804–8814.
19. Hersel, U., Dahmen, C., & Kessler, H. (2003). RGD modified polymers: Biomaterials for stimulated cell adhesion and beyond. *Biomaterials, 24*(24), 4385–4415.
20. Li, J., & Kao, W. J. (2003). Synthesis of polyethylene glycol (PEG) derivatives and PEGylated−peptide biopolymer conjugates. *Biomacromolecules, 4*(4), 1055–1067.
21. Torgbo, S., & Sukyai, P. (2018). Bacterial cellulose-based scaffold materials for bone tissue engineering. *Applied Materials Today, 11*, 34–49.
22. Dobos, A. M., Onofrei, M.-D., & Ioan, S. (2015). Cellulose acetate nanocomposites with antimicrobial properties. *Eco-friendly Polymer Nanocomposites* (pp. 367–398). Springer.
23. Park, S. M., & Kim, H. S. (2017). Preparation of acetylated chitosan/carbonated hydroxyapatite composite barriers for guided bone regeneration. *Macromolecular Research, 25*(2), 158–164.
24. Dumont, V. C., et al. (2016). Chitosan and carboxymethyl-chitosan capping ligands: Effects on the nucleation and growth of hydroxyapatite nanoparticles for producing biocomposite membranes. *Materials Science and Engineering: C, 59*, 265–277.
25. Zhang, K., et al. (2010). Preparation of chitosan/hydroxyapatite guided membrane used for periodontal tissue regeneration. *Chinese Journal of Polymer Science, 28*(4), 555–561.
26. Gooneh-Farahani, S., et al. (2021). A pH-sensitive nanocarrier based on BSA-stabilized graphene-chitosan nanocomposite for sustained and prolonged release of anticancer agents. *Scientific Reports, 11*(1), 17404.
27. Gooneh-Farahani, S., Naghib, S. M., & Naimi-Jamal, M. R. (2020). A novel and inexpensive method based on modified ionic gelation for pH-responsive controlled drug release of homogeneously distributed chitosan nanoparticles with a high encapsulation efficiency. *Fibers and Polymers, 21*(9), 1917–1926.

28. Gooneh-Farahani, S., Naimi-Jamal, M. R., & Naghib, S. M. (2019). Stimuli-responsive graphene-incorporated multifunctional chitosan for drug delivery applications: A review. *Expert Opinion on Drug Delivery, 16*(1), 79–99.
29. Sheikh, Z., Hasanpour, S., & Glogauer, M. (2018). Bone grafting. *Mandibular Implant Prostheses* (pp. 155–174). Springer.
30. Rakhmatia, Y. D., et al. (2013). Current barrier membranes: Titanium mesh and other membranes for guided bone regeneration in dental applications. *Journal of Prosthodontic Research, 57*(1), 3–14.
31. Naenni, N., et al. (2017). Randomized clinical study assessing two membranes for guided bone regeneration of peri-implant bone defects: Clinical and histological outcomes at 6 months. *Clinical Oral Implants Research, 28*(10), 1309–1317.
32. Thoma, D. S., et al. (2017). Bone augmentation at peri-implant dehiscence defects comparing a synthetic polyethylene glycol hydrogel matrix vs. standard guided bone regeneration techniques. *Clinical Oral Implants Research, 28*(7), e76–e83.
33. Benic, G. I., et al. (2016). Guided bone regeneration of peri-implant defects with particulated and block xenogenic bone substitutes. *Clinical Oral Implants Research, 27*(5), 567–576.
34. Stuart, B. W., et al. (2018). Gallium incorporation into phosphate based glasses: Bulk and thin film properties. *Journal of the Mechanical Behavior of Biomedical Materials, 82*, 371–382.
35. Naghib, S. M., et al. (2012). Bioactivation of 304 stainless steel surface through 45S5 bioglass coating for biomedical applications. *International Journal of Electrochemical Science, 7*(4), 2890–2903.
36. Morteza, N. S., et al. (2012). Development and characterization of 316 L stainless steel coated by melt-derived and sol-gel derived 45S5 bioglass for orthopedic applications. *Ceramics-Silikáty, 56*(1), 89–93.
37. Dascălu, C.-A., et al. (2019). Synthesis and characterization of biocompatible polymer-ceramic film structures as favorable interface in guided bone regeneration. *Applied Surface Science, 494*, 335–352.
38. Duta, L., et al. (2017). Comparative physical, chemical and biological assessment of simple and titanium-doped ovine dentine-derived hydroxyapatite coatings fabricated by pulsed laser deposition. *Applied Surface Science, 413*, 129–139.
39. Fu, L., et al. (2017). Bilayer poly (lactic-co-glycolic acid)/nano-hydroxyapatite membrane with barrier function and osteogenesis promotion for guided bone regeneration. *Materials, 10*(3), 257.
40. Lee, C. H., Singla, A., & Lee, Y. (2001). Biomedical applications of collagen. *International Journal of Pharmaceutics, 221*(1–2), 1–22.
41. Young, S., et al. (2005). Gelatin as a delivery vehicle for the controlled release of bioactive molecules. *Journal of Controlled Release, 109*(1–3), 256–274.
42. Puppi, D., et al. (2010). Polymeric materials for bone and cartilage repair. *Progress in Polymer Science, 35*(4), 403–440.
43. Nair, L. S., & Laurencin, C. T. (2007). Biodegradable polymers as biomaterials. *Progress in Polymer Science, 32*(8–9), 762–798.
44. Sultana, N., et al. (2015). Chitosan-based nanocomposite scaffolds for tissue engineering applications. *Materials and Manufacturing Processes, 30*(3), 273–278.
45. Qin, J., Zhong, Z., & Ma, J. (2016). Biomimetic synthesis of hybrid hydroxyapatite nanoparticles using nanogel template for controlled release of bovine serum albumin. *Materials Science and Engineering: C, 62*, 377–383.
46. Przekora, A., Palka, K., & Ginalska, G. (2016). Biomedical potential of chitosan/HA and chitosan/β-1, 3-glucan/HA biomaterials as scaffolds for bone regeneration—A comparative study. *Materials Science and Engineering: C, 58*, 891–899.
47. López-Noriega, A., et al. (2015). Incorporation of polymeric microparticles into collagen-hydroxyapatite scaffolds for the delivery of a pro-osteogenic peptide for bone tissue engineering. *APL Materials, 3*(1), 014910.
48. Palmero, P. (2016). Ceramic–polymer nanocomposites for bone-tissue regeneration. *Nanocomposites for Musculoskeletal Tissue Regeneration*, 331–367.

49. Moheb Afzali, A., Kheradmand, M. A., & Naghib, S. M. (2024). Bioreactor design-assisted bioprinting of stimuli-responsive materials for tissue engineering and drug delivery applications. *Bioprinting, 37*, e00325.
50. Askari, E., et al. (2021). Reduced graphene oxide-grafted bovine serum albumin/bredigite nanocomposites with high mechanical properties and excellent osteogenic bioactivity for bone tissue engineering. *Bio-Design and Manufacturing, 4*(2), 243–257.
51. Wahl, D., & Czernuszka, J. (2006). Collagen-hydroxyapatite composites for hard tissue repair. *European Cells & Materials, 11*(1), 43–56.
52. Clarke, K., et al. (1993). Investigation into the formation and mechanical properties of a bioactive material based on collagen and calcium phosphate. *Journal of Materials Science: Materials in Medicine, 4*(2), 107–110.
53. Kikuchi, M., et al. (2004). Glutaraldehyde cross-linked hydroxyapatite/collagen self-organized nanocomposites. *Biomaterials, 25*(1), 63–69.
54. O'Brien, F. J., et al. (2004). Influence of freezing rate on pore structure in freeze-dried collagen-GAG scaffolds. *Biomaterials, 25*(6), 1077–1086.
55. Carlson, G. A., et al. (2004). Bacteriostatic properties of biomatrices against common orthopaedic pathogens. *Biochemical and Biophysical Research Communications, 321*(2), 472–478.
56. Xie, J., Baumann, M. J., & McCabe, L. R. (2004). Osteoblasts respond to hydroxyapatite surfaces with immediate changes in gene expression. *Journal of Biomedical Materials Research Part A: An Official Journal of The Society for Biomaterials, The Japanese Society for Biomaterials, and The Australian Society for Biomaterials and the Korean Society for Biomaterials, 71*(1), 108–117.
57. Wang, R., et al. (1995). Synthesis of nanophase hydroxyapatite/collagen composite. *Journal of Materials Science Letters, 14*(7), 490–492.
58. Serre, C., et al. (1993). In vitro induction of a calcifying matrix by biomaterials constituted of collagen and/or hydroxyapatite: An ultrastructural comparison of three types of biomaterials. *Biomaterials, 14*(2), 97–106.
59. Yamauchi, K., et al. (2004). Preparation of collagen/calcium phosphate multilayer sheet using enzymatic mineralization. *Biomaterials, 25*(24), 5481–5489.
60. Lawson, A. C., & Czernuszka, J. (1998). Collagen-calcium phosphate composites. *Proceedings of the Institution of Mechanical Engineers, Part H: Journal of Engineering in Medicine, 212*(6), 413–425.
61. Johnson, K. D., et al. (1996). Porous ceramics as bone graft substitutes in long bone defects: A biomechanical, histological, and radiographic analysis. *Journal of Orthopaedic Research, 14*(3), 351–369.
62. El-Amin, S., et al. (2003). Extracellular matrix production by human osteoblasts cultured on biodegradable polymers applicable for tissue engineering. *Biomaterials, 24*(7), 1213–1221.
63. Wang, X., et al. (2004). Tissue engineering of biphasic cartilage constructs using various biodegradable scaffolds: An in vitro study. *Biomaterials, 25*(17), 3681–3688.
64. Scabbia, A., & Trombelli, L. (2004). A comparative study on the use of a HA/collagen/chondroitin sulphate biomaterial (Biostite®) and a bovine-derived HA xenograft (Bio-Oss®) in the treatment of deep intra-osseous defects. *Journal of Clinical Periodontology, 31*(5), 348–355.
65. Maté-Sánchez de Val, J. E., et al. (2014). Comparison of three hydroxyapatite/β-tricalcium phosphate/collagen ceramic scaffolds: An in vivo study. *Journal of Biomedical Materials Research Part A, 102*(4), 1037–1046.
66. Li, W., et al. (2015). Antibacterial 45S5 Bioglass®-based scaffolds reinforced with genipin cross-linked gelatin for bone tissue engineering. *Journal of Materials Chemistry B, 3*(16), 3367–3378.
67. Dreßler, M., et al. (2011). Influence of gelatin coatings on compressive strength of porous hydroxyapatite ceramics. *Journal of the European Ceramic Society, 31*(4), 523–529.
68. Lien, S.-M., Ko, L.-Y., & Huang, T.-J. (2010). Effect of crosslinking temperature on compression strength of gelatin scaffold for articular cartilage tissue engineering. *Materials Science and Engineering: C, 30*(4), 631–635.

69. Thomas, A., & Bera, J. (2019). Preparation and characterization of gelatin-bioactive glass ceramic scaffolds for bone tissue engineering. *Journal of Biomaterials Science, Polymer Edition, 30*(7), 561–579.
70. Migneault, I., et al. (2004). Glutaraldehyde: Behavior in aqueous solution, reaction with proteins, and application to enzyme crosslinking. *BioTechniques, 37*(5), 790–802.
71. Mozafari, M., et al. (2010). Development of macroporous nanocomposite scaffolds of gelatin/bioactive glass prepared through layer solvent casting combined with lamination technique for bone tissue engineering. *Ceramics International, 36*(8), 2431–2439.
72. Daemi, H., & Barikani, M. (2012). Synthesis and characterization of calcium alginate nanoparticles, sodium homopolymannuronate salt and its calcium nanoparticles. *Scientia Iranica, 19*(6), 2023–2028.
73. Daemi, H., et al. (2016). A robust super-tough biodegradable elastomer engineered by supramolecular ionic interactions. *Biomaterials, 84*, 54–63.
74. Sisson, K., et al. (2009). Evaluation of cross-linking methods for electrospun gelatin on cell growth and viability. *Biomacromolecules, 10*(7), 1675–1680.
75. Kailasanathan, C., & Selvakumar, N. (2012). Comparative study of hydroxyapatite/gelatin composites reinforced with bio-inert ceramic particles. *Ceramics International, 38*(5), 3569–3582.
76. Damink, L. O., et al. (1995). Glutaraldehyde as a crosslinking agent for collagen-based biomaterials. *Journal of Materials Science: Materials in Medicine, 6*(8), 460–472.
77. Murugan, R., & Ramakrishna, S. (2005). Development of nanocomposites for bone grafting. *Composites Science and Technology, 65*(15–16), 2385–2406.
78. Pietak, A. M., et al. (2007). Silicon substitution in the calcium phosphate bioceramics. *Biomaterials, 28*(28), 4023–4032.
79. Li, H., Khor, K., & Cheang, P. (2002). Titanium dioxide reinforced hydroxyapatite coatings deposited by high velocity oxy-fuel (HVOF) spray. *Biomaterials, 23*(1), 85–91.
80. Jayakumar, R., et al. (2011). Biomaterials based on chitin and chitosan in wound dressing applications. *Biotechnology Advances, 29*(3), 322–337.
81. Anitha, A., et al. (2014). Chitin and chitosan in selected biomedical applications. *Progress in Polymer Science, 39*(9), 1644–1667.
82. Venkatesan, J., & Kim, S.-K. (2010). Chitosan composites for bone tissue engineering—An overview. *Marine drugs, 8*(8), 2252–2266.
83. Younes, I., & Rinaudo, M. (2015). Chitin and chitosan preparation from marine sources. Structure, properties and applications. *Marine Drugs, 13*(3), 1133–1174.
84. Shahidi, F., & Abuzaytoun, R. (2005). Chitin, chitosan, and co-products: Chemistry, production, applications, and health effects. *Advances in Food and Nutrition Research, 49*, 93–137.
85. Wysokowski, M., et al. (2015). Poriferan chitin as a versatile template for extreme biomimetics. *Polymers, 7*(2), 235–265.
86. Ehrlich, H. (2013). Biomimetic potential of chitin-based composite biomaterials of poriferan origin. *Biomimetic Biomaterials* (pp. 46–66). Elsevier.
87. Di Martino, A., Sittinger, M., & Risbud, M. V. (2005). Chitosan: A versatile biopolymer for orthopaedic tissue-engineering. *Biomaterials, 26*(30), 5983–5990.
88. Naghib, S. M., Rabiee, M., & Omidinia, E. (2014). Electrochemical biosensor for L-phenylalanine based on a gold electrode modified with graphene oxide nanosheets and chitosan. *International Journal of Electrochemical Science, 9*(5), 2341–2353.
89. Dehghani, N., et al. (2023). Chitosan/silk fibroin/nitrogen-doped carbon quantum dot/α-tricalcium phosphate nanocomposite electrospinned as a scaffold for wound healing application: In vitro and in vivo studies. *International Journal of Biological Macromolecules, 238*, 124078.
90. Saberi, J., et al. (2018). Chitosan-polyacrylic acid hybrid nanoparticles as novel tissue adhesive: Synthesis and characterization. *Fibers and Polymers, 19*(12), 2458–2464.
91. Jayakumar, R., et al. (2011). Chitin scaffolds in tissue engineering. *International Journal of Molecular Sciences, 12*(3), 1876–1887.

92. Pina, S., Oliveira, J. M., & Reis, R. L. (2015). Natural-based nanocomposites for bone tissue engineering and regenerative medicine: A review. *Advanced Materials, 27*(7), 1143–1169.
93. Deepthi, S., et al. (2016). An overview of chitin or chitosan/nano ceramic composite scaffolds for bone tissue engineering. *International Journal of Biological Macromolecules, 93*, 1338–1353.
94. Dutta, P. K. (2016). *Chitin and chitosan for regenerative medicine*. Springer.
95. Suh, J.-K.F., & Matthew, H. W. (2000). Application of chitosan-based polysaccharide biomaterials in cartilage tissue engineering: A review. *Biomaterials, 21*(24), 2589–2598.
96. VandeVord, P. J., et al. (2002). Evaluation of the biocompatibility of a chitosan scaffold in mice. *Journal of Biomedical Materials Research: An Official Journal of The Society for Biomaterials, The Japanese Society for Biomaterials, and The Australian Society for Biomaterials and the Korean Society for Biomaterials, 59*(3), 585–590.
97. Madihally, S. V., & Matthew, H. W. (1999). Porous chitosan scaffolds for tissue engineering. *Biomaterials, 20*(12), 1133–1142.
98. Shi, C., et al. (2006). Therapeutic potential of chitosan and its derivatives in regenerative medicine. *Journal of Surgical Research, 133*(2), 185–192.
99. Ladet, S., et al. (2011). Multi-membrane chitosan hydrogels as chondrocytic cell bioreactors. *Biomaterials, 32*(23), 5354–5364.
100. Chedly, J., et al. (2017). Physical chitosan microhydrogels as scaffolds for spinal cord injury restoration and axon regeneration. *Biomaterials, 138*, 91–107.
101. Malaise, S., et al. (2014). Bioresorption mechanisms of chitosan physical hydrogels: A scanning electron microscopy study. *Materials Science and Engineering: C, 42*, 374–384.
102. Sartipzadeh, O., et al. (2022). Microfluidic-assisted synthesis and modeling of stimuli-responsive monodispersed chitosan microgels for drug delivery applications. *Scientific Reports, 12*(1), 8382.
103. Garshasbi, H., et al. (2023). Stimuli-responsive injectable chitosan-based hydrogels for controlled drug delivery systems. *Frontiers in Bioengineering and Biotechnology, 10*, 1126774.
104. Naghib, S. M., Amiri, S., & Mozafari, M. R. (2024). Stimuli-responsive chitosan-based nanocarriers for drug delivery in wound dressing applications: A review. *Carbohydrate Polymer Technologies and Applications, 7*, 100497.
105. Naghib, S. M., Ahmadi, B., & Mozafari, M. R. (2024). Smart physicochemical-triggered chitosan-based nanogels for siRNA delivery and gene therapy: A focus on emerging strategies and paradigms for cancer therapy. *Current Medicinal Chemistry*.
106. Montembault, A., Viton, C., & Domard, A. (2005). Rheometric study of the gelation of chitosan in aqueous solution without cross-linking agent. *Biomacromolecules, 6*(2), 653–662.
107. Naghib, S. M., Ahmadi, B., & Mozafari, M. R. (2024). Stimuli-sensitive chitosan-based nanosystems-immobilized nucleic acids for gene therapy in breast cancer and hepatocellular carcinoma. *Current Topics in Medicinal Chemistry*.
108. Lampin, M., et al. (1997). Correlation between substratum roughness and wettability, cell adhesion, and cell migration. *Journal of Biomedical Materials Research: An Official Journal of The Society for Biomaterials and The Japanese Society for Biomaterials, 36*(1), 99–108.
109. Eppley, B. L. (2005). Biomechanical testing of alloplastic PMMA cranioplasty materials. *Journal of Craniofacial Surgery, 16*(1), 140–143.
110. Van de Velde, K., & Kiekens, P. (2002). Biopolymers: Overview of several properties and consequences on their applications. *Polymer Testing, 21*(4), 433–442.
111. Todo, M., et al. (2006). Relationship between microstructure and fracture behavior of bioabsorbable HA/PLLA composites. *Composites Part A: Applied Science and Manufacturing, 37*(12), 2221–2225.
112. DeStefano, V., Khan, S., & Tabada, A. (2020). Applications of PLA in modern medicine. *Engineered Regeneration, 1*, 76–87.
113. Farah, S., Anderson, D. G., & Langer, R. (2016). Physical and mechanical properties of PLA, and their functions in widespread applications—A comprehensive review. *Advanced Drug Delivery Reviews, 107*, 367–392.

114. Zimina, A., et al. (2020). Biocompatibility and physico-chemical properties of highly porous PLA/HA scaffolds for bone reconstruction. *Polymers, 12*(12), 2938.
115. Valantin, M.-A., et al. (2003). Polylactic acid implants (New-Fill)® to correct facial lipoatrophy in HIV-infected patients: Results of the open-label study VEGA. *AIDS, 17*(17), 2471–2477.
116. Fitzgerald, R., & Vleggaar, D. (2011). Facial volume restoration of the aging face with poly-l-lactic acid. *Dermatologic Therapy, 24*(1), 2–27.
117. Fitzgerald, R., et al. (2018). Physiochemical characteristics of poly-L-lactic acid (PLLA). *Aesthetic Surgery Journal, 38*(suppl_1), S13–S17.
118. Jin, S., et al. (2021). Recent advances in PLGA-based biomaterials for bone tissue regeneration. *Acta Biomaterialia, 127*, 56–79.
119. Singh, L., Kumar, V., & Ratner, B. D. (2004). Generation of porous microcellular 85/15 poly (DL-lactide-co-glycolide) foams for biomedical applications. *Biomaterials, 25*(13), 2611–2617.
120. Torres, F., et al. (2007). Mechanical properties and bioactivity of porous PLGA/TiO_2 nanoparticle-filled composites for tissue engineering scaffolds. *Composites Science and Technology, 67*(6), 1139–1147.
121. Boccaccini, A. R., et al. (2005). Preparation and characterisation of poly (lactide-co-glycolide)(PLGA) and PLGA/Bioglass® composite tubular foam scaffolds for tissue engineering applications. *Materials Science and Engineering: C, 25*(1), 23–31.
122. Zhang, Y., et al. (2005). Characterization of the surface biocompatibility of the electrospun PCL-collagen nanofibers using fibroblasts. *Biomacromolecules, 6*(5), 2583–2589.
123. Kiran, A. S. K., et al. (2019). Drug loaded electrospun polymer/ceramic composite nanofibrous coatings on titanium for implant related infections. *Ceramics International, 45*(15), 18710–18720.
124. Banimohamad-Shotorbani, B., et al. (2021). The efficiency of PCL/HAp electrospun nanofibers in bone regeneration: A review. *Journal of Medical Engineering & Technology, 45*(7), 511–531.
125. Russias, J., et al. (2006). Fabrication and mechanical properties of PLA/HA composites: A study of in vitro degradation. *Materials Science and Engineering: C, 26*(8), 1289–1295.
126. Aboudzadeh, N., et al. (2010). Fabrication and characterization of poly (D, L-lactide-co-glycolide)/hydroxyapatite nanocomposite scaffolds for bone tissue regeneration. *Journal of Biomedical Materials Research Part A: An Official Journal of The Society for Biomaterials, The Japanese Society for Biomaterials, and The Australian Society for Biomaterials and the Korean Society for Biomaterials, 94*(1), 137–145.
127. Rogel, M. R., Qiu, H., & Ameer, G. A. (2008). The role of nanocomposites in bone regeneration. *Journal of Materials Chemistry, 18*(36), 4233–4241.
128. Sun, T.-W., et al. (2017). Hydroxyapatite Nanowire@ magnesium silicate core-shell hierarchical nanocomposite: Synthesis and application in bone regeneration. *ACS Applied Materials & Interfaces, 9*(19), 16435–16447.
129. Du, X., Fu, S., & Zhu, Y. (2018). 3D printing of ceramic-based scaffolds for bone tissue engineering: An overview. *Journal of Materials Chemistry B, 6*(27), 4397–4412.
130. Scholz, M.-S., et al. (2011). The use of composite materials in modern orthopaedic medicine and prosthetic devices: A review. *Composites Science and Technology, 71*(16), 1791–1803.
131. Ribas, R. G., et al. (2019). Current advances in bone tissue engineering concerning ceramic and bioglass scaffolds: A review. *Ceramics International, 45*(17), 21051–21061.
132. Jiang, L., et al. (2009). Preparation and characterization of a novel composite containing carboxymethyl cellulose used for bone repair. *Materials Science and Engineering: C, 29*(1), 193–198.
133. Hing, K. A., et al. (2004). Mediation of bone ingrowth in porous hydroxyapatite bone graft substitutes. *Journal of Biomedical Materials Research Part A: An Official Journal of The Society for Biomaterials, The Japanese Society for Biomaterials, and The Australian Society for Biomaterials and the Korean Society for Biomaterials, 68*(1), 187–200.

134. Bharadwaz, A., & Jayasuriya, A. C. (2020). Recent trends in the application of widely used natural and synthetic polymer nanocomposites in bone tissue regeneration. *Materials Science and Engineering: C, 110*, 110698.
135. Peroglio, M., et al. (2010). Mechanical properties and cytocompatibility of poly (ε-caprolactone)-infiltrated biphasic calcium phosphate scaffolds with bimodal pore distribution. *Acta Biomaterialia, 6*(11), 4369–4379.
136. Martínez-Vázquez, F. J., et al. (2010). Improving the compressive strength of bioceramic robocast scaffolds by polymer infiltration. *Acta Biomaterialia, 6*(11), 4361–4368.
137. Amini, A. R., Laurencin, C. T., & Nukavarapu, S. P. (2012). Bone tissue engineering: Recent advances and challenges. *Critical Reviews™ in Biomedical Engineering, 40*(5).
138. Sun, F., Zhou, H., & Lee, J. (2011). Various preparation methods of highly porous hydroxyapatite/polymer nanoscale biocomposites for bone regeneration. *Acta Biomaterialia, 7*(11), 3813–3828.
139. Barone, D. J., Raquez, J. M., & Dubois, P. (2011). Bone-guided regeneration: From inert biomaterials to bioactive polymer (nano) composites. *Polymers for Advanced Technologies, 22*(5), 463–475.
140. Pina, S., Oliveira, J. M., & Reis, R. L. (2016). Biomimetic strategies to engineer mineralized human tissues. In I. V. Antoniac (Ed.), *Handbook of Bioceramics and Biocomposites* (pp. 503–519). Springer International Publishing.
141. Prabhakaran, M. P., Vatankhah, E., & Ramakrishna, S. (2013). Electrospun aligned PHBV/collagen nanofibers as substrates for nerve tissue engineering. *Biotechnology and Bioengineering, 110*(10), 2775–2784.
142. Pastorino, L., et al. (2014). Oriented collagen nanocoatings for tissue engineering. *Colloids and Surfaces B: Biointerfaces, 114*, 372–378.
143. Kijeńska, E., et al. (2012). Electrospun bio-composite P (LLA-CL)/collagen I/collagen III scaffolds for nerve tissue engineering. *Journal of Biomedical Materials Research Part B: Applied Biomaterials, 100*(4), 1093–1102.
144. Bhowmick, A., et al. (2015). Novel magnetic antimicrobial nanocomposites for bone tissue engineering applications. *RSC Advances, 5*(32), 25437–25445.
145. Langer, R., & Tirrell, D. A. (2004). Designing materials for biology and medicine. *Nature, 428*(6982), 487–492.
146. Thorrez, L., et al. (2008). Growth, differentiation, transplantation and survival of human skeletal myofibers on biodegradable scaffolds. *Biomaterials, 29*(1), 75–84.
147. Chae, T., et al. (2013). Novel biomimetic hydroxyapatite/alginate nanocomposite fibrous scaffolds for bone tissue regeneration. *Journal of Materials Science: Materials in Medicine, 24*(8), 1885–1894.
148. Abdelaal, O. A., & Darwish, S. M. (2011). Fabrication of tissue engineering scaffolds using rapid prototyping techniques. *International Journal of Industrial and Manufacturing Engineering, 5*(11), 2310–2318.
149. Dehghani, F., & Annabi, N. (2011). Engineering porous scaffolds using gas-based techniques. *Current Opinion in Biotechnology, 22*(5), 661–666.
150. Van de Witte, P., et al. (1996). Phase separation processes in polymer solutions in relation to membrane formation. *Journal of Membrane Science, 117*(1–2), 1–31.
151. Liapis, A., Pim, M., & Bruttini, R. (1996). Research and development needs and opportunities in freeze drying. *Drying Technology, 14*(6), 1265–1300.
152. Hou, Q., Grijpma, D. W., & Feijen, J. (2003). Porous polymeric structures for tissue engineering prepared by a coagulation, compression moulding and salt leaching technique. *Biomaterials, 24*(11), 1937–1947.
153. Oliveira, J. M., et al. (2009). Macroporous hydroxyapatite scaffolds for bone tissue engineering applications: Physicochemical characterization and assessment of rat bone marrow stromal cell viability. *Journal of Biomedical Materials Research Part A: An Official Journal of The Society for Biomaterials, The Japanese Society for Biomaterials, and The Australian Society for Biomaterials and the Korean Society for Biomaterials, 91*(1), 175–186.

154. Zhang, J., et al. (2014). Preparation and characterization of bionic bone structure chitosan/hydroxyapatite scaffold for bone tissue engineering. *Journal of Biomaterials Science, Polymer Edition, 25*(1), 61–74.
155. Venkatesan, J., & Kim, S.-K. (2014). Nano-hydroxyapatite composite biomaterials for bone tissue engineering—A review. *Journal of Biomedical Nanotechnology, 10*(10), 3124–3140.
156. Kong, L., et al. (2006). A study on the bioactivity of chitosan/nano-hydroxyapatite composite scaffolds for bone tissue engineering. *European Polymer Journal, 42*(12), 3171–3179.
157. Kong, L., et al. (2005). Preparation and characterization of nano-hydroxyapatite/chitosan composite scaffolds. *Journal of Biomedical Materials Research Part A: An Official Journal of The Society for Biomaterials, The Japanese Society for Biomaterials, and The Australian Society for Biomaterials and the Korean Society for Biomaterials, 75*(2), 275–282.
158. Liu, Y., et al. (2009). Segmental bone regeneration using an rhBMP-2-loaded gelatin/nanohydroxyapatite/fibrin scaffold in a rabbit model. *Biomaterials, 30*(31), 6276–6285.
159. Oliveira, J. M., et al. (2006). Novel hydroxyapatite/chitosan bilayered scaffold for osteochondral tissue-engineering applications: Scaffold design and its performance when seeded with goat bone marrow stromal cells. *Biomaterials, 27*(36), 6123–6137.
160. Yan, L.-P., et al. (2013). Bioactive macro/micro porous silk fibroin/nano-sized calcium phosphate scaffolds with potential for bone-tissue-engineering applications. *Nanomedicine, 8*(3), 359–378.
161. Leeuwenburgh, S., Ana, I., & Jansen, J. (2010). Sodium citrate as an effective dispersant for the synthesis of inorganic–organic composites with a nanodispersed mineral phase. *Acta Biomaterialia, 6*(3), 836–844.
162. Slaughter, B. V., et al. (2009). Hydrogels in regenerative medicine. *Advanced Materials, 21*(32–33), 3307–3329.
163. Hoffman, A. S. (2012). Hydrogels for biomedical applications. *Advanced Drug Delivery Reviews, 64*, 18–23.
164. Annabi, N., et al. (2010). Controlling the porosity and microarchitecture of hydrogels for tissue engineering. *Tissue Engineering Part B: Reviews, 16*(4), 371–383.
165. Haraguchi, K. (2007) Nanocomposite gels: New advanced functional soft materials. In *Macromolecular Symposia*. Wiley Online Library.
166. Drury, J. L., & Mooney, D. J. (2003). Hydrogels for tissue engineering: Scaffold design variables and applications. *Biomaterials, 24*(24), 4337–4351.
167. Gaharwar, A. K., Peppas, N. A., & Khademhosseini, A. (2014). Nanocomposite hydrogels for biomedical applications. *Biotechnology and Bioengineering, 111*(3), 441–453.
168. Skelton, S., et al. (2013). Biomimetic adhesive containing nanocomposite hydrogel with enhanced materials properties. *Soft Matter, 9*(14), 3825–3833.
169. Wei, G., & Ma, P. X. (2004). Structure and properties of nano-hydroxyapatite/polymer composite scaffolds for bone tissue engineering. *Biomaterials, 25*(19), 4749–4757.
170. Wen, Y., et al. (2017). 3D printed porous ceramic scaffolds for bone tissue engineering: A review. *Biomaterials Science, 5*(9), 1690–1698.
171. Detsch, R., et al. (2011). In vitro-osteoclastic activity studies on surfaces of 3D printed calcium phosphate scaffolds. *Journal of Biomaterials Applications, 26*(3), 359–380.
172. Niu, L.-N., et al. (2014). A review of the bioactivity of hydraulic calcium silicate cements. *Journal of Dentistry, 42*(5), 517–533.
173. Pei, P., et al. (2017). The effect of calcium sulfate incorporation on physiochemical and biological properties of 3D-printed mesoporous calcium silicate cement scaffolds. *Microporous and Mesoporous Materials, 241*, 11–20.
174. Shen, Y., et al. (2014). Engineering scaffolds integrated with calcium sulfate and oyster shell for enhanced bone tissue regeneration. *ACS Applied Materials & Interfaces, 6*(15), 12177–12188.
175. Fu, S., et al. (2018). 3D printed porous β-Ca_2SiO_4 scaffolds derived from preceramic resin and their physicochemical and biological properties. *Science and Technology Of Advanced Materials, 19*(1), 495–506.

176. Ginebra, M.-P., et al. (2010). New processing approaches in calcium phosphate cements and their applications in regenerative medicine. *Acta Biomaterialia, 6*(8), 2863–2873.
177. Shakor, P., et al. (2017). Modified 3D printed powder to cement-based material and mechanical properties of cement scaffold used in 3D printing. *Construction and Building Materials, 138*, 398–409.
178. Bertol, L. S., Schabbach, R., & Dos Santos, L. A. L. (2017). Dimensional evaluation of patient-specific 3D printing using calcium phosphate cement for craniofacial bone reconstruction. *Journal of Biomaterials Applications, 31*(6), 799–806.
179. Asadi-Eydivand, M., et al. (2016). Structure, properties, and in vitro behavior of heat-treated calcium sulfate scaffolds fabricated by 3D printing. *PLoS ONE, 11*(3), e0151216.
180. Shekhawat, D., et al. (2021). A short review on polymer, metal and ceramic based implant materials. In *IOP Conference Series: Materials Science and Engineering*. IOP Publishing.
181. Liu, Y.-T., Lee, T.-M., & Lui, T.-S. (2013). Enhanced osteoblastic cell response on zirconia by bio-inspired surface modification. *Colloids and Surfaces B: Biointerfaces, 106*, 37–45.
182. Okamoto, M., & John, B. (2013). Synthetic biopolymer nanocomposites for tissue engineering scaffolds. *Progress in Polymer Science, 38*(10–11), 1487–1503.
183. Marsich, E., et al. (2013). Nano-composite scaffolds for bone tissue engineering containing silver nanoparticles: Preparation, characterization and biological properties. *Journal of Materials Science: Materials in Medicine, 24*(7), 1799–1807.
184. Zhang, Y., et al. (2013). Biomedical applications of zinc oxide nanomaterials. *Current Molecular Medicine, 13*(10), 1633–1645.
185. Idumah, C. I. (2020). Progress in polymer nanocomposites for bone regeneration and engineering. *Polymers and Polymer Composites*, 0967391120913658.
186. Chen, Y., et al. (2014). Recent advances on the development of magnesium alloys for biodegradable implants. *Acta Biomaterialia, 10*(11), 4561–4573.
187. Li, N., & Zheng, Y. (2013). Novel magnesium alloys developed for biomedical application: A review. *Journal of Materials Science & Technology, 29*(6), 489–502.
188. Dayaghi, E., et al. (2019). Magnesium-zinc scaffold loaded with tetracycline for tissue engineering application: In vitro cell biology and antibacterial activity assessment. *Materials Science and Engineering: C, 102*, 53–65.
189. Zheng, Y. F., Gu, X. N., & Witte, F. (2014). Biodegradable metals. *Materials Science and Engineering: R: Reports, 77*, 1–34.
190. Sanchez, A. H. M., et al. (2015). Mg and Mg alloys: How comparable are in vitro and in vivo corrosion rates? A review. *Acta Biomaterialia, 13*, 16–31.
191. Vial, S., Reis, R. L., & Oliveira, J. M. (2017). Recent advances using gold nanoparticles as a promising multimodal tool for tissue engineering and regenerative medicine. *Current Opinion in Solid State and Materials Science, 21*(2), 92–112.
192. Matsuno, H., et al. (2001). Biocompatibility and osteogenesis of refractory metal implants, titanium, hafnium, niobium, tantalum and rhenium. *Biomaterials, 22*(11), 1253–1262.
193. Jansen, J. H., et al. (2010). Stimulation of osteogenic differentiation in human osteoprogenitor cells by pulsed electromagnetic fields: An in vitro study. *BMC Musculoskeletal Disorders, 11*(1), 1–11.
194. Santini, M. T., et al. (2003). Effects of a 50 Hz sinusoidal magnetic field on cell adhesion molecule expression in two human osteosarcoma cell lines (MG-63 and Saos-2). *Bioelectromagnetics, 24*(5), 327–338.
195. Taylor, K. F., et al. (2006). Effect of pulsed electromagnetic fields on maturation of regenerate bone in a rabbit limb lengthening model. *Journal of Orthopaedic Research, 24*(1), 2–10.
196. Chang, K., & Chang, W. H. S. (2003). Pulsed electromagnetic fields prevent osteoporosis in an ovariectomized female rat model: A prostaglandin E2-associated process. *Bioelectromagnetics, 24*(3), 189–198.
197. Zhang, X.y., Xue, Y., & Zhang, Y. (2006). Effects of 0.4 T rotating magnetic field exposure on density, strength, calcium and metabolism of rat thigh bones. *Bioelectromagnetics: Journal of the Bioelectromagnetics Society, The Society for Physical Regulation in Biology and Medicine, The European Bioelectromagnetics Association, 27*(1), 1–9.

198. Fini, M., et al. (2002). The effect of pulsed electromagnetic fields on the osteointegration of hydroxyapatite implants in cancellous bone: A morphologic and microstructural in vivo study. *Journal of Orthopaedic Research, 20*(4), 756–763.
199. Sun, S., et al. (2004). Monodisperse mfe2o4 (m= fe, co, mn) nanoparticles. *Journal of the American Chemical Society, 126*(1), 273–279.
200. Bretcanu, O., et al. (2006). Magnetic properties of the ferrimagnetic glass-ceramics for hyperthermia. *Journal of Magnetism and Magnetic Materials, 305*(2), 529–533.
201. Jun, Y.-W., et al. (2005). Nanoscale size effect of magnetic nanocrystals and their utilization for cancer diagnosis via magnetic resonance imaging. *Journal of the American Chemical Society, 127*(16), 5732–5733.
202. Kim, J., et al. (2006). Magnetic fluorescent delivery vehicle using uniform mesoporous silica spheres embedded with monodisperse magnetic and semiconductor nanocrystals. *Journal of the American Chemical Society, 128*(3), 688–689.
203. Wang, L., et al. (2006). A biocompatible method of decorporation: Bisphosphonate-modified magnetite nanoparticles to remove uranyl ions from blood. *Journal of the American Chemical Society, 128*(41), 13358–13359.
204. Campbell, R. B. (2007). Battling tumors with magnetic nanotherapeutics and hyperthermia: turning up the heat. *Nanomedicine*, 649–652.
205. Gupta, A. K., & Gupta, M. (2005). Synthesis and surface engineering of iron oxide nanoparticles for biomedical applications. *Biomaterials, 26*(18), 3995–4021.
206. Jeong, U., et al. (2007). Superparamagnetic colloids: Controlled synthesis and niche applications. *Advanced Materials, 19*(1), 33–60.
207. Jun, Y.-W., Seo, J.-W., & Cheon, J. (2008). Nanoscaling laws of magnetic nanoparticles and their applicabilities in biomedical sciences. *Accounts of Chemical Research, 41*(2), 179–189.
208. Bock, N., et al. (2010). A novel route in bone tissue engineering: Magnetic biomimetic scaffolds. *Acta Biomaterialia, 6*(3), 786–796.
209. Tampieri, A., et al. (2012). Intrinsic magnetism and hyperthermia in bioactive Fe-doped hydroxyapatite. *Acta Biomaterialia, 8*(2), 843–851.
210. Wu, C., et al. (2011). Multifunctional magnetic mesoporous bioactive glass scaffolds with a hierarchical pore structure. *Acta Biomaterialia, 7*(10), 3563–3572.
211. Meng, J., et al. (2010). Paramagnetic nanofibrous composite films enhance the osteogenic responses of pre-osteoblast cells. *Nanoscale, 2*(12), 2565–2569.
212. De Santis, R., et al. (2011). A basic approach toward the development of nanocomposite magnetic scaffolds for advanced bone tissue engineering. *Journal of Applied Polymer Science, 122*(6), 3599–3605.
213. Tampieri, A., et al. (2010). A conceptually new type of bio-hybrid scaffold for bone regeneration. *Nanotechnology, 22*(1), 015104.
214. Bañobre-López, M., et al. (2011). Poly (caprolactone) based magnetic scaffolds for bone tissue engineering. *Journal of Applied Physics, 109*(7), 07B313.
215. Wei, Y., et al. (2011). Magnetic biodegradable Fe3O4/CS/PVA nanofibrous membranes for bone regeneration. *Biomedical Materials, 6*(5), 055008.
216. Nicolini, C. (2016). High performance lightweight magnesium nanocomposites for engineering and biomedical applications. *NanoWorld Journal, 1*(2).
217. Moravej, M., & Mantovani, D. (2011). Biodegradable metals for cardiovascular stent application: Interests and new opportunities. *International Journal of Molecular Sciences, 12*(7), 4250–4270.
218. Abazari, S., et al. (2020). Graphene family nanomaterial reinforced magnesium-based matrix composites for biomedical application: A comprehensive review. *Metals, 10*(8), 1002.
219. Kumar, P. S., et al. (2011). Synthesis, characterization and cytocompatibility studies of α-chitin hydrogel/nano hydroxyapatite composite scaffolds. *International Journal of Biological Macromolecules, 49*(1), 20–31.
220. Li, M., et al. (2013). In situ synthesis and biocompatibility of nano hydroxyapatite on pristine and chitosan functionalized graphene oxide. *Journal of Materials Chemistry B, 1*(4), 475–484.

221. Lian, X.-J., et al. (2013). Structural and biomedical properties of zirconia-hydroxyapatite nano-crystal ceramics. *Journal of Biomaterials and Tissue Engineering, 3*(3), 330–334.
222. Tripathi, A., et al. (2012). Bio-composite scaffolds containing chitosan/nano-hydroxyapatite/nano-copper–zinc for bone tissue engineering. *International Journal of Biological Macromolecules, 50*(1), 294–299.
223. Saravanan, S., et al. (2011). Preparation, characterization and antimicrobial activity of a bio-composite scaffold containing chitosan/nano-hydroxyapatite/nano-silver for bone tissue engineering. *International Journal of Biological Macromolecules, 49*(2), 188–193.

The manufacturer's authorised representative in the EU is Springer Nature Customer Service Centre GmbH, Europaplatz 3, 69115 Heidelberg, Germany. If you have any concerns regarding our products, please contact ProductSafety@springernature.com

Printed and bound by CPI Group (UK) Ltd, Croydon, CR0 4YY
01/10/2026
02225931-0002